QUANTUM CHAOS

This book introduces the quantum mechanics of classically chaotic systems, or Quantum Chaos for short. The basic concepts of quantum chaos can be grasped easily by any student of physics, but the underlying physical principles tend to be obscured by the mathematical apparatus used to describe it. The author's philosophy, therefore, has been to keep the discussion simple and to illustrate theory, wherever possible, with experimental or numerical examples. The microwave billiard experiments, initiated by the author and his group, play a major role in this respect. A basic knowledge of quantum mechanics is assumed.

Beginning with a presentation of the various types of billiard experiments, random matrix theory is described, including an introduction to supersymmetry techniques. Systems with periodic time dependences are treated, with special emphasis on dynamical localization. Another topic is the analogy between the dynamics of a one-dimensional gas with a repulsive interaction and spectral level dynamics. Distribution and fluctuation properties of scattering matrix elements are treated within the framework of scattering theory. Finally semiclassical quantum mechanics and periodic orbit theory are presented, including a derivation of the Gutzwiller trace formula together with a number of applications.

This book addresses both students and specialists working in the field.

H.-J. Stöckmann was born in 1945 in Göttingen, Germany. He started his studies in physics and mathematics in 1964 at the University of Heidelberg. He performed his diploma work in experimental physics on optical spectroscopy, which he finished in 1969. For his doctoral work he changed to nuclear solid state physics, with experiments at the research reactor of the Kernforschungszentrum Karlsruhe. After finishing his thesis in 1972 he continued this type of experiments at different sites, among others the Institute Laue-Langevin in Grenoble, France. He joined the institute as a guest for one year, and for several short-term visits. In 1978 he acquired his habilitation. In 1979 the author was appointed a university professor at the University of Marburg, where he still works. He extended his previous research activities with experiments at the Berlin synchrotron facility, BESSY. In 1988 he started the microwave experiments on quantum chaotic questions which frequently appear in the present monograph. Meanwhile, these experiments have become his main field of interest.

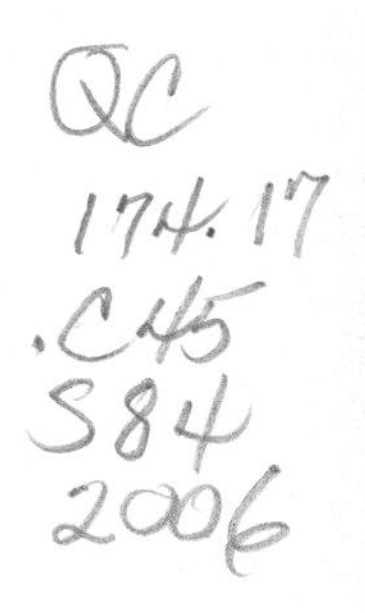

QUANTUM CHAOS

An Introduction

HANS-JÜRGEN STÖCKMANN

Philipps Universität Marburg

CAMBRIDGE UNIVERSITY PRESS

CAMBRIDGE UNIVERSITY PRESS
Cambridge, New York, Melbourne, Madrid, Cape Town, Singapore, São Paulo

Cambridge University Press
The Edinburgh Building, Cambridge CB2 2RU, UK

Published in the United States of America by Cambridge University Press, New York

www.cambridge.org
Information on this title: www.cambridge.org/9780521592840

First published 1999
Reprinted 2000
This digitally printed first paperback version (with corrections) 2006

A catalogue record for this publication is available from the British Library

Library of Congress Cataloguing in Publication data
Stöckmann, Hans-Jürgen.
Quantum chaos : an introduction / Hans-Jürgen Stöckmann.
p. cm.
ISBN 0 521 59284 4 (hb)
1. Quantum chaos. I. Title.
QC174.17.C45S84 1999
530.12–dc21 98-45454 CIP

ISBN-13 978-0-521-59284-0 hardback
ISBN-10 0-521-59284-4 hardback

ISBN-13 978-0-521-02715-1 paperback
ISBN-10 0-521-02715-2 paperback

Contents

Preface

This monograph is based on the script of a lecture series on the quantum mechanics of classically chaotic systems given by the author at the University of Marburg during the summer term 1995. The lectures were attended by students with basic knowledge in quantum mechanics, including members of the author's own group working on microwave analogous experiments on quantum chaotic questions.

When preparing the lectures the author became aware that a comprehensive textbook, covering both the theoretical and the experimental aspects, was not available. The present monograph is intended to fill this gap.

The basic concepts of the quantum mechanics of classically chaotic systems, termed 'quantum chaos' for short, are easy to grasp by any student of physics. The mathematical apparatus needed, however, often tends to obscure the physical background. That is why the theoretical results will be illustrated by real experimental or numerical data whenever possible.

Chapter 1 gives a short introduction on the essential ideas of semiclassical quantum mechanics, which is illustrated by two examples taken from the microwave billiards and the kicked rotator.

Chapter 2 treats the different types of billiard experiments. Methods to study vibrating solids and liquids are presented. The main part of this chapter deals with microwave techniques, as by far the most experiments have been performed in microwave billiards. The chapter ends with a discussion of mesoscopic billiards including quantum corrals.

Chapter 3 introduces random matrix theory. It was developed in the sixties, and is treated in several reviews and monographs. Therefore only the fundamental concepts are discussed, such as level spacing distributions and spectral correlation functions. The last part of the chapter introduces supersymmetry techniques, which have become more and more important in recent years.

In Chapter 4 systems with periodic time dependences are discussed, with special emphasis on dynamical localization, i.e. the suppression of classical chaos by quantization. Since dynamical localization is a special case of Anderson localization, which is observed for electrons in disordered lattices, the chapter includes a short discussion of the latter systems.

Chapter 5 deals with the analogy between the dynamics of the eigenvalues of a chaotic system as a function of an external parameter and the dynamics of a one-dimensional gas, a model introduced by Pechukas and Yukawa. When varying the parameter, the phases of the wave functions change as well. These so-called geometrical phases, also called 'Berry's phases', are treated in the last section of this chapter.

In Chapter 6 scattering theory is introduced to describe the influence of coupled antennas on the spectrum, with special emphasis on microwave billiards. For depths and widths of the resonances Porter–Thomas distributions are found, these have been well-known in nuclear physics for many years. The chapter ends with a discussion of the fluctuations of scattering matrix elements, known as Ericson fluctuations in nuclear physics, and as universal conductance fluctuations in mesoscopic systems.

In Chapter 7 semiclassical quantum mechanics is developed. Starting with the Feynman path integral for the quantum mechanical propagator, stationary phase approximations are applied to obtain semiclassical expressions for the propagator and the Green function. The main result of the chapter is the Gutzwiller trace formula expressing the quantum mechanical spectrum in terms of the classical periodic orbits.

In Chapter 8 several applications of periodic orbit theory are presented. It starts with a discussion of Fourier transform techniques to extract the contributions of periodic orbits from the spectra. The semiclassical theory of spectral rigidity establishes a link between periodic orbit and random matrix theory. Subsequently resummation schemes are developed allowing the calculation of quantum mechanical spectra from the periodic orbits under favourable conditions. The chapter ends with a discussion of billiards on a metric with constant negative curvature, and of the Selberg trace formula, the non-Euclidean equivalent of the Gutzwiller trace formula.

From the very beginning, when starting the microwave experiments in the late eighties, I have been supported and encouraged by many colleagues from theory. Prof. S. Großmann, Marburg, roused my interest in nonlinear dynamics. In the following years Prof. B. Eckhardt, Marburg, and Prof. F. Haake, Essen, have been my main interlocutors in theoretical problems. Moreover, I want to thank all my coworkers, especially J. Stein, with whom I started the experiments, and U. Kuhl, my senior coworker. The experiments have been supported by the Sonderforschungsbereich 'Nichtlineare Dynamik' of the Deutsche Forschungsgemeinschaft, both financially and scientifically. Here I want to mention in particular the groups of Prof. T. Geisel, Göttingen, and Prof. A. Richter, Darmstadt. The Fachbereich Physik at the University of Marburg has provided me with the local support necessary to perform the experiments.

I further want to thank the authors and the publishers who gave permission to reproduce figures from their publications. In addition most of the authors have provided me with good copies of the figures. Some of the figures have been prepared by U. Kuhl. My coworker M. Barth has cared for the regular updating of our internal quantum chaos bibliography used for this monograph. Finally I am indebted to my wife E.-B. Stöckmann for her critical reading of the whole manuscript and stylistic corrections.

Hans-Jürgen Stöckmann

Preface to the paperback edition

As stated in the preface of the hard cover edition, it was the idea of this book to illustrate that the basic concepts of quantum chaos are easy to grasp by any student of physics. To this end the theoretical results were illustrated by real experimental or numerical data whenever possible. The positive response of many colleagues and students gave me the impression that I had not been completely unsuccessful with this concept. Thanks to the publisher there is now a paperback edition available at a reasonable price for every student. Apart from correcting some errors the text remains unchanged. As to the most recent developments in quantum chaos, only a short survey shall be given in the following material.

There has been a considerable increase in the number of groups using classical waves to study quantum mechanical questions, by far the most of them using microwaves. Noticeable exceptions are the still small number of experiments in the regime of visible light [1]. The main field of interest meanwhile changed from closed to open i.e. scattering systems.

The number of universality classes, traditionally restricted to three (see Section 3.1), has increased to ten [2]. The new classes are of particular relevance whenever the creation and annihilation of electron-hole pairs or of particles and antiparticles is involved. The main fields of application are consequently superconductivity and quantum electrodynamics. Random matrix theory as a whole has been reviewed in a special volume of Journal of Physics A [3], and its application to scattering theory in another special edition of the same journal [4].

Really important progress has been made in semiclassical quantum mechanics. Twenty years ago Berry showed that the leading term of the spectral form factor in random matrix theory could be obtained within the so-called diagonal approximation from the periodic orbits of a chaotic system (see Section 8.2). Sieber and Richter were able to get the next term in the series expansion by going beyond the diagonal approximation [5]. The main ingredients are pairs of orbits, one with a self-intersection, the other with an avoided crossing. Meanwhile all terms of the series expansion could be reproduced by Müller et al. [6] by an extension of the ideas of Sieber and Richter. Thus a decisive step in the proof of the BGS conjecture (see Section 3.2) has been achieved.

Marburg, October 2005 Hans-Jürgen Stöckmann

References

[1] V. Doya et al., Phys. Rev. Lett. 88, 014102 (2002); J. Dingjan et al., Phys. Rev. Lett. 88, 064101 (2002); T. Gensty et al., Phys. Rev. Lett. 94, 233901 (2005).

[2] M. R. Zirnbauer, J. Math. Phys. 37, 4986 (1996); A. Altland, M. R. Zirnbauer, Phys. Rev. B 55, 1142 (1997).

[3] P. J. Forrester, N. C. Snaith, and J. J. M. Verbaarschot (guest editors), J. Phys. A: Math. Gen. 36 (2003).

[4] Y. V. Fyodorov, T. Kottos, and H.-J. Stöckmann (guest editors), J. Phys. A: Math. Gen. 38 (2005).

[5] M. Sieber and K. Richter, Phys. Scr. T90, 128 (2001).

[6] S. Müller et al., Phys. Rev. Lett. 93, 014103 (2004).

1
Introduction

From the very beginning classical nonlinear dynamics has enjoyed much popularity even among the noneducated public as is documented by numerous articles in well-renowned magazines, including nonscientific papers. For its nonclassical counterpart, the quantum mechanics of chaotic systems, termed in short 'quantum chaos', the situation is completely different. It has always been considered as a more or less mysterious topic, reserved to a small exclusive circle of theoreticians. Whereas the applicability of classical nonlinear dynamics to daily life is comprehensible for a complete outsider, quantum chaos, on the other hand, seems to be of no practical relevance at all. Moreover, in classical nonlinear dynamics the theory is supported by numerous experiments, mainly in hydrodynamics and laser physics, whereas quantum chaos at first sight seems to be the exclusive domain of theoreticians. In the beginning the only experimental contributions came from nuclear physics [Por65]. This preponderance of theory seems to have suppressed any experimental effort for nearly two decades. The situation gradually changed in the middle of the eighties, since when numerous experiments have been performed. An introductory presentation also suited to the experimentalist with no or only little basic knowledge is still missing.

It is the intention of this monograph to demonstrate that there is no reason to be afraid of quantum chaos. The underlying ideas are very simple. It is essentially the mathematical apparatus that makes things difficult and often tends to obscure the physical background. Therefore the philosophy adopted in this presentation is to illustrate theory by experimental results whenever possible, which leads to a strong accentuation of billiard systems for which a large number of experiments now exist. Consequently, results on microwave billiards obtained by the author's own group will be frequently represented. This should not be misunderstood as an unappropriate preference given to their own work. The billiard, though being conceptually simple, nevertheless exhibits the full complexity of nonlinear dynamics, including its quantum mechanical aspects. Probably there is no essential aspect of quantum chaos which cannot be found in chaotic billiards.

The nonexpert for whom this book is mainly written may ask whether quantum chaos is really an interesting topic in its own right. After all, quantum mechanics has now existed for more than 60 years and has probably become the best tested physical theory ever conceived. Quantum mechanics can handle not only the hydrogen atom which is classically integrable but also the classically nonintegrable helium atom. We may even ask whether there is anything like quantum chaos at all. The Schrödinger equation is a linear equation leaving no room for chaos. The correspondence principle, on the other hand, demands that in the semiclassical region, i.e. at length scales large compared to the de Broglie wavelength, quantum mechanics continuously develops into classical mechanics.

That is why there has even been a debate whether the term 'quantum chaos' should be used at all. In 1989 the leading scientists in the field came together to discuss these questions at a summer school in Les Houches [Gia89]. The proceedings are titled 'chaos and quantum physics' thus avoiding the dubious term. Berry [Ber89] once again proposed the term 'quantum chaology', introduced by him previously [Ber87]. This would obviously have been a much better choice than 'quantum chaos', but was not generally accepted. In the following years the debate ceased. Today the term 'quantum chaos' is generally understood to comprise all problems concerning the quantum mechanical behaviour of classically chaotic systems. This view will also be adopted in this book. For billiard experiments another aspect has to be considered. Most of them are analogue experiments using the equivalence of the Helmholtz equation with the stationary Schrödinger equation. That is why the term 'wave chaos' is sometimes preferred in this context. Most of the phenomena discussed in this book indeed apply to all waves and are not primarily of quantum mechanical origin.

The problems with the proper definition of the term 'quantum chaos' have their origin in the concept of the trajectory, which completely loses its significance in quantum mechanics. Only in the semiclassical region do the trajectories eventually reappear, an aspect of immense significance in the context of semiclassical theories. For purposes of illustration, let us consider the evolution of a classical system with N dynamical variables $x_1, \ldots, x_N$ under the influence of an interaction. Typically the x_n comprise all components of the positions and the momenta of the particles. Consequently the number of dynamical variables is $N = 6M$ for a three-dimensional M particle system.

Let $\mathbf{x}(0) = [x_1(0), \ldots, x_N(0)]$ be the vector of the dynamical variables at the time $t = 0$. At any later time t we may write $\boldsymbol{x}(t)$ as a function of the initial conditions and the time as

$$\boldsymbol{x}(t) = \boldsymbol{F}[\boldsymbol{x}(0),\ t]. \tag{1.1}$$

If the initial conditions are infinitesimally changed to

$$\boldsymbol{x}_1(0) = \boldsymbol{x}(0) + \boldsymbol{\xi}(0), \tag{1.2}$$

then at a later time t the dynamical variables develop according to

$$\boldsymbol{x}_1(t) = \boldsymbol{F}[\boldsymbol{x}(0) + \boldsymbol{\xi}(0),\ t]. \tag{1.3}$$

The distance $\boldsymbol{\xi}(t) = \boldsymbol{x}_n(t) - \boldsymbol{x}(t)$ between the two trajectories is obtained from Eqs. (1.1) and (1.3) in linear approximation as

$$\boldsymbol{\xi}(t) = (\boldsymbol{\xi}(0)\boldsymbol{\nabla})\boldsymbol{F}[\boldsymbol{x}(0),\ t], \tag{1.4}$$

where $\boldsymbol{\nabla}$ is the gradient of $\boldsymbol{F}$ with respect to the initial values. Written in components Eq. (1.4) reads

$$\xi_n(t) = \sum_m \frac{\partial F_n}{\partial x_m} \xi_m(0). \tag{1.5}$$

The eigenvalues of the matrix $M = (\partial F_n / \partial x_m)$ determine the stability properties of the trajectory. If the moduli of all eigenvalues are smaller than one, the trajectory is stable, and all deviations from the initial trajectory will rapidly approach zero. If the modulus of at least one eigenvalue is larger than one, both trajectories will exponentially depart from each other even for infinitesimally small initial deviations $\boldsymbol{\xi}(0)$. Details can be found in every textbook on nonlinear dynamics (see Refs. [Sch84, Ott93]).

In quantum mechanics this definition of chaos becomes obsolete, since the uncertainty relation

$$\Delta x \Delta p \geqslant \tfrac{1}{2}\hbar \tag{1.6}$$

prevents a precise determination of the initial conditions. This can best be illustrated for the propagation of a point-like particle in a box with infinitely high walls. For obvious reasons these systems are called billiards. They will accompany us throughout this book. For the quantum mechanical treatment two steps are necessary. First the Schrödinger equation

$$-\frac{\hbar^2}{2m}\Delta\psi = \imath\hbar\frac{\partial\psi}{\partial t} \tag{1.7}$$

has to be solved with the Dirichlet boundary condition

$$\psi|_S = 0, \tag{1.8}$$

where S denotes the walls of the box. Stationary solutions of the Schrödinger equation are obtained by separating the time dependence,

$$\psi_n(x,\ t) = \psi_n(x)e^{\imath\omega_n t}. \tag{1.9}$$

Insertion into Eq. (1.7) yields

$$(\Delta + k_n^2)\psi_n(x) = 0 \tag{1.10}$$

where ω_n and k_n are connected via the dispersion relation

$$\omega_n = \frac{\hbar}{2m} k_n^2. \tag{1.11}$$

Equation (1.10) is also obtained if we start with the wave equation

$$\left(\Delta - \frac{1}{c^2}\frac{\partial^2}{\partial t^2}\right)\psi = 0, \tag{1.12}$$

where c is the wave velocity, and if we separate again the time dependence by means of the ansatz (1.9). In contrast to the quadratic dispersion relation (1.11) for the quantum mechanical case we now have the linear relation

$$\omega_n = ck_n \tag{1.13}$$

between ω_n and k_n. It is exactly this correspondence between the stationary Schrödinger equation and the stationary wave equation, also called the Helmholtz equation, which has been used in many billiard experiments to study quantum chaotic problems using wave analogue systems (see Chapter 2).

As soon as the stationary solutions of the Schrödinger equation are known, a wave packet can be constructed by a superposition of eigenfunctions,

$$\psi(x, t) = \sum_n a_n \psi_n(x) e^{-\imath\omega_n t}. \tag{1.14}$$

For a Gaussian shaped packet centred at a wave number $\overline{k}$ and of width Δk the coefficients a_n are given by

$$a_n = a \exp\left[-\frac{1}{2}\left(\frac{k_n - \overline{k}}{\Delta k}\right)^2\right], \tag{1.15}$$

where a is chosen in such a way that the total probability of finding the particle in the packet is normalized to one. If the a_n are known at time $t = 0$, e.g. by a measurement of the momentum with an uncertainty of $\Delta p = \hbar\Delta k$, the quantum mechanical evolution of the packet can be calculated for any later time with arbitrary precision. Moreover, to construct wave packets with a given width, the sum in Eq. (1.14) can be restricted to a finite number of terms. Apart from untypical exceptions, the resulting function is not periodic, since in general the ω_n are not commensurable, but *quasi-periodic*. Thus the wave packet will always reconstruct itself, possibly after a long period of time. The exponential departure of neighbouring trajectories known from classical nonlinear dynamics has completely disappeared.

The wave properties of matter do not provoke an additional spreading of the probability density as we might intuitively think. On the contrary, in systems where the classical probability density continuously diffuses with time, e.g. by a random walk process, quantum mechanics tends to freeze the diffusion and to localize the wave packet [Cas79]. This has been established in numerous calculations and has even been demonstrated experimentally [Gal88, Bay89, Moo94]. The phenomenon of quantum mechanical localization will be discussed in detail in Chapter 4.

To demonstrate how the wave packet just constructed evolves with time, we now take the simplest of all possible billiards, a particle in a one-dimensional box with infinitely high walls. Taking the walls at the positions $x = 0$ and $x = l$, the eigenfunctions of the system are given by

$$\psi_n(x) = \sqrt{\frac{2}{l}} \sin k_n x, \qquad n = 1, 2, 3, \ldots \tag{1.16}$$

with the wave numbers

$$k_n = \frac{\pi n}{l}. \tag{1.17}$$

Insertion into Eq. (1.14) yields

$$\psi(x, t) = a\sqrt{\frac{2}{l}} \sum_{n=1}^{\infty} \exp\left[-\frac{1}{2}\left(\frac{k_n - \bar{k}}{\Delta k}\right)^2\right] \sin k_n x \, e^{-\imath \omega_n t}. \tag{1.18}$$

This equation holds for the propagation of both particle packets and ordinary waves, provided that the respective dispersion relations (1.11) or (1.13) are obeyed. The calculation is somewhat easier for ordinary waves. Therefore this situation will now be considered by putting $\omega_n = c k_n$. For particle waves the calculation follows exactly the same scheme. To simplify the calculation it will be further assumed that the average momentum is large compared to the width of the distribution, i.e. $\bar{k} \gg \Delta k$. Then the sum can be extended from $-\infty$ to $+\infty$, and we can apply the *Poisson sum relation*

$$\sum_{n=-\infty}^{\infty} f(n) = \sum_{n=-\infty}^{\infty} g(n), \tag{1.19}$$

where

$$g(n) = \int_{-\infty}^{\infty} f(n) e^{2\pi \imath n m} \, dn \tag{1.20}$$

is the Fourier transform of $f(n)$. Application to Eq. (1.18) yields

$$\psi(x, t) = a\sqrt{\frac{2}{l}} \sum_{m=-\infty}^{\infty} \frac{l}{\pi} \int_{-\infty}^{\infty} \exp\left[-\frac{1}{2}\left(\frac{k - \bar{k}}{\Delta k}\right)^2\right] \sin kx \, e^{\imath(2lm - ct)k} \, dk, \tag{1.21}$$

where the integration variable n has been substituted by $k = n\pi/l$. The integration is easily carried out using the well-known relation

$$\int_{-\infty}^{\infty} \exp[-(ak^2 + 2bk + c)]\, dk = \sqrt{\frac{\pi}{a}} \exp\left(\frac{b^2}{a} - c\right) \tag{1.22}$$

for Gaussian integrals which also holds for complex values of a, b, c provided that $\mathrm{Re}(a) > 0$. The result is

$$\psi(x, t) = \sum_{m=-\infty}^{\infty} [\phi(x - ct_m) - \phi(l - x - ct_{m+\frac{1}{2}})], \tag{1.23}$$

where

$$t_m = t - m\frac{2l}{c}, \tag{1.24}$$

and

$$\phi(x) = 2a\sqrt{\frac{l}{\pi}}\Delta k \exp\left[\imath \bar{k} x - \frac{1}{2}(x\Delta k)^2\right]. \tag{1.25}$$

Equation (1.23) allows a straightforward interpretation. It describes the propagation of a Gaussian pulse with width $\Delta x = 1/\Delta k$ and velocity c, passing to and fro between the two walls and changing sign upon every reflection. For the propagation of particle waves the situation is qualitatively similar, but now the quadratic dispersion relation (1.11) leads to a spreading of the pulse with time and a pulse width $\Delta x(t)$ given by

$$\Delta x(t) = \frac{1}{\Delta k}\left[1 + \left(\frac{\hbar(\Delta k)^2 t}{m}\right)^2\right]^{1/4}. \tag{1.26}$$

For time $t = 0$ we obtain $\Delta x \Delta k = 1$ as for ordinary waves. This is just the quantum mechanical uncertainty relation.

By means of the Poisson sum relation two different expressions for $\psi(x, t)$ have been obtained. First, in Eq. (1.18), it is expressed in terms of a sum over the eigenfunctions of the systems, second, in Eq. (1.23), it is written as a pulse propagating with the velocity c being periodically reflected at the walls. This reciprocity, with the quantum mechanical spectrum on the one side and the classical trajectories on the other, will become one of the main ingredients of the semiclassical theory, in particular of the Gutzwiller trace formula. In the special example presented here the applied procedure worked especially well, since the set $\{k_n\}$ of eigenvalues was equidistant, leading to a perfect pulse reconstruction after every reflection. In the general case the pulse will be destroyed after a small number of reflections, but pulse reconstructions are still possible. The correspondence between classical and quantum mechanics will be demonstrated by two examples.

Figure 1.1 shows the propagation of a microwave pulse in a cavity in the shape of a quarter stadium [Ste95]. The measuring technique will be described in detail in Section 2.2.1. A circular wave is emitted from an antenna, propagates through the billiard, and is eventually reflected by the walls, thereby undergoing a change of sign (this can be seen especially well in Fig. 1.1(d) for the reflection of the pulse at the top and the bottom walls). After a number of additional reflections the pulse amplitude is distributed more or less equally

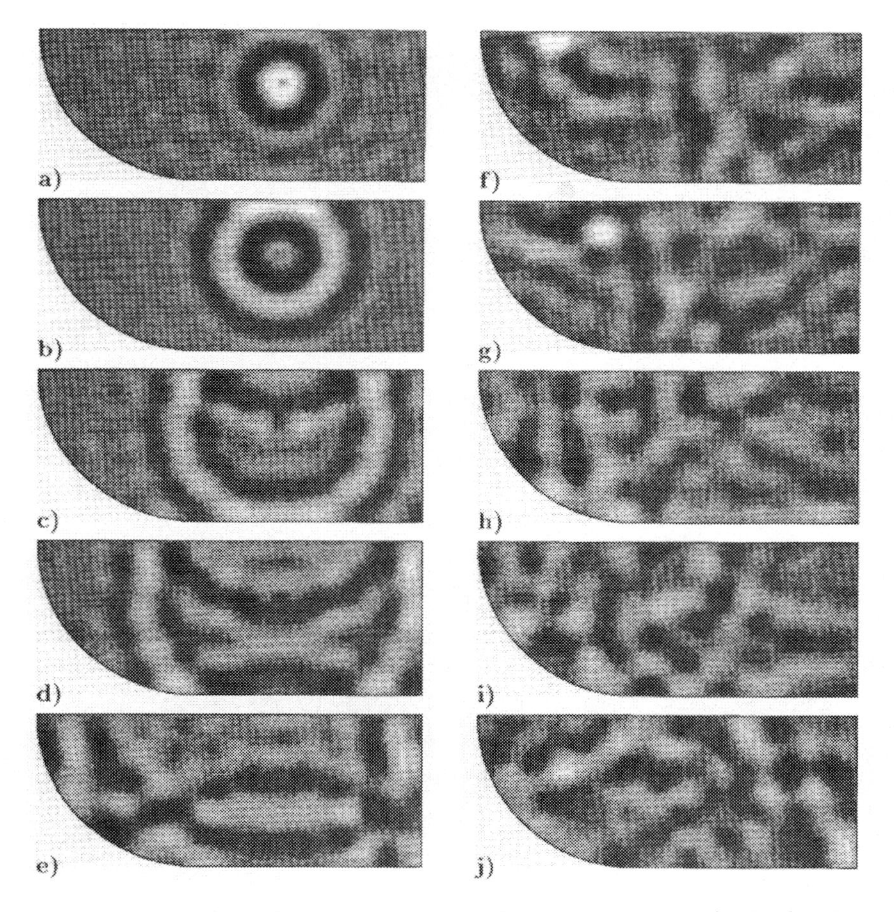

Figure 1.1. Propagation of a microwave pulse in a microwave cavity in the shape of a quarter stadium (length of the straight part $l = 18$ cm, radius $r = 13.5$ cm, height $h = 0.8$ cm) for different times $t/10^{-10}$ s: 0.36 (a), 1.60 (b), 2.90 (c), 3.80 (d), 5.63 (e), 9.01 (f), 10.21 (g), 12.0 (h), 14.18 (i), 19.09 (j) [Ste95] (Copyright 1995 by the American Physical Society).

over the billiard. But after some time the pulse suddenly reappears (see Fig. 1.1(f)). This is even more evident in Fig. 1.2 where the pulses are shown in a three-dimensional representation for two snapshots corresponding to Figs. 1.1(a) and (f). This reconstruction has nothing to do with the quantum

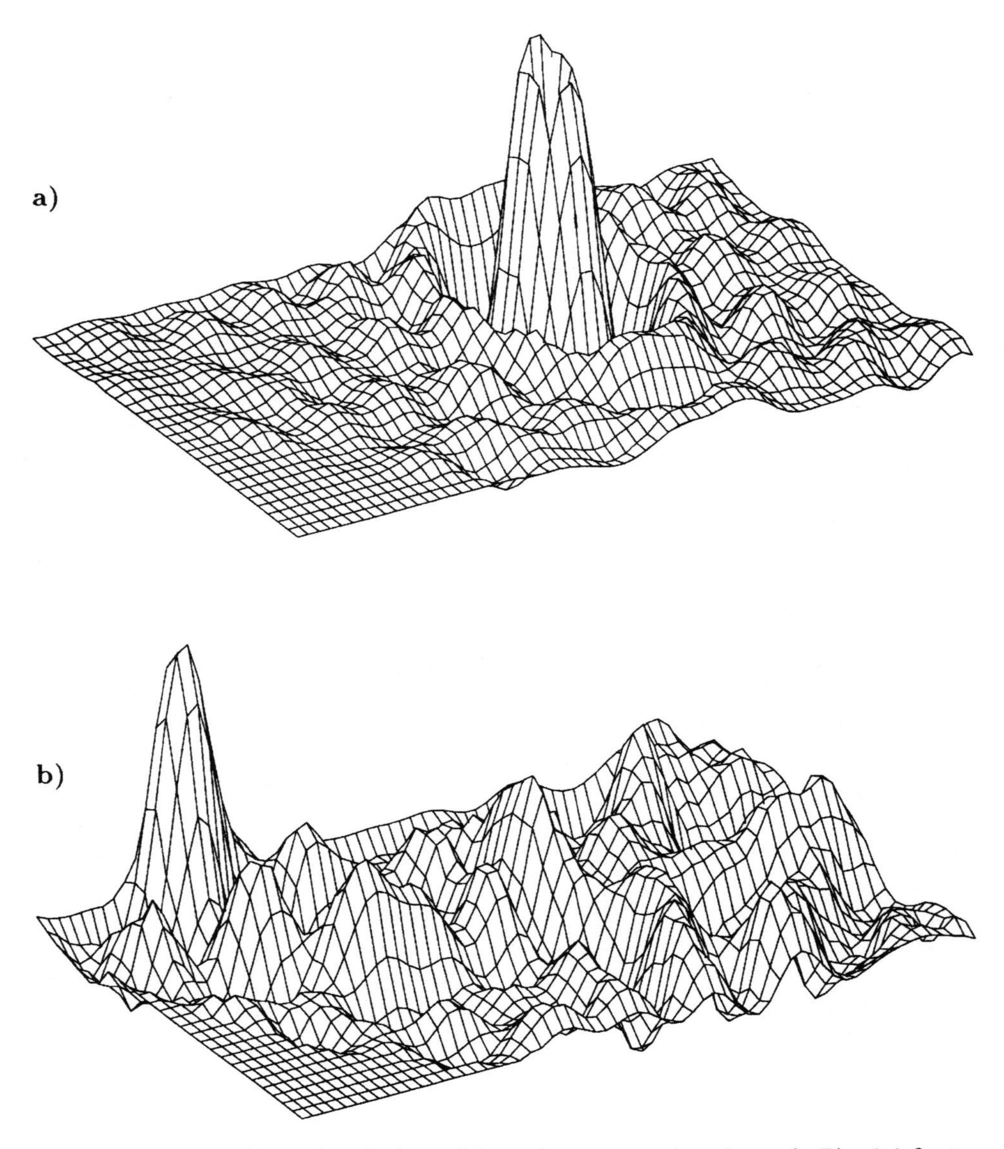

Figure 1.2. Three-dimensional view of the pulse propagation shown in Fig. 1.1 for two times, corresponding to Figs. 1.1(a) and (f), respectively [Ste95] (Copyright 1995 by the American Physical Society).

mechanical revival discussed above, it is just a manifestation of the focussing properties of the circular boundary. All classical paths reflected by these parts of the stadium will be simultaneously focussed later in the image point of the antenna position after reflection at the circular mirror. This pulse reconstruction again demonstrates the fact that wave properties and classical trajectories are just two sides of the same coin.

The second example is taken from the kicked rotator. It is one of the most thoroughly studied chaotic systems, both classically and quantum mechanically, and will be discussed in detail in Section 4.2.1. Its Hamiltonian is given by

$$\mathscr{H}(t) = \frac{L^2}{2I} + k \cos\theta \sum_n \delta(t - nT). \tag{1.27}$$

The first term describes the rotation of a pendulum with angular momentum L, and a moment of inertia I. The second term describes periodic kicks with a period T by a pulsed gravitational potential of strength $k = mgh$. (This example convincingly demonstrates the superiority of theory over experiment. It is not so easy to realize such a situation experimentally.) The kicked rotator belongs to the Hamiltonian systems. The equations of motion are obtained from the canonical equations

$$\dot{\theta} = \frac{\partial \mathscr{H}}{\partial L}, \qquad \dot{L} = -\frac{\partial \mathscr{H}}{\partial \theta}, \tag{1.28}$$

whence follows, with the Hamiltonian (1.27)

$$\dot{\theta} = L, \qquad \dot{L} = k \sin\theta \sum_n \delta(t - n), \tag{1.29}$$

where now I and T have been normalized to one (readers not too familiar with classical mechanics will find a short recapitulation in Section 7.2.2).

As L changes discontinuously, the equations of motion define a map for the dynamical variables θ, L. Let θ_n and L_n be the values of the variables just before the $(n+1)$th kick. Immediately after the kick L_n takes the value $L_n + k \sin\theta_n$ whereas θ_n is not changed. Between the kicks L remains constant, and θ increases linearly with t. Just before the next kick the dynamical variables take the values

$$\theta_{n+1} = \theta_n + L_{n+1} \tag{1.30}$$

$$L_{n+1} = L_n + k \sin\theta_n. \tag{1.31}$$

This is Chirikov's *standard map* [Chi79]. It is especially well suited to studying the transition of a Hamiltonian system from regular to chaotic behaviour when changing one control parameter. Figure 1.3 shows Poincaré sections of the map

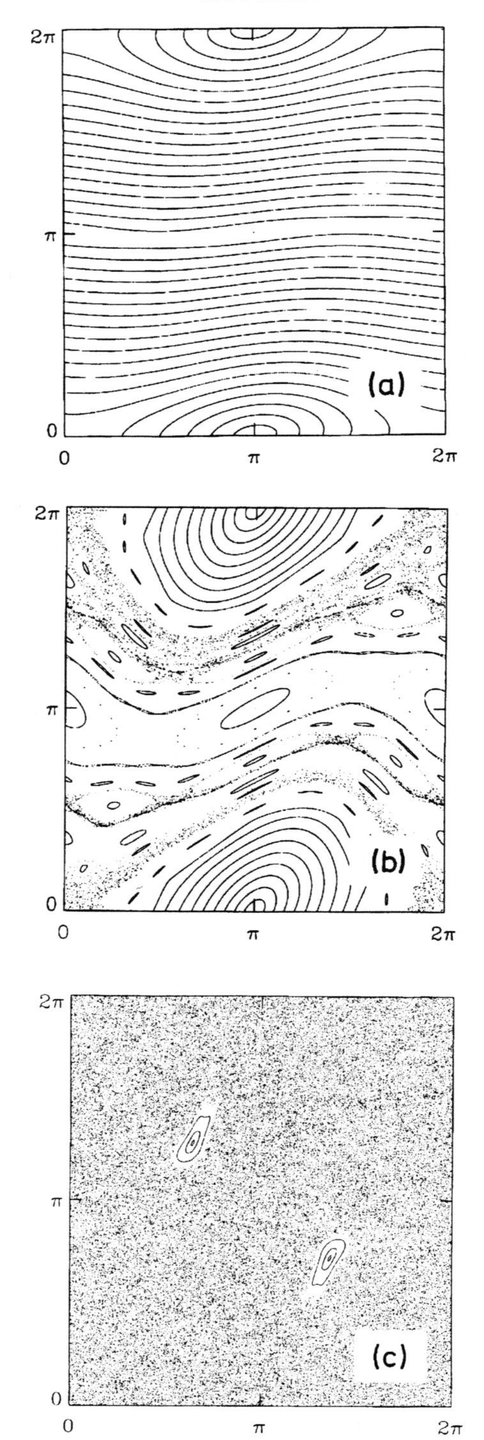

2π
π
0
0
π
2π
(a)
2π
π
0
0
π
2π
(b)
2π
π
0
0
π
2π
(c)

for different values of k. Each point corresponds to a pair (θ, L) immediately after a kick. For small k values of about 0.2 the rotator behaves regularly for most of the initial values of θ and L. The motion in phase space proceeds on the so-called invariant tori, where L varies only little, while θ takes all values between 0 and 2π with about equal probability. With increasing k more and more tori break up, until above a critical value of $k = k_c = 0.9716\ldots$ the last invariant torus will have been destroyed. There are, however, still regular regions in the phase space. For $k \geqslant 5$ finally, the major part of the phase space has become chaotic.

At the critical value $k = k_c$ the motion in phase space becomes diffusive. If the initial conditions are chosen within the chaotic sea, the rotator will eventually explore the complete accessible phase space. The still existing tori represent inpenetrable diffusion barriers which will never be passed. However, the chains of cantori, found in Fig. 1.3(b) for L values in the ranges of $1 \leqslant L \leqslant 2.5$ and $3.5 \leqslant L \leqslant 5$, also act as dynamical barriers. The cantori are the remnants of destroyed tori. If the coupling strength exceeds a critical value, a torus dissolves into a chain of subtori, which, after a further increase of k, breaks up into sub-subtori, and so on. A detailed description of this scenario can be found in Chapter 7 of Ref. [Ott93]. If the rotator is trapped in the fractal structure of the cantori, it may last for a very long time until an eventual escape.

Since the Hamiltonian (1.27) is time-dependent, it is not possible to separate the time dependence from the Schrödinger equation, with the consequence that the energy is no longer conserved. In Chapter 4 we shall discuss techniques to treat systems with periodic time dependences. In kicked systems the time evolution of the quantum mechanical state vector is obtained from

$$\psi_n = F^n \psi_0, \tag{1.32}$$

where ψ_0, ψ_n are the state vectors at the beginning and after the nth kick, respectively. F is called the Floquet operator. For the kicked rotator it is given by

$$F = \exp\left(-\frac{\imath}{\hbar} k \cos\theta\right) \exp\left(\frac{\imath T}{\hbar} \frac{L^2}{2I}\right), \tag{1.33}$$

Figure 1.3. Poincaré map of the standard map for different kick strengths. Angle θ is plotted versus angular momentum L. For $k = 0.2$ (a) the motion is on invariant tori for most initial conditions, for $k = 0.9716\ldots$ (b) the last invariant torus is just destroyed, and unlimited diffusion becomes possible, for $k = 5$ (c) the phase space has become chaotic, apart from some surviving stable islands.

where now L has to be interpreted as the quantum mechanical operator $-\imath\hbar\partial/\partial\theta$. The matrix elements of F are easily determined in the basis of the eigenfunctions of L. The calculation of the time evolution thus reduces to a straightforward matrix multiplication.

In this way Geisel and coworkers [Gei86] calculated the quantum mechanical evolution of the angular momentum distribution exactly at the critical border $k = k_c$. They started with an angular momentum eigenstate $L = 3.2\hbar$. Figure 1.4 shows the asymptotic angular momentum distribution. We do not find a more or less Gaussian distribution as we might have expected for a diffusive process, but three sharp borders both on the low and the high angular momentum side. The first is found at the L values corresponding to the last surviving torus which breaks up just at k_c (see Fig. 1.3(b)). Although in quantum mechanics impenetrable barriers no longer exist because of the uncertainty relation, the influence of the classical border is nevertheless seen in the quantum mechanical angular momentum distribution, too. Perhaps the next two borders are even more surprising. They can be attributed to two successive

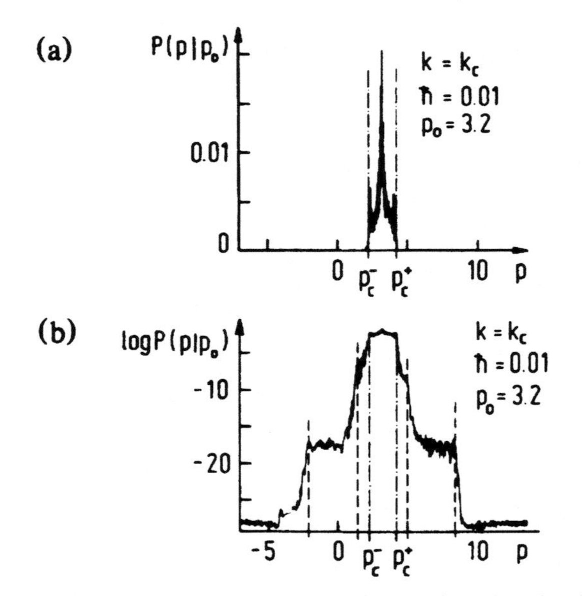

Figure 1.4. Asymptotic angular momentum distribution for the kicked rotator with $k = k_c = 0.9716\ldots$ on a linear (a) and on a logarithmic (b) scale. The initial value for the angular momentum was $L = 3.2\hbar$. The Planck constant is given in units of I/T, where I is the moment of inertia, and T is the period. Three pairs of borders are seen. The inner one, denoted by p_c^+, p_c^-, and marked by dashed-dotted lines, can be associated with the last surviving invariant torus, the two outer ones, marked by dashed lines, are caused by chains of cantori, see Fig. 1.3(b) [Gei86] (Copyright 1986 by the American Physical Society).

chains of cantori. In quantum mechanics their role as barriers is even more impressive than in classical mechanics.

Both examples have shown that there is a strong connection between classical dynamics and wave mechanics or quantum mechanics, respectively. To explore this connection, especially to look for the traces of classical chaos in the corresponding quantum mechanical system, is the main concern of this monograph.

2
Billiard experiments

Until about 1990 only a very small number of experiments on the quantum mechanics of chaotic systems existed, apart from the early studies of nuclear spectra [Por65]. In this context the experiments on hydrogen atoms in strong microwave fields by Bayfield and Koch [Bay74], and in strong magnetic fields by Welge and his group [Hol86, Mai86] have to be mentioned in particular. The studies of irregularly shaped microwave cavities by Stöckmann and Stein [Stö90] have introduced a new type of quantum chaos research. The microwave billiards and a number of variants to be discussed in this chapter are analogue systems, as they use the equivalence of the Helmholtz equation and the time independent Schrödinger equation. Whether there is a complete correspondence with quantum mechanics or not depends on the respective boundary conditions. As most of the phenomena discussed in the following are common to all types of waves, this does not reduce the conclusiveness of the analogue experiments.

Starting with a historical review, the state of the art in billiard experiments is presented with emphasis on a general survey and the technical background. The results and their quantum mechanical implications will be presented later. The hydrogen experiments, too, will be described in the proper context. The discussion of mesoscopic systems is restricted to billiard-like structures such as antidot lattices [Wei91], quantum dots [Mar92], and tunnelling devices [Fro94]. A more profound discussion of mesoscopic systems would go far beyond the intended scope of this monograph, which is rather regrettable, since many concepts developed in quantum chaos can be directly applied to mesoscopic systems [Guh98]. The chapter ends with a presentation of the experiments by Eigler and his group [Cro93a] on quantum corrals which allow a direct visualization of electronic wave functions in billiard systems.

2.1 Wave propagation in solids and liquids

2.1.1 Chladni figures

At the end of the eighteenth century Ernst Florens Chladni (see Fig. 2.1) conceived an experiment which can be considered as a precursor of all the billiard set-ups to be described in the following. He noticed that dust, randomly distributed on glass or metal plates, arranges itself in characteristic figures, if the plates are set to vibrate by means of a violin bow. This possibility of 'making the sound visible' excited not only the leading scientists of that time such as Laplace, Poisson and others, but also the scientifically interested public. In 1809, when Chladni stayed in Paris for a longer period, he was invited to give a demonstration of his experiments in the presence of Napoleon and his court at the Tuileries. Chladni himself gave a vivid report of this visit reprinted in a booklet on Chladni's life by Melde [Mel88]. As it is probably inaccessible to most readers, a short passage shall be cited. After describing his arrival at the Tuileries, Chladni continues:

When I entered, he welcomed me, standing in the centre of the room, with the expressions of his favour. Napoleon showed much interest in my experiments and

Figure 2.1. Ernst Florens Friedrich Chladni (1756–1827), inventor of the sound figures of vibrating plates bearing his name [Mel88].

explanations and asked me, as an expert in mathematical questions, to explain all topics thoroughly, so that I could not take the matter too easy. He was well informed that one is not yet able to apply a calculation to irregularly shaped areas, and that, if one were successful in this respect, it could be useful for applications to other subjects as well. (Abridged translation by the author from the German original.[1])

The emperor was impressed by the performance, it seems. The next day Chladni received a gratuity of 6000 francs, and a prize of 3000 francs was granted for the correct mathematical foundation of the Chladni figures. The prize was taken by Sophie Germain in 1816 (she belonged to the fairly large number of women who had extreme difficulties in making their way in a scientific community dominated by men; in her letters to Gauss she took a male pseudonym as she feared not to be taken seriously as a woman [Dal91]). Her solution was not yet complete, however. The correct explanation for circularly shaped plates was not found until 1850 by Kirchhoff, and in a handbook of physics of 1891 [Mel91] we still find the statement that, apart from a small number of simply shaped plates, the problem of the Chladni figures is still unsolved. In the same article a very irregularly shaped plate is depicted with the remark that any plate of whatever shape shows a characteristic sequence of Chladni figures.

The irregularly shaped plates especially renewed the interest in these nearly 200 year old experiments for the following reason: the vibrations of stiff plates are described by the equation (see §25 of Ref. [Lan59])

$$(\Delta)^2 \psi_n = k_n^4 \psi_n, \tag{2.1.1}$$

where

$$\Delta = \frac{\partial^2}{\partial x^2} + \frac{\partial^2}{\partial y^2} \tag{2.1.2}$$

is the two-dimensional Laplace operator, $\psi_n(x, y)$ is the amplitude function of the nth resonance, and k_n the associated wave number. Note that in contrast to the vibrations of membranes without internal stiffness it is the square of the Laplace operator which enters Eq. (2.1.1). Thus the solutions belong to two classes obeying the equations

$$\Delta \psi_n = -k_n^2 \psi_n, \tag{2.1.3}$$

[1] 'Als ich eintrat, empfing er mich, in der Mitte des Zimmers stehend, mit Aeusserungen des Wohlwollens [...]. Napoléon bezeigte meinen Experimenten und Erklärungen viele Aufmerksamkeiten und verlangte, als Kenner mathematischer Gegenstaende, dass ich ihm Alles recht von Grund aus erklären sollte, so dass ich also die Sache nicht eben von der leichten Seite nehmen durfte. Er wusste auch recht wohl, dass man noch nicht im Stande ist, Flächen, die nach mehr als einer Richtung auf verschiedene Art gekrümmt sind,[...] dem Calcul zu unterwerfen [...], dass wenn man hierin weitere Fortschritte machen könnte, es auch zur Anwendung auf manche andere Gegenstaende nützlich sein würde [...].'

and

$$\Delta\psi_n = k_n^2 \psi_n, \tag{2.1.4}$$

respectively. Equation (2.1.3) is the ordinary Helmholtz equation. If the plates are clamped along the outer rim (which, however, is usually not the case), the problem is completely equivalent to the quantum mechanics of a particle in a box with infinitely high walls. Such an analogy does not exist for Eq. (2.1.4). For regularly shaped plates the corresponding classical dynamics is integrable, i.e. the number of degrees of freedom equals the number of constants of motion. For circular plates these are the total energy E and the angular momentum L. For rectangular plates the squares of p_x and p_y, the components of the momentum parallel to the sides, are conserved. For irregularly shaped plates the only constant of motion is the total energy E, and the corresponding classical motion is nonintegrable, i.e. chaotic. The interpretation of Chladni figures of irregularly shaped plates is thus intimately connected with the quantum mechanics of chaotic billiards.

Chladni figures are not appropriate for quantitative measurements because of the rather strong damping of the resonances. They are, however, very well suited for didactic purposes as they are easy to perform and inexpensive. Figure 2.2 shows Chladni figures for differently shaped glass plates [Stö95a]. The plates have been fixed in the centre, placed on an overhead projector and set to vibrate by means of a loudspeaker. Typical vibration frequencies were in the range from 300 to 2000 Hz. As the outer rim is not fixed, Neumann boundary conditions hold. Thus there is no direct equivalence with quantum mechanics. Figure 2.2(a) shows one of the nodal patterns for a circular plate observed by Chladni himself. We find a regular net of intersecting circles and straight lines typical for integrable systems. The central mounting does not disturb the integrability since the rotational invariance is not broken. The situation is different for rectangular plates (see Fig. 2.2(b)). Here the mounting reduces the symmetry, and the billiard becomes pseudointegrable (such systems will be discussed in Section 3.2.2). For low frequencies the mounting has little influence, for higher frequencies it induces avoidance of some of the crossings, as is shown by the figure. Figure 2.2(c) finally displays a Chladni figure for a nonintegrable plate in the shape of a quarter Sinai billiard. A Sinai billiard is formed by a rectangle or square with a central circle. For these systems the Russian mathematician Sinai has been able to prove the ergodicity of the classical motion, one of the rare cases where such a proof has succeeded [Sin63]. Figure 2.2(c) shows a pattern of meandering lines as observed for the first time by McDonald and Kaufman [McD79] in calculations for the stadium billiard (see Section 6.2.1).

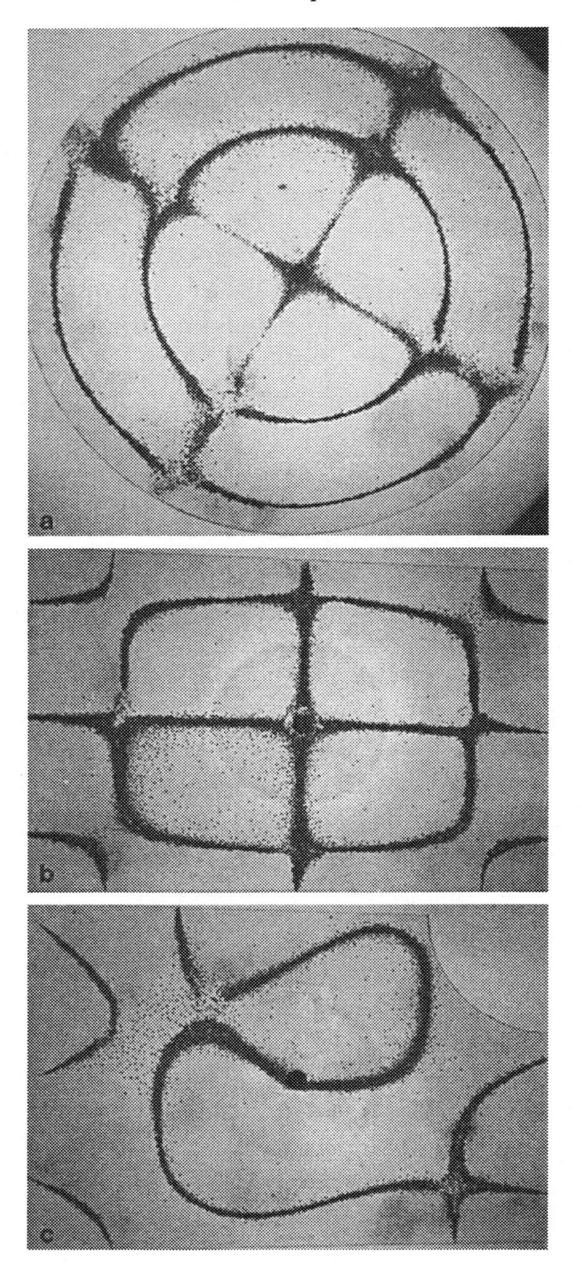

Figure 2.2. Chladni figures of glass plates of circular ($r = 10.5$ cm), rectangular ($a = 21$ cm, $b = 14$ cm), and quartered Sinai billiard ($a = 21$ cm, $b = 14$ cm, $r = 5$ cm) shapes. All plates are fixed in the centre and have been excited to vibrate by a loudspeaker, which has been removed for the photograph. The plates correspond to integrable (a), pseudointegrable (b), and nonintegrable (c) billiards [Stö95a].

Although the quality of these measurements is rather poor, they nevertheless are very well suited to illustrate the difference between integrable and nonintegrable systems to a nonexpert audience. A modern variant of the technique yielding high quality Chladni figures will be presented in Section 2.1.3. Up to the present day Chladni figures have been used routinely to study the vibrations of the resonance boards of pianos and string instruments [Fle91]. In most cases, however, the dust is now replaced by holographic illumination. Chladni figures are used in particular for the fine-tuning of high quality violins [Hut81].

2.1.2 Water surface waves

Another classroom demonstration of wave chaos in chaotic billiards can be performed with water surface waves in suitably formed vessels. Lindelof *et al.* [Lin86] have studied the propagation of water waves in different arrangements of circular scatterers to demonstrate localization by disorder. Their set-up is very simple. A point source of light was reflected from the water surface and projected to the ceiling. The whole water bath could be set to vibrate within the frequency range of 10 to 100 Hz. For a regular arrangement of scattering bodies we expect allowed and forbidden transmission bands corresponding to the Bloch states of solid state physics. Indeed, in a square array of scattering bodies the water surface is more or less homogeneously excited at the appropriate frequencies (see Fig. 2.3(a)), whereas for other frequencies no transmission is found. In an irregular arrangement on the other hand, only localized excitations are observed (see Figs. 2.3(b) and (c)) thus nicely demonstrating the effect of localization by destructive interference. A completely analogous experiment with microwaves will be discussed in Section 2.2.2.

With a similar arrangement Blümel *et al.* have studied standing waves on the surface of circular and stadium-shaped water vessels [Blü92], which have been placed on an overhead projector and shaken as a whole with frequencies up to somewhat above 1000 Hz. For water surface waves the velocity υ depends on the wavelength λ via (see Section 15.52 of Ref. [Mil68])

$$v^2 = \left(\frac{g\lambda}{2\pi} + \frac{2\pi\sigma}{\rho\lambda}\right)\tanh\left(\frac{2\pi h}{\lambda}\right), \qquad (2.1.5)$$

where $g = 9.81\ \mathrm{m\,s^{-2}}$ is the acceleration of gravity, h the depth of the vessel, $\rho = 10^3\ \mathrm{kg\,m^{-3}}$ the density of water, and $\sigma = 0.074\ \mathrm{N\,m^{-1}}$ its surface tension. Using $k = 2\pi/\lambda$ and $v = \omega/k$ this can be expressed in terms of a dispersion relation as

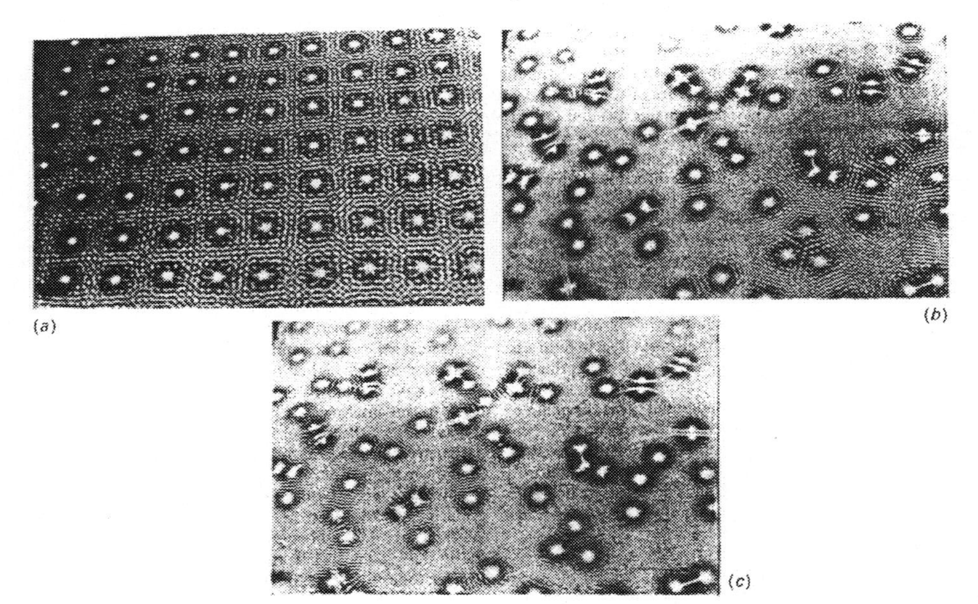

Figure 2.3. Propagation of water waves in different arrangements of circular scatterers. In a regular grid of scatterers the waves can propagate if the frequency is chosen appropriately (a). In an irregular arrangement the waves are localized due to destructive interference (b, c) [Lin86].

$$\omega^2 = \left(gk + \frac{\delta}{\sigma} k^3 \right) \tanh kh. \tag{2.1.6}$$

For wavelengths $\lambda \leqslant 1.7$ cm, used in the experiments of Blümel *et al.* [Blü92], the second term on the right hand side of the equation dominates. This is the range of the so-called capillary waves.

The propagation of waves in liquids is governed by the nonlinear Navier–Stokes equation. But as long as the deviation $f(x, y)$ of the surface from the equilibrium level is small, the equations can be linearized and lead again to a Helmholtz equation

$$(\Delta + k^2) f = 0 \tag{2.1.7}$$

with a Neumann boundary condition for $f(x, y)$. For the circular vessel highly symmetric standing wave patterns are observed, two of which are shown in Fig. 2.4. The situation changes drastically for the stadium-shaped vessel. Figure 2.5 shows three typical situations. The pattern in Fig. 2.5(a) corresponds to a standing wave between the long sides of the stadium. Its classical counterpart is a particle reflected to and fro between the two straight sides of the stadium. This periodic motion has been termed suggestively ‘bouncing ball’ and will

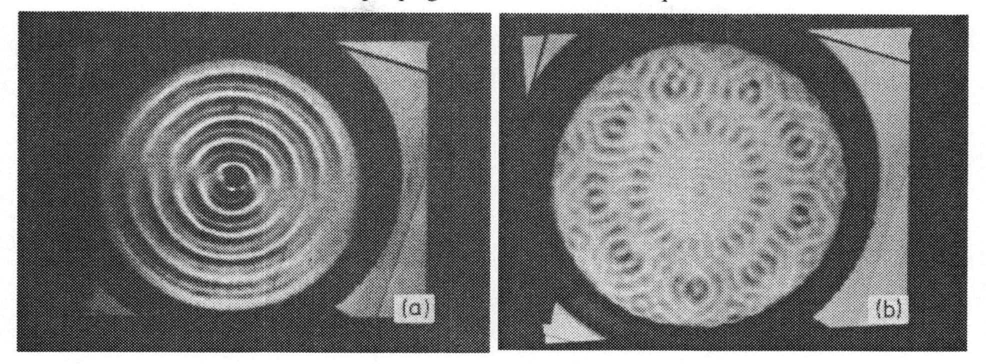

Figure 2.4. Water surface waves on a circular water vessel ($r = 5.3$ cm) placed on an overhead projector and set to vibrate by an electromechanical shaker [Blü92] (Copyright 1992 by the American Physical Society). R. Blümel is thanked for kind permission to reprint a number of unpublished photographs.

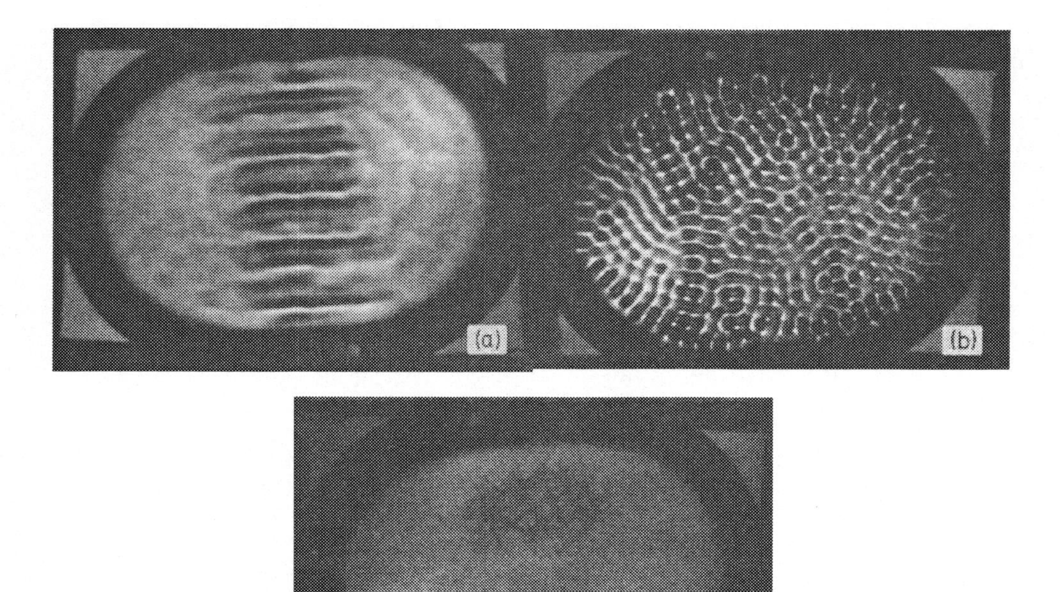

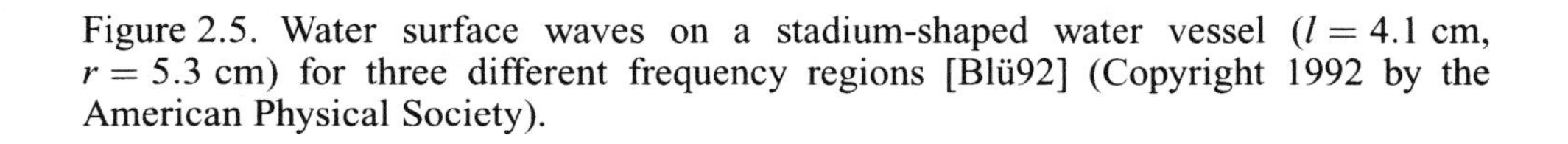

Figure 2.5. Water surface waves on a stadium-shaped water vessel ($l = 4.1$ cm, $r = 5.3$ cm) for three different frequency regions [Blü92] (Copyright 1992 by the American Physical Society).

accompany us throughout this book. The regions near the two endcaps of the stadium remain nearly uninfluenced. This is a first demonstration of the so-called 'scarring' phenomenon discovered by Heller [Hel84]. Eigenfunctions in chaotic systems often show a tendency to large amplitudes close to classical periodic orbits and small amplitudes in the regions in between. Intuitively we may look on scars as standing waves along periodic orbits. This phenomenon will be discussed in detail in Section 8.1.3. At higher frequencies speckle patterns are observed (see Fig. 2.5(b)) similar to those found with laser light. This is not just a coincidence, since such patterns build up if waves of the same frequency but with different directions are superimposed randomly. At very high frequencies above 700 Hz odd scratch-like structures are found (Fig. 2.5(c)). Although they no longer have the appearance of standing waves, they nevertheless are. A superposition of a very large number of waves with random directions yields exactly such structures [O'Con87]. The detailed discussion of these findings is postponed to Section 6.2.1.

For $\lambda \geqslant 1.7$ cm the first term on the right hand side of Eq. (2.1.6) dominates. This is the range of the gravity waves. Moreover, if the wavelengths are large compared to the depth of the vessel, the equation reduces to

$$\omega^2 = ghk^2, \tag{2.1.8}$$

or

$$v = \sqrt{gh}. \tag{2.1.9}$$

In this limiting case the wave propagation is nondispersive. By an appropriate modelling of the vessel depth profile it is possible to simulate billiards with position-dependent propagation velocities, or, expressed in other words, of billiards on a non-Euclidean metric! The range of validity of Eq. (2.1.9) is unfortunately rather limited. To obtain a precision of 10 per cent the frequency must be below 5 Hz and the depth must be smaller than 0.5 cm. Nevertheless it has been recently demonstrated in the author's group [Sch96] that this technique can actually reproduce wave propagation on a metric with constant negative curvature. Details will be given in Section 8.4.1.

2.1.3 Vibrating blocks

A number of new aspects are involved in the case of sound waves in solids. They are again described by the Helmholtz equation, but now three differently polarized modes exist, two transverse and one longitudinal mode. A further complication arises from the fact that these modes are coupled by the reflection at the surface of the solid. In general a purely longitudinally or transversely polarized wave is split into two differently polarized waves after the reflection.

Thus we are rather far from the corresponding quantum mechanical system, and we may ask whether the techniques to be discussed in the following chapters, especially random matrix and periodic orbit theory, can be applied to these systems at all.

The billiard sound experiments were started by Weaver [Wea89] and continued by Ellegaard and his group [Ell95]. The technique applied is surprisingly simple. In the original experiments a brick-shaped aluminum block with linear dimensions in the cm region was set to ultrasound vibrations by dropping a small steel ball onto the block. The block was supported by a foam pad, and an ultrasound transducer was glued to one side to register the vibrations. Both the pad and the transducer proved to have a negligible effect on the experiment. The resonance spectrum of the block was obtained from a Fourier analysis of the transducer output. Thus a spectral range between 0 and 250 kHz could be registered. The insert of Fig. 2.6 shows part of such a spectrum. The quality factor of the system, also called Q value, can be determined from the widths of the resonances. It is defined as $Q = \nu/\Delta\nu$ where

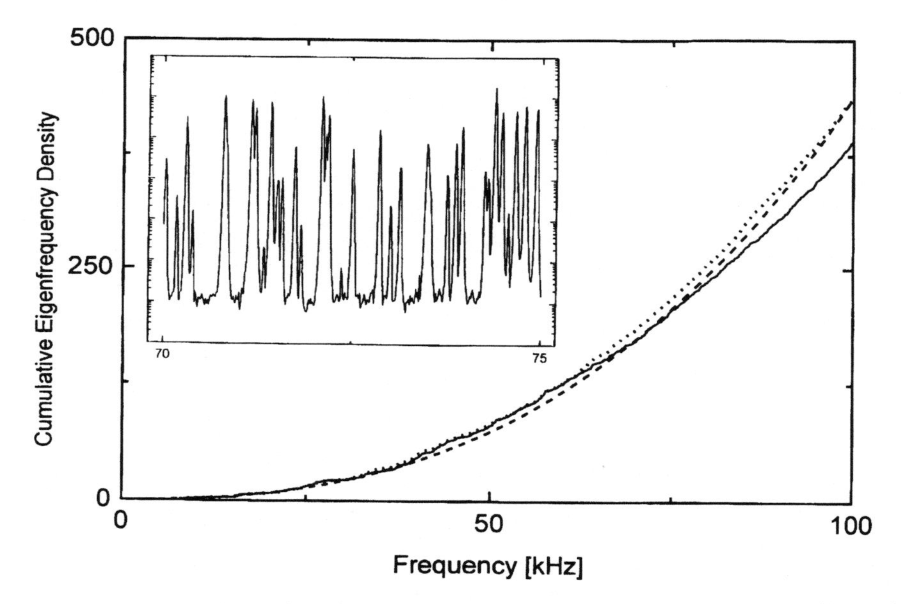

Figure 2.6. Integrated density of states for a rectangular Al block with the dimensions $a = 60.6$ mm, $b = 98.0$ mm, $c = 158.6$ mm (solid line) and a Sinai billiard of the same dimensions, obtained by cutting the octant of a sphere with $r = 20$ mm out of one corner (dotted line). The dashed line corresponds to the theoretical expectation (see Eq. (2.1.12)). In the insert part of the spectrum of the rectangular block is shown on a linear-logarithmic plot [Ell95] (Copyright 1995 by the American Physical Society).

ν is the frequency of the resonance, and $\Delta\nu$ its width. Q values between 5000 and 10 000 have been obtained from the measurements shown in Fig. 2.6. This is the same order of magnitude as found in normally conducting microwave cavities (see Section 2.2.1).

In a modified set-up Ellegaard and coworkers used quartz blocks supported by three sapphire needles such as those used for record-players [Ell96]. Two of the needles were coupled to piezo-electric transducers to excite and record the vibrations. For further reduction of damping the measurements were performed at a reduced pressure of 10^{-2} Torr. Due to these improvements even quality factors of the order of 10^5 have been obtained, close to the values found in superconducting microwave cavities.

In all analogue experiments the figure of merit is the fraction of registered resonances. To determine this quantity, usually the integrated density of states $N(\nu)$, i.e. the total number of resonances up to a given frequency ν, is considered. For three-dimensional billiards the number of resonances up to a given frequency increases asymptotically according to [Bal70, Bal71, Bal76]

$$N_{\rm av}(\nu) = aV\nu^3 + bS\nu^2, \tag{2.1.10}$$

where V and S are volume and surface area of the resonator, respectively, and a and b are constants depending on the type of the waves and the boundary conditions (see Section 7.3.1). In two-dimensional systems the corresponding expression reads

$$N_{\rm av}(\nu) = aA\nu^2 + bL\nu, \tag{2.1.11}$$

where now A and L are area and circumference of the billiard, respectively. For sound waves with periodic boundary conditions it follows in particular [Ell95]

$$\begin{aligned} N_{\rm av}(\nu) = & \frac{4\pi V}{3}\left(\frac{2}{c_T^3} + \frac{1}{c_L^3}\right)\nu^3 \\ & + \frac{\pi S}{4}\frac{2 - 3(c_L/c_T)^2 + 3(c_L/c_T)^4}{c_L^2[(c_L/c_T)^2 - 1]}\nu^2, \end{aligned} \tag{2.1.12}$$

where c_T and c_L are the sound velocities of the transverse and the longitudinal waves, respectively. In the high frequency range Eqs. (2.1.10) to (2.1.12) reduce to the *Weyl formula*: for $\nu \to \infty$ the integrated density of states becomes proportional to $V_d\nu^d$, where d is the dimension of the system, and V_d its d-dimensional volume (see Section 7.3.1).

Figure 2.6 shows the integrated density of states for a rectangular block and a Sinai shaped block (a rectangular block where the octant of a sphere has been removed from one of the corners) together with the theoretical curve obtained from Eq. (2.1.12). Though the boundary conditions are obviously not periodic in the present case, the results for the Sinai block nevertheless coincide closely with the theoretical curve. This shows that first the boundary conditions are apparently not really important here, and second no resonances are missed. To come to this result three measurements with different impact positions of the steel ball have to be combined. This is mandatory since a vibration cannot be excited if the impact is incidentally close to a node line. This probability is reduced by combining different measurements. We shall see that similar techniques are also applied in microwave billiards. Nevertheless, for the rectangular resonator a missing fraction of about 15% remains, which is caused by the fact that in integrable systems there is a tendency to resonance clustering. Consequently adjacent resonances can no longer be separated if their distance becomes smaller than the resonance width.

Recently the technique was extended to the spatial mapping of sound figures of vibrating plates [Sch97]. The experiments were performed with aluminum plates of a thickness of 3 mm and the shape of a quarter Sinai-stadium billiard. This is a billiard with an outer boundary in the shape of a stadium and an additional circular boundary in the centre. The vibrations were excited by a transducer from underneath, the receiver was placed on top at the end of a pick-up arm, the position of which was scanned in the x and the y directions, thus registering the vibration amplitude as a function of position. Figure 2.7 shows the sound figures for two eigenfrequencies obtained in this way. On the right the probability distributions for the squared amplitudes are shown. The solid lines correspond to the so-called Porter–Thomas distribution. We shall come back to this in Section 6.2.2.

The measurements performed so far were aimed at testing random matrix theory. In one accompanying theoretical study the spectra have been analysed using periodic orbit theory [Boh91]. In Chapter 3 these results will be discussed in detail. However, it can already be stated that deviations from the theoretical expectations have not been found anywhere. Astonishing or not, the techniques originally developed for the quantum mechanics of chaotic systems also work perfectly well in cases where their applicability could be doubted at first sight.

2.1.4 Ultrasonic fields in water-filled cavities

As we have already seen, the coexistence of sound waves with transverse and longitudinal polarization in solids leads to considerable complications. These are avoided in fluids where only longitudinal waves can exist. Fluids have the further advantage that standing wave patterns can be made visible by a special illumination technique. Figure 2.8 shows a sketch of the set-up used by

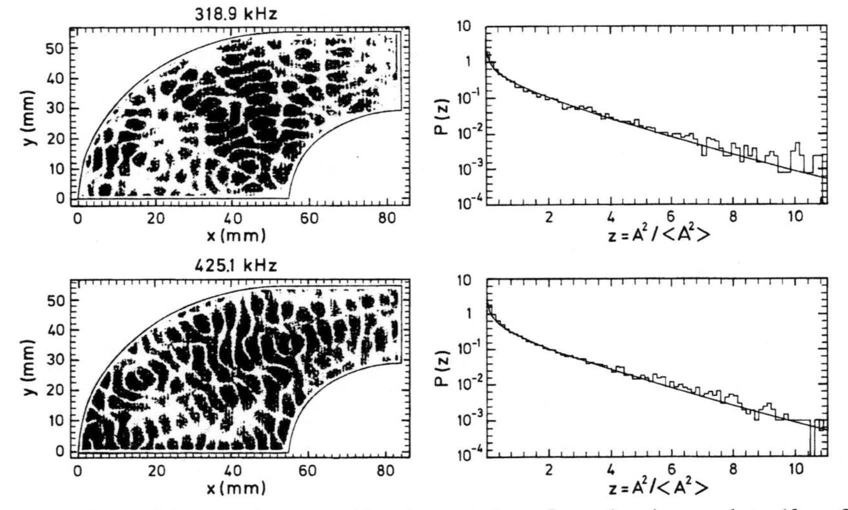

Figure 2.7. Sound figures for two vibration modes of an aluminum plate ($h = 3$ mm) in the shape of a quarter Sinai-stadium billiard (left), and probability distributions for the squared vibration amplitudes (right). The solid lines correspond to Porter–Thomas distributions [Sch97].

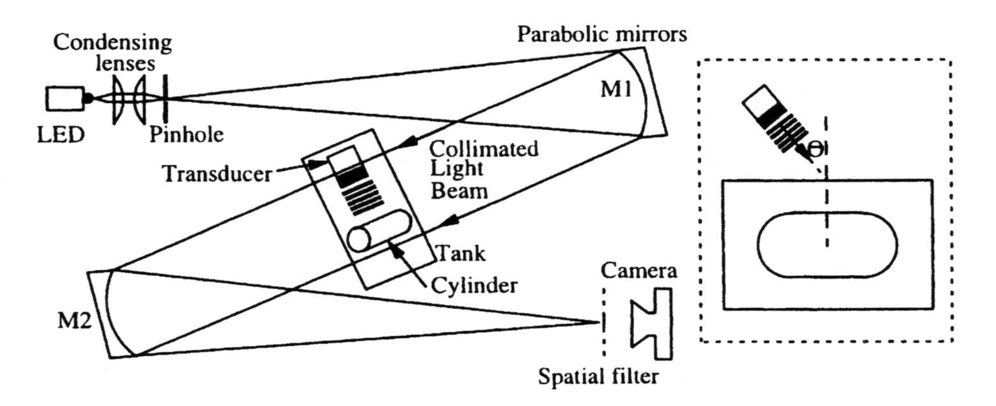

Figure 2.8. Illumination system for the visualization of the sound pressure field in a cylinder of circular or stadium shape filled with water. The sound is produced by an ultrasonic transducer, and could be transmitted from different directions into the cylinder to allow the selective excitation of different eigenmodes [Chi97].

Chinnery *et al.* [Chi97]. A hollow cylinder with a circular or stadium-shaped cross-section is suspended within a tank filled with water. Sound waves are excited by means of an ultrasonic transducer with frequencies up to some MHz. The pressure field p within the fluid column again satisfies the Helmholtz equation

$$(\Delta + k^2)p = 0 \tag{2.1.13}$$

with a Neumann boundary condition at the cylinder walls. Light is passed through the cylinder parallel to its axis. The schlieren technique is used to make the sound field visible. From a high power light emitting diode light is focussed on a pinhole or, alternatively, on a random array of pinholes (see Fig. 2.8). Reflected by a parabolic mirror the collimated beam enters the tank through a window and travels through the cylinder parallel to the axis. Here part of the light is diffracted by the density variations produced by the sound field. After leaving the tank the light is focussed again by a second parabolic mirror. In its focal plane the zero order light is suppressed by an appropriately formed mask leaving only the diffracted components. According to the authors the optical distribution approximates the square of the pressure distribution for low acoustic pressures, thus mapping the field distribution.

In contrast to the sound experiments discussed in the last section and to the microwave experiments presented in the following, the technique has the advantage that the whole pattern is obtained immediately without any scanning. Due to a rather low quality of $Q \approx 2000$ only some hundred resonances in the low frequency range can be resolved. In the MHz region therefore a large number of overlapping resonances have always been excited simultaneously. Figure 2.9 shows a selection of eigenmode patterns obtained for circular and stadium-shaped cylinders [Chi97, Chi96]. Again eigenmodes corresponding to the bouncing ball and other scarred structures are found. The ridge structures we already know from the water surface waves (see Fig. 2.5) appear once more at very high frequencies.

2.2 Microwave billiards

2.2.1 Basic principles

Since most of the billiard experiments have been performed with differently shaped microwave cavities, known as microwave billiards, these systems will be discussed in detail. The starting point is the Maxwell equations

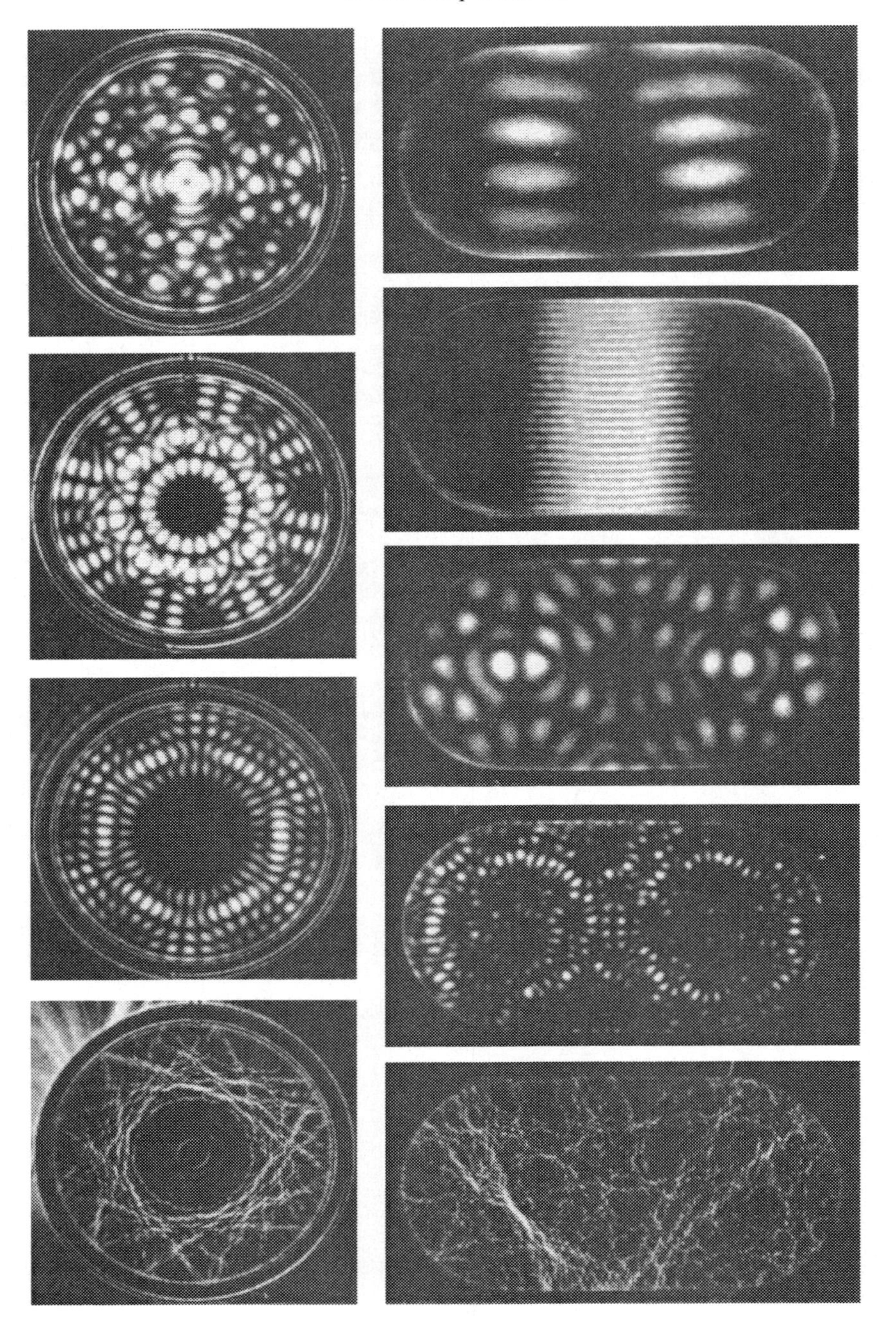

Figure 2.9. Standing sound waves in water-filled cylinders of circular ($r = 14.3$ mm) and stadium shapes. The eigenfrequencies in kHz are 526.2, 607.9, 663.8, 3500 (from top to bottom) for the circle [Chi97], and 150.9, 720.2, 261.1, 664.8, 2500 for the stadium ($l = 25.4$ mm, $r = 12.7$ mm) [Chi96].

$$\nabla \times \boldsymbol{E} = -\frac{\partial \boldsymbol{B}}{\partial t}, \tag{2.2.1}$$

$$\nabla \times \boldsymbol{H} = \frac{\partial \boldsymbol{D}}{\partial t}, \tag{2.2.2}$$

$$\nabla \boldsymbol{D} = 0, \tag{2.2.3}$$

$$\nabla \boldsymbol{B} = 0, \tag{2.2.4}$$

where, in a vacuum, displacement $\boldsymbol{D}$ and induction $\boldsymbol{B}$ are related to the electric and magnetic fields $\boldsymbol{E}$ and $\boldsymbol{H}$, respectively, by

$$\boldsymbol{D} = \epsilon_0 \boldsymbol{E}, \tag{2.2.5}$$

$$\boldsymbol{B} = \mu_0 \boldsymbol{H}. \tag{2.2.6}$$

Here ϵ_0 and μ_0 are dielectric constant and permeability of the vacuum, respectively. Assuming periodic time dependences for the electromagnetic fields, we obtain Helmholtz equations for $\boldsymbol{E}$ and $\boldsymbol{B}$ by means of a standard procedure which can be found in every textbook on electrodynamics (see Ref. [Jac62]):

$$(\Delta + k^2)\boldsymbol{E} = 0, \tag{2.2.7}$$

$$(\Delta + k^2)\boldsymbol{B} = 0, \tag{2.2.8}$$

where $k = \omega/c$ is the wavenumber and ω is the angular frequency. Additionally the boundary conditions

$$\boldsymbol{n} \times \boldsymbol{E} = 0, \qquad \boldsymbol{n}\boldsymbol{B} = 0, \tag{2.2.9}$$

have to be met where $\boldsymbol{n}$ is the unit vector normal to the surface. In most of the experiments resonators with a cylindrical geometry and different cross-sections have been used. In this case the equations are considerably simplified. Taking the z axis parallel to the axis of the cylinder, the boundary conditions (2.2.9) reduce to

$$E_z|_S = 0, \qquad \nabla_\perp B_z|_S = 0, \tag{2.2.10}$$

on the cylinder surface, where $\nabla_\perp$ denotes the normal derivative. For E_z this follows immediately from Eq. (2.2.9), for B_z it is a straightforward consequence of Eq. (2.2.9) and the second Maxwell equation. As the details of the further derivation can be found elsewhere, e.g. in Chapter 8 of Ref. [Jac62], only the results will be given in the following. There are two possibilities to obey the boundary conditions (2.2.10). For the so-called transverse magnetic (TM) modes we have

$$E_z(x, y, z) = E(x, y)\cos\left(\frac{n\pi z}{d}\right), \qquad n = 0, 1, 2, \ldots, \tag{2.2.11}$$

$$B_z(x, y, z) = 0, \tag{2.2.12}$$

where $E(x, y)$ obeys the two-dimensional Helmholtz equation

$$\left[\Delta + k^2 - \left(\frac{n\pi}{d}\right)^2\right] E = 0 \tag{2.2.13}$$

with the Dirichlet boundary condition

$$E(x, y)|_S = 0 \tag{2.2.14}$$

on the cylinder surface. The x and y components of $\boldsymbol{E}$ and $\boldsymbol{B}$ can be calculated from $E(x, y)$ but are not relevant here. For the transverse electric (TE) modes we analogically obtain

$$E_z(x, y, z) = 0, \tag{2.2.15}$$

$$B_z(x, y, z) = B(x, y)\sin\left(\frac{n\pi z}{d}\right), \qquad n = 1, 2, 3, \ldots \tag{2.2.16}$$

where now $B(x, y)$ obeys the two-dimensional Helmholtz equation

$$\left[\Delta + k^2 - \left(\frac{n\pi}{d}\right)^2\right] B = 0, \tag{2.2.17}$$

but with the Neumann boundary condition

$$\nabla_\perp B(x, y)|_S = 0. \tag{2.2.18}$$

For frequencies $\nu < c/2d$ corresponding to wavenumbers $k < \pi/d$ only TM modes with $n = 0$ are possible, and Eq. (2.2.13) reduces to

$$(\Delta + k^2)E = 0. \tag{2.2.19}$$

In the following we shall denote billiards of this type as quasi-two-dimensional. There is a complete equivalence to the two-dimensional Schrödinger equation for a particle in a box with infinitely high walls, including the boundary conditions. In this analogy $E(x, y)$ corresponds to the wavefunction, and k^2 to the eigenenergy.

To measure the eigenmodes, microwaves are fed into the resonator by means of an antenna, which typically consists of a small wire with a diameter of some 0.1 mm introduced into the resonator through a small hole. In most cases the reflected microwave power is measured using a microwave bridge to separate the incoming and outcoming waves, but transmission measurements between two or more antennas are also possible.

Even in the fifties the equivalence between microwaves and sound waves was used to simulate room acoustics [Sch87]. The first microwave experiment treating quantum chaotic questions, however, was not performed until 1990 [Stö90]. The authors studied the eigenmodes of a resonator of quarter stadium shape by measuring the reflected microwave power as a function of frequency. Figure 2.10 shows one part of the reflection spectra. Each minimum in the reflected microwave power corresponds to an eigenmode. There are good reasons to study systems without symmetries. The complete stadium possesses two mirror reflection symmetries resulting in a superposition of four independent spectra belonging to different parity classes. This would unnecessarily complicate the analysis of the data in terms of random matrix theory, as will become clear in Section 3.1.1.

Confronted with a chaotic looking spectrum such as displayed in Fig. 2.10 we may wonder whether there is any relevant information in it. In this section only an idea will be given of what can be done. In the sixties similar problems had to be faced with the spectra of complex nuclei. It proved to be useful to plot the distribution function $P(s)$ of the spacings $s_n = E_n - E_{n-1}$ of adjacent eigenenergies [Por65]. Figure 2.11 shows the distributions for different microwave billiards. The mean spacing $\langle s \rangle$ has been normalized to one. Figure 2.11(a) shows $P(s)$ for a rectangular billiard [Haa91]. The corresponding classical dynamics is integrable. In this case a Poissonian nearest neighbour spacing distribution

$$P(s) = \exp(-s) \tag{2.2.20}$$

is expected, as we shall see later. The experiment follows the theoretical curve rather closely, but there are some significant deviations. The experimental distribution shows a pronounced hole at small s values, whereas a maximum

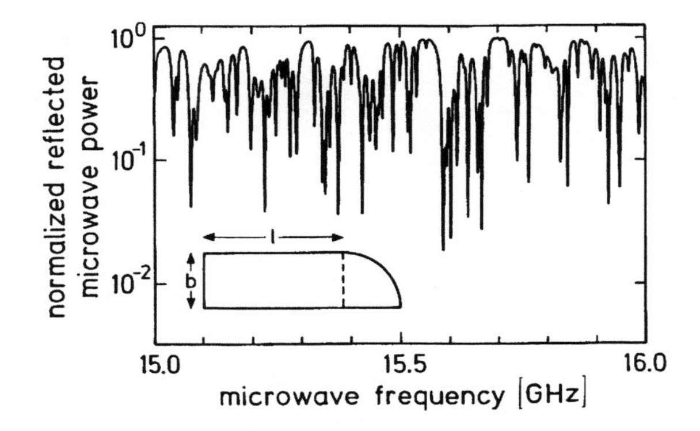

Figure 2.10. Part of a microwave reflection spectrum of a cavity in the shape of a quarter stadium billiard ($b = 20$ cm, $l = 36$ cm) with a height of $h = 0.8$ cm. Each minimum in the reflected microwave power corresponds to a resonator eigenfrequency [Stö90] (Copyright 1990 by the American Physical Society).

should be expected according to Eq. (2.2.20). The cause for this discrepancy is the finite width of the resonance curves. If two resonances are separated by a distance smaller than the experimental line width, they are registered as one. It should be remembered that this fact was equally responsible for the loss of eigenfrequencies observed in the sound experiments in rectangular blocks (see Fig. 2.6). Further deviations show up at larger distances. Here the influence of the coupling wire becomes manifest. A comparable situation has been found for the rectangular glass plate fixed in the centre (see Fig. 2.2). Due to the existence of the antenna the system becomes pseudointegrable. As with the glass plate the influence is small in the low frequency region but eventually becomes pronounced. Figure 2.11(b) shows $P(s)$ for the same rectangle as above but now for a higher frequency region. There is no longer any similarity with the Poisson distribution (2.2.20). The solid line has been obtained by a

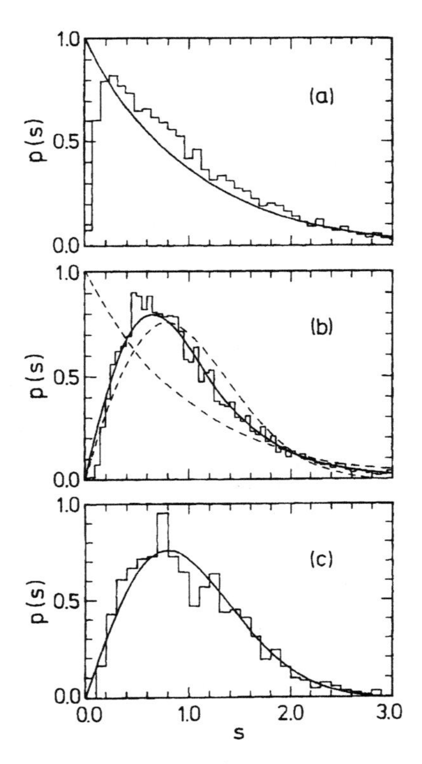

Figure 2.11. Nearest neighbour distance histograms collected from a number of rectangularly shaped microwave resonators with side lengths $a = 16.5 \ldots 51.0$ cm, $b = 20$ cm in the two frequency ranges 5 to 10 GHz (a) and 15 to 18 GHz (b) [Haa91], and for the quarter stadium billiard shown in Fig. 2.10 (c) [Stö90] (Copyright 1990+91 by the American Physical Society).

calculation method explicitly considering the influence of the antenna [Haa91, Šeb90]. An almost perfect correspondence with the experiment is found. Figure 2.11(c) finally shows $P(s)$ for the quarter stadium billiard the spectrum of which has been presented in Fig. 2.10. The solid line corresponds to the Wigner distribution

$$P(s) = \frac{\pi}{2} s \exp\left(-\frac{\pi}{4} s^2\right). \tag{2.2.21}$$

This distribution is omnipresent in the spectra of chaotic systems, and will accompany us throughout this monograph.

We shall now discuss the widths of the resonances. The boundary conditions (2.2.10) hold for ideally conducting walls. In reality, however, we always find a finite conductivity. As a result the microwaves partially penetrate into the walls. The penetration depth, called skin depth, is given by

$$\delta = \sqrt{\frac{2}{\mu_0 \omega \sigma}} \tag{2.2.22}$$

(see Chapter 7 of Ref. [Jac62]), where σ is the conductivity of the walls. For a good conductor like brass with $\sigma = 2 \times 10^7\ \Omega^{-1}\,\mathrm{m}^{-1}$ and a typical microwave frequency $\nu = \omega/2\pi$ of 10 GHz, typical skin depths are in the order of 1 μm.

The dissipation of the microwaves in the wall leads to an exponential damping of the electromagnetic energy in the resonator,

$$W(t) = W_0 \exp(-t/\tau), \tag{2.2.23}$$

and a corresponding damping of the electric fields,

$$\boldsymbol{E}(t) = \boldsymbol{E}_0 \exp(-t/2\tau)\cos(\omega_0 t). \tag{2.2.24}$$

An analogous relation holds for $\boldsymbol{B}$. The spectral distribution of the electromagnetic energy, the so-called power spectrum, is given by

$$S(\omega) = |\hat{E}(\omega)|^2, \tag{2.2.25}$$

where $\hat{E}(\omega)$ is the Fourier transform of $\boldsymbol{E}(t)$,

$$\hat{E}(\omega) = \frac{1}{\sqrt{2\pi}} \int_{-\infty}^{\infty} E(t) e^{\imath\omega t}\, dt. \tag{2.2.26}$$

The elementary integration yields

$$S(\omega) \sim \frac{1}{(\omega - \omega_0)^2 + \left(\frac{1}{2\tau}\right)^2}. \tag{2.2.27}$$

Thus the energy loss in the walls leads to a Lorentzian broadening of the resonance with a full width at half maximum of

$$\Delta\omega = \frac{1}{\tau}. \tag{2.2.28}$$

The resonator quality $Q = \omega_0/\Delta\omega$ introduced in Section 2.1.3 is related to the decay time τ via

$$\tau = \frac{Q}{\omega_0}. \tag{2.2.29}$$

Typical qualities of normally conducting cavities are in the range of 10^3 to 10^4. To calculate the quality, the boundary conditions have to be modified to take into account the finite conductivity of the wall. Thus the expression

$$Q = \alpha \frac{V}{S\delta} \tag{2.2.30}$$

is obtained for the quality, where V and S are volume and surface area of the resonator, respectively, and α is a factor of the order of one, depending on the resonator geometry. For details see Chapter 8 of Ref. [Jac62]. Up to a constant factor the quality is thus given by the energy stored in the resonator divided by the energy stored in the walls.

The quality limits the total number of eigenfrequencies which can be determined experimentally. For reasons of simplicity we assume a rectangular resonator with the side lengths a, b, c. The wavenumbers associated with the eigenfrequencies are then given by

$$k_{lmn} = \sqrt{\left(\frac{l\pi}{a}\right)^2 + \left(\frac{m\pi}{b}\right)^2 + \left(\frac{n\pi}{c}\right)^2} \qquad l,\ m,\ n = 0,\ 1,\ 2,\ \ldots. \tag{2.2.31}$$

Apart from cases in which one of the quantum numbers is zero, every resonance has to be counted twice, since there are transversal electric *and* magnetic modes. The number of resonances in the range $\langle k,\ k + dk\rangle$ is thus given by

$$\begin{aligned} \rho(k)\, dk &= 2 \sum_{k < k_{lmn} < k+dk} 1 \\ &= 2 \sum_{l,m,n} \int_k^{k+dk} \delta\left(k - \sqrt{\left(\frac{l\pi}{a}\right)^2 + \left(\frac{m\pi}{b}\right)^2 + \left(\frac{n\pi}{c}\right)^2}\right) dk. \end{aligned} \tag{2.2.32}$$

Replacing the sums by integrals we get asymptotically

$$\rho(k)\, dk = \frac{2}{\pi^2} V k^2 dk, \tag{2.2.33}$$

where $V = abc$ is the volume of the resonator. This is again the Weyl formula which was used by Planck in the derivation of his blackbody radiation formula. Though the result has been obtained for a rectangular resonator it holds asymptotically for resonators of arbitrary shapes as will be shown in Section 7.3.1. For the number of resonances below a given k we obtain from Eq. (2.2.33)

$$N(k) = \int_0^k \rho(k)\, dk$$
$$= \frac{2}{3\pi^2} V k^3. \qquad (2.2.34)$$

This corresponds to the leading term of Eq. (2.1.10). The next term which is proportional to the surface of the resonator has been calculated by Balian and Bloch [Bal71] for electromagnetic fields. The experimental limits of resolution are obtained at a value k_{max} where the mean distance between neighbouring resonances, i.e. the reciprocal density of states, equals the width Δk of a resonance,

$$\Delta k = \left(\frac{2}{\pi^2} V k_{max}^2 \right)^{-1}, \qquad (2.2.35)$$

whence follows for the quality, using $\omega = ck$,

$$Q = \frac{\omega_{max}}{\Delta\omega} = \frac{k_{max}}{\Delta k} = \frac{2}{\pi^2} V k_{max}^3. \qquad (2.2.36)$$

Using Eq. (2.2.34) this can be expressed as

$$N_{max} = \frac{1}{3} Q, \qquad (2.2.37)$$

where N_{max} is the maximum number of resolvable resonances. For qualities of some 1000 this number is of the order of 1000 in accordance with the experiments.

The experiments have been extended by Richter and his group [Grä92] to superconducting cavities where quality factors of 10^5 to even 10^7 were obtained, resulting in extraordinary sharp resonance lines. Figure 2.12 shows part of the spectrum of a stadium billiard. According to Eq. (2.2.37) millions of resonances should be obtainable, exceeding the best calculations by magnitudes. These measurements have not yet been realized, however. With a Q value of 10^6 and a maximum frequency of 40 GHz (the actually available limit), we learn from Eq. (2.2.36) that a volume of 10^4 cm^3 would be needed to have all resolvable resonances within the technically accessible range. Volumes of this size have not yet been studied although they are probably not out of reach. The extremely low damping not only allows one to register a spectrum with an unprecedented resolution, it is also possible to study the decay of the resonances to temporal depths unaccessible by other methods [Alt95]. The latter experiments will be

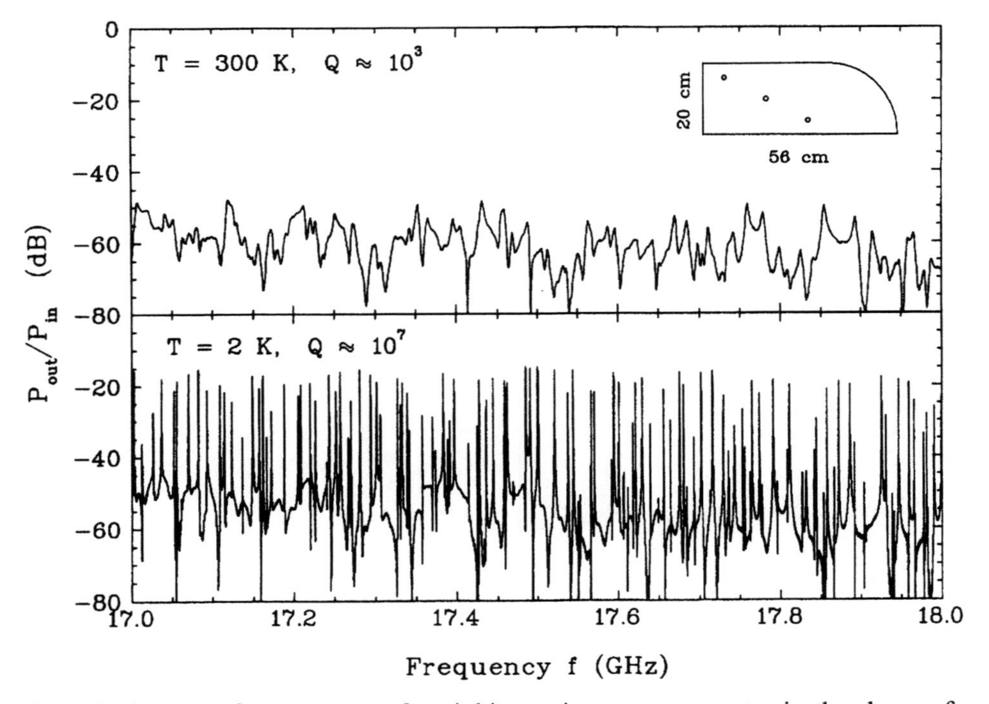

Figure 2.12. Part of a spectrum of a niobium microwave resonator in the shape of a quarter stadium billiard. The spectrum shown at the top was taken at room temperature, for the bottom part the resonator was in the superconducting state. In the insert the resonator and the positions of the three coupled antennas are shown [Grä92] (Copyright 1992 by the American Physical Society).

described in Section 6.3.1. A survey of the experiments with superconducting microwave cavities is given by Richter in Ref. [Ric98].

The number of microwave investigations of spectra of chaotic billiards has increased considerably in recent years. The questions addressed ranged from tests of random matrix and periodic orbit theory [Stö90, Grä92, Kud94] to scattering matrix approaches [Dor90, Alt95, Ste95] and spectral level dynamics [Kol94]. Billiards with ray-splitting properties have been studied by loading microwave cavities with dielectric material [Sir97].

A number of experiments on three-dimensional billiards have been published as well, both using normal [Deu95] and superconducting [Alt96, Alt97] cavities. Here the equivalence between quantum mechanics and electrodynamics is lost. In general a separation into transverse electric and magnetic modes is no longer possible. Nevertheless, the statistical properties of the spectra did not differ significantly from those obtained in quasi-two-dimensional billiards and quantum mechanical systems as we shall see in Section 3.2.1.

2.2.2 Field distributions in microwave cavities

Microwave experiments also allow one to study billiard wave functions. We have seen that in quasi-two-dimensional billiards the electric field strength E_z is equivalent to the wave function of the corresponding quantum billiard. To determine E_z two experimental techniques have been applied. First we can use the fact that the depths of the resonances in the reflection spectrum are proportional to the square of the electric field strength at the position of the coupling antenna. Qualitatively we know that in the direct neighbourhood of a node line a resonance cannot be excited at all. Close to a maximum, on the other hand, this will be especially easy. Thus the different depths of the resonances in Fig. 2.10 can be understood. A two-dimensional scan of the antenna position accordingly yields a direct image of the field distribution in the resonator. For a quantitative understanding the interaction between the coupling antenna and the billiard has to be analysed by a scattering matrix approach. We shall come back to this aspect in Section 6.1.2.

Figure 2.13 shows a number of wave functions for a stadium billiard, obtained in this way [Ste92]. All wave functions are strongly scarred, i.e. show high amplitudes close to classical periodic orbits. In Fig. 2.13(a) this is the bouncing-ball orbit we have already met, in Fig. 2.13(b) it is an orbit in the shape of a figure eight. The orbit associated with the wave function in Fig. 2.13(c) is called whispering gallery. The analogy to the acoustic phenomenon is obvious.

An alternative way for the determination of wave functions is the perturbing bead method. If a small metallic bead is introduced into the resonator, a positive frequency shift of the resonance is observed which is proportional to the square of the electric field strength at the bead position. Qualitatively this is again easily understood: due to the presence of the bead the volume of the resonator is somewhat reduced, and in consequence the density of states is reduced as well (see Eq. (2.2.33)). Alternatively using a dielectric bead, a negative frequency shift will be observed. Measurement of the frequency shift as a function of the bead position again yields a mapping of the field distribution. Wave functions in a number of differently shaped billiards have thus been determined by Sridhar and coworkers [Sri91, Sri92]. Kudrolli *et al.* [Kud95] have studied the localization of electromagnetic fields in a random arrangement of scatterers, bearing great resemblance to the water wave results shown in Fig. 2.3.

Let us discuss one of these experiments in a bit more detail. For many years it has been debated whether there is a one-to-one correspondence between the boundary of a billiard and its spectrum. This induced M. Kac to put the popular

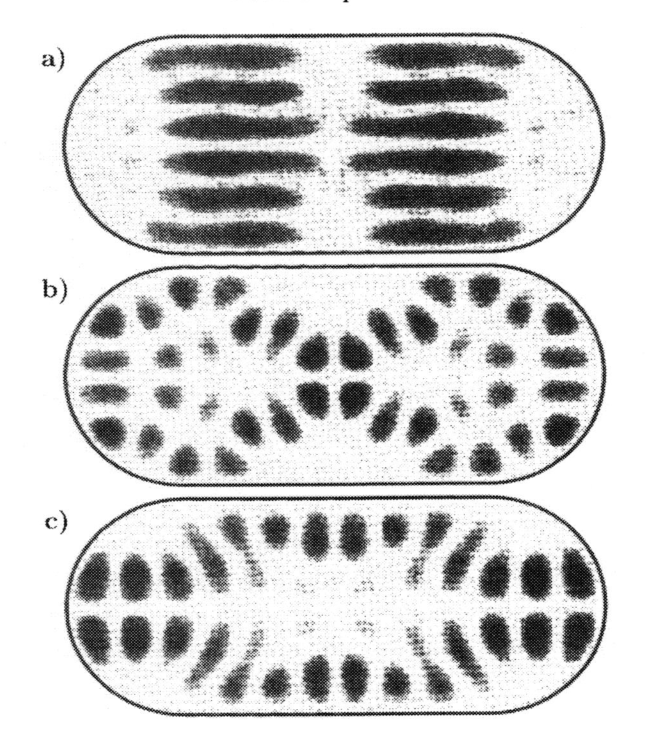

Figure 2.13. Experimental eigenfunctions in a microwave resonator of the shape of a stadium billiard ($l = 18$ cm, $r = 13.5$ cm) at three frequencies 3.384 GHz (a), 3.865 GHz (b), and 4.056 GHz (c). For the display of the wave functions the stadium has been completed by a twofold reflection. All wave functions show strong scarring close to classical periodic orbits [Ste92].

question: 'Can you hear the shape of a drum?' [Kac66]. Probably most experts would have affirmed this, though Kac himself was not so positive about this possibility. Therefore the recent presentation of a counter-example has been a surprise [Gor92]: there are two differently shaped billiards with identical spectra. They are made up of a number of right-angled triangles and somewhat resemble the pieces of a pentomino puzzle. We shall see later that these billiards belong to the pseudointegrable systems. In Fig. 2.14 experimental eigenfunctions for three pairs of eigenfrequencies are displayed, which are identical within the limits of error [Sri94].

Both methods of studying wave functions measure $|E_z|^2$ and are thus insensitive to the sign. If the sign is wanted, too, transmission measurements have to be performed. It will be shown in Section 6.1.2 that the transmission amplitude is proportional to the product of the electric field strengths at the points of entrance and exit, respectively. By fixing the entrance antenna and

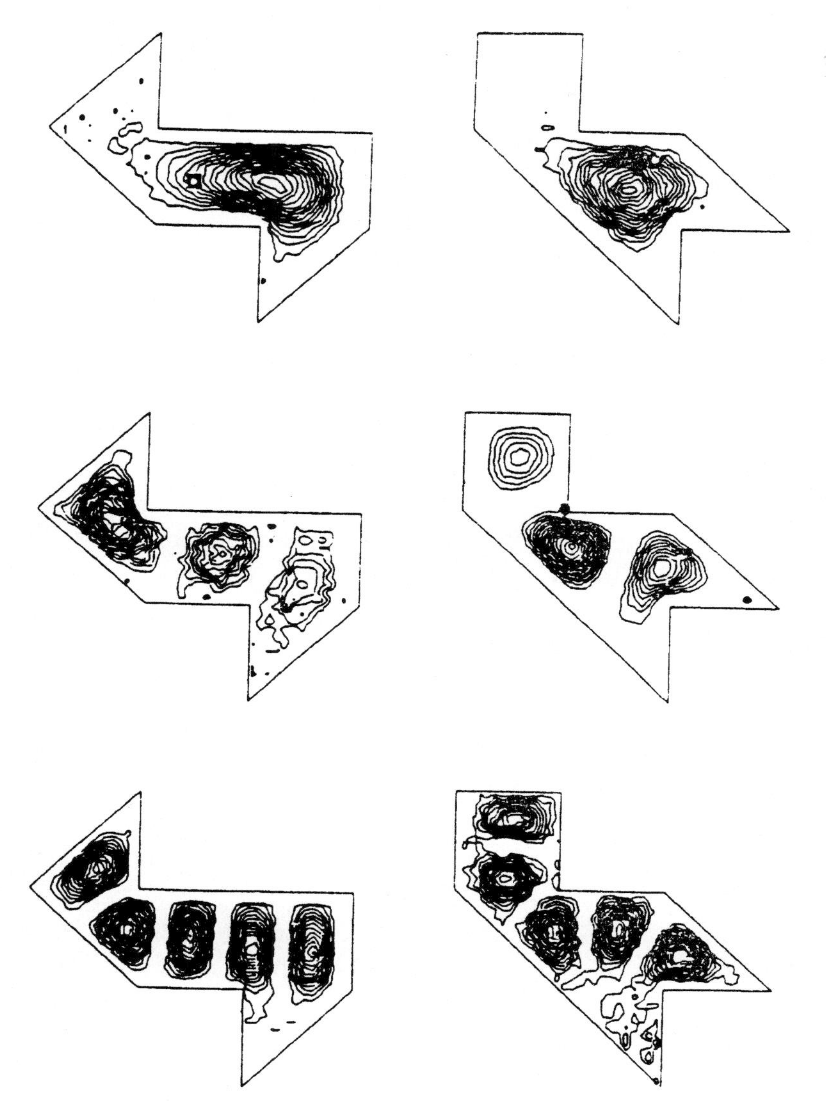

Figure 2.14. Three pairs of eigenfunctions for two isospectral microwave billiards. The cavities have a unit length of 7.6 cm. The eigenfrequencies in GHz are 1.9907, 1.9908 (top), 2.8413, 2.8418 (middle), 3.7964, 3.7924 (bottom) [Sri94] (Copyright 1994 by the American Physical Society).

varying the position of the exit antenna thus the wave functions can be obtained including the sign [Ste95].

If only measuring the modulus of either reflection or transmission coefficient, a scalar network analyser is sufficient. If the phase is wanted, too, a vector network analyser is needed, meaning a considerable additional effort, both technically and financially. The advantages of a vector network analyser, on the other hand, are considerable. If both amplitude and phase can be measured as a function of frequency, the study of pulse propagation also becomes possible by a simple Fourier transform to the time domain [Ste95]. An example has already been given in the introduction.

Recently the perturbing bead method has been extended to a three-dimensional Sinai billiard [Dör98]. In three dimensions the frequency shift observed depends on the shape of the perturber. For spherical objects it is proportional to $-2|\boldsymbol{E}|^2 + |\boldsymbol{B}|^2$ [Mai52]. We shall discuss these experiments in detail in Sections 6.1.4 and 6.2.2.

We have seen that resonances are lost if the coupling antenna is close to a node line. Additional losses result, especially in integrable systems, if the distance between two resonances falls below the line width. The total number of missing resonances is typically of the order of 10 to 15%. There are two ways to eliminate these problems. The loss due to nodal lines can be effectively reduced by combining measurements from different antenna positions. The positions of the antennas not used have to be closed by 50 Ω loads, otherwise the resonances will be somewhat shifted if another coupling antenna is selected. The loss due to degeneracies, however, cannot be reduced in this way. Both sources of loss can be avoided by level dynamics measurements. If the length of the billiard is varied, resonances disappear and reappear while a node line passes the antenna position. In the same way resonances being degenerate at one length are well separated at others (Fig. 5.2 in Section 5.2.1 gives an experimental example). By applying either of these two methods, it is possible to reduce the loss to zero.

The usual microwave cavities correspond to billiards with hard wall reflections, the potential energy being zero within and infinite outside the billiards. But it is also possible to simulate soft potentials, as has been demonstrated by Lauter in his Ph.D. thesis [Lau94]. To illustrate the idea, we go back to the Helmholtz equation (2.2.13) for $\boldsymbol{E}$, but now with $n = 1$,

$$\left[\Delta + k^2 - \left(\frac{\pi}{d}\right)^2\right] E = 0. \tag{2.2.38}$$

A comparison with the Schrödinger equation

$$\left[\Delta + \frac{2mE}{\hbar^2} - \frac{2mV(x, y)}{\hbar^2}\right] \psi = 0 \tag{2.2.39}$$

shows a complete equivalence, if the term $(\pi/d)^2$ is interpreted as a potential energy. By varying the height d with the position, soft potentials may likewise be simulated. If the outer wall of the billiard is removed, the modes with $n = 0$

disappear, and only $n = 1$ and higher modes survive. For a position dependent d, however, the separation of the z dependence of $\boldsymbol{E}$ (see Eq. (2.2.11)) is only approximately correct, since there are now additional terms depending on the gradient of d. Only for adiabatic changes, where $d(x, y)$ varies little over one wavelength, can these terms be discarded. The technique has been tested with a two-dimensional harmonic oscillator, and there has been good agreement between the measured and the calculated wave functions. As another example eigenfunctions of the Hénon–Heiles potential

$$V(x, y) = \frac{x^2 + y^2}{2} + xy^2 + \frac{y^3}{3}, \tag{2.2.40}$$

an often studied model system for nonlinear dynamics (see Ref. [Gut90]), have been measured. As the potential has a threefold symmetry axis and three reflection planes, the measurement could be restricted to one sixth of the xy plane, the remaining parts being obtained by reflection and rotation. Figure 2.15

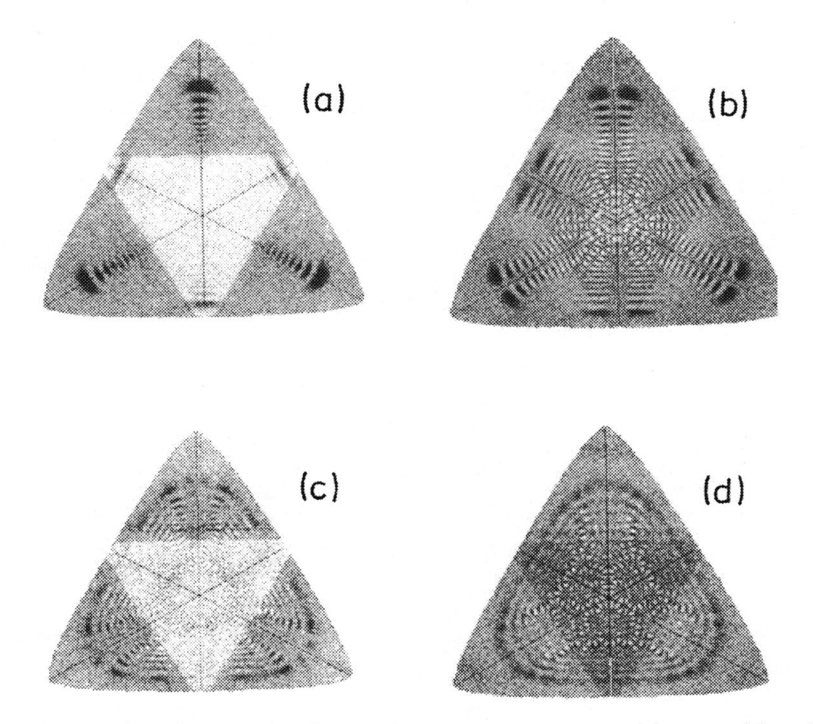

Figure 2.15. Eigenfrequencies in a microwave resonator with a position-dependent height to simulate the Hénon–Heiles potential. The total side length of the triangular region was 130 cm, the eigenfrequencies in GHz were 15.190 (a), 15.093 (b), 15.246 (c), 15.067 (d) [Lau94].

shows four of these wave functions, nicely demonstrating the feasibility of the method.

2.2.3 Billiards with broken time-reversal symmetry

The statistical properties of the spectra of chaotic systems with and without time-reversal symmetry show characteristic differences, as we shall see later. In ordinary microwave billiards time-reversal symmetry is obeyed. The propagation direction of a wave will just be reversed if the sign of time is changed. The same is valid for an electron in the absence of magnetic fields. But as soon as the field is switched on, the situation changes. Now the electron performs a circular or spiral motion under the influence of the Lorentz force, and time-reversal symmetry is broken. For microwaves this cannot be achieved so easily. One possibility to break time-reversal symmetry uses the unidirectional properties of ferrites, and will be described in the following.

A ferrite is a material with a spontaneous internal electronic magnetization $\boldsymbol{M}_0$. A typical representative is Fe_3O_4. Ferrites are neither ferromagnetic nor antiferromagnetic, they are ferrimagnetic. Here the electronic spins are aligned partly parallel, partly antiparallel, resulting in a net magnetization. For details see the special literature such as the monographs by Soohoo [Soo60] and Lax and Button [Lax62].

Under the influence of an external magnetic field $\boldsymbol{H}$ the magnetization obeys the equation of motion

$$\frac{d\boldsymbol{M}}{dt} = \gamma \boldsymbol{M} \times \boldsymbol{H}. \tag{2.2.41}$$

The term on the right hand side describes the precession of the magnetization under the influence of the torque exerted by $\boldsymbol{H}$, where $\gamma = -2.21 \times 10^5\ \mathrm{Am^{-1}\,s^{-1}}$ is the gyromagnetic ratio of the electron. In Eq. (2.2.41) the damping is neglected, which has to be accounted for in a more realistic calculation by an additional term on the right hand side.

Now consider the situation that $\boldsymbol{H}$ is a superposition of a static magnetic field H_0 along the z axis and an additional radiofrequency field:

$$\boldsymbol{H} = H_0 \boldsymbol{e}_z + \boldsymbol{h} e^{\imath\omega t}, \tag{2.2.42}$$

where $\boldsymbol{e}_z$ is the unit vector in the z direction. For the magnetization we make a corresponding ansatz

$$\boldsymbol{M} = M_0 \boldsymbol{e}_z + \boldsymbol{m} e^{\imath\omega t}, \tag{2.2.43}$$

where $M_0 = \chi_0 H_0$ is the static magnetization, and χ_0 the static susceptibility. Entering expressions (2.2.42) and (2.2.43) into the equation of motion, and taking only terms linear in $\boldsymbol{h}$ and $\boldsymbol{m}$ we get

$$\imath\omega m_x = -\gamma M_0 h_y + \gamma H_0 m_y, \tag{2.2.44}$$

$$\imath\omega m_y = \gamma M_0 h_x - \gamma H_0 m_x, \tag{2.2.45}$$

$$\imath\omega m_z = 0. \tag{2.2.46}$$

We have obtained a linear equation system for m_x and m_y, which is easily solved:

$$\begin{pmatrix} m_x \\ m_y \end{pmatrix} = \begin{pmatrix} \chi & -\imath\kappa \\ i\kappa & \chi \end{pmatrix} \begin{pmatrix} h_x \\ h_y \end{pmatrix}, \tag{2.2.47}$$

where

$$\chi = \frac{\omega_0 \omega_M}{\omega_0^2 - \omega^2}, \tag{2.2.48}$$

$$\kappa = -\frac{\omega \omega_M}{\omega_0^2 - \omega^2}. \tag{2.2.49}$$

Here $\omega_0 = -\gamma H_0$ and $\omega_M = -\gamma M_0$ are the precession angular frequencies about the external field H_0 and the magnetization M_0, respectively.

The relation obtained between the components of $\boldsymbol{M}$ and $\boldsymbol{H}$ may be written in compact matrix notation as

$$\begin{pmatrix} M_x \\ M_y \\ M_z \end{pmatrix} = \begin{pmatrix} \chi & -\imath\kappa & \cdot \\ \imath\kappa & \chi & \cdot \\ \cdot & \cdot & \chi_0 \end{pmatrix} \begin{pmatrix} H_x \\ H_y \\ H_z \end{pmatrix}, \tag{2.2.50}$$

or

$$\boldsymbol{M} = \overleftrightarrow{\chi} \boldsymbol{H}. \tag{2.2.51}$$

To derive the wave equation for the propagation of microwaves in billiards containing ferritic material, we start again with the Maxwell equations. The relation (2.2.6) between $\boldsymbol{B}$ and $\boldsymbol{H}$, however, has now to be replaced by

$$\boldsymbol{B} = \mu_0(\boldsymbol{H} + \boldsymbol{M}), \tag{2.2.52}$$

which, using Eq. (2.2.51), can be expressed as

$$\begin{aligned} \boldsymbol{B} &= \mu_0(1 + \overleftrightarrow{\chi})\boldsymbol{H} \\ &= \overleftrightarrow{\mu}\boldsymbol{H}, \end{aligned} \tag{2.2.53}$$

where $\overleftrightarrow{\mu}$ is the permeability tensor. Additionally the second and the third Maxwell equations (2.2.2) and (2.2.3) should be modified by adding a current density term $\boldsymbol{j}$ and a charge density ρ on the right hand sides, respectively. But in ferrites with their high resistivities of the order of 10^4 to $10^7\ \Omega\,\mathrm{m}$ these terms can be neglected. In this respect ferrites show similar behaviour to a lossless dielectric. Proceeding in an analogous way as in Section 2.2.1 we again get a wave equation for $\boldsymbol{E}$, which now reads

$$\nabla \times \left[(1+\overleftrightarrow{\chi})^{-1}\nabla \times \boldsymbol{E}\right] = \left(\frac{\omega}{c}\right)^2 \boldsymbol{E}. \tag{2.2.54}$$

For the inverse of $(1+\overleftrightarrow{\chi})$ we get from Eq. (2.2.50)

$$(1+\overleftrightarrow{\chi})^{-1} = \frac{1}{\delta}\begin{pmatrix} 1+\chi & \imath\kappa & \cdot \\ -\imath\kappa & 1+\chi & \cdot \\ \cdot & \cdot & \delta(1+\chi_0)^{-1} \end{pmatrix}, \tag{2.2.55}$$

where

$$\delta = (1+\chi)^2 - \kappa^2. \tag{2.2.56}$$

Inserting this into Eq. (2.2.54) and specializing to two-dimensional billiards, we obtain for the z component of the electric field

$$\left\{\nabla\frac{1+\chi}{\delta}\nabla + i\left[\boldsymbol{e}_z \times \nabla\frac{\kappa}{\delta}\right]\nabla + \left(\frac{\omega}{c}\right)^2\right\}E_z = 0. \tag{2.2.57}$$

This has to be compared with the stationary Schrödinger equation for a free particle in a magnetic field,

$$\frac{1}{2m}\left(\frac{\hbar}{i}\nabla - \frac{e}{c}\boldsymbol{A}\right)^2\psi = E\psi, \tag{2.2.58}$$

which may be alternatively written as

$$\left(\nabla^2 - i\frac{2e}{\hbar c}\boldsymbol{A}\nabla + \frac{2mE}{\hbar^2}\right)\psi = 0, \tag{2.2.59}$$

where the Lorentz gauge $\nabla\boldsymbol{A} = 0$ has been applied. In the last equation the term quadratic in $\boldsymbol{A}$ has been neglected. Apart from the first term the equations (2.2.57) and (2.2.59) are equivalent if $[\boldsymbol{e}_z \times \nabla(\kappa/\delta)]$ is identified with $\boldsymbol{A}$. In the first term a complete correspondence is merely obtained if χ is constant. This is only realized if the resonator is homogeneously filled with the ferrite. But even if this is not the case, the existence of the imaginary term in Eq. (2.2.57) is completely sufficient to break the time-reversal symmetry as we shall see in Chapter 3.

The equivalence of the two equations (2.2.57) and (2.2.59) has been used by So and coworkers [So95] to study the spectra of micowave billiards with broken time-reversal symmetry. They did not fill the billiard completely with ferrites, but only coated one wall with a stripe of ferritic material. The details of the experiment will be discussed later in Section 3.2.1, as some knowledge on random matrix theory is necessary for understanding this.

In another experiment the unidirectional properties of a microwave isolator have been used [Sto95b]. An isolator is made of a rectangular waveguide with

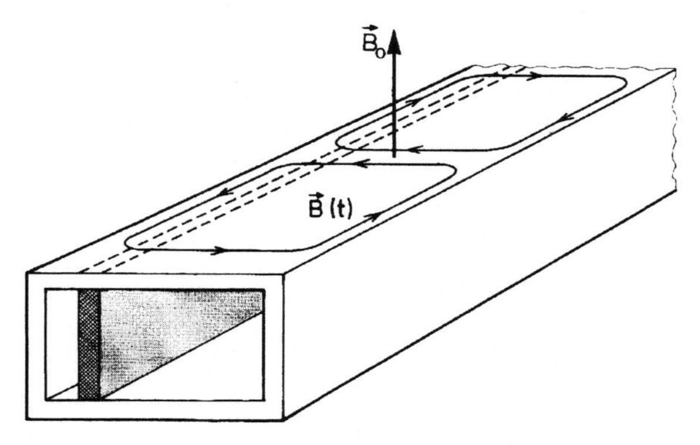

Figure 2.16. Microwave isolator consisting of a rectangular waveguide with a ferrite slab in an off-centre position. The electron spins in the ferrite perform a precession in a magnetic field B_0 applied perpendicularly. Travelling waves produce a time-varying magnetic field $B(t)$ in the ferrite which rotates clock- or counterclock-wise depending on the propagation direction. The component which rotates synchronously with the electron spins induces an electron spin resonance and is absorbed. The other one propagates unattenuated.

a ferrite slab in an off-centre position and a magnetic field applied perpendicularly. In the isolator the microwaves can propagate in one direction, but are absorbed in the other. Figure 2.16 gives a qualitative description of the underlying principle. More details can be found in the monographs cited above. Billiards with attached absorbing channels differ from the systems considered above, as they are no longer conservative. They will be discussed in Section 6.1.3.

2.2.4 Josephson junctions

An interesting alternative to study microwave propagation in billiard-like structures has been developed by Krülle *et al.* [Krü94]. They have used appropriately formed Josephson tunnel junctions, made of superconducting top and bottom electrodes with an AlO_x layer in between. Figure 2.17 displays the cross-section of a junction of a quarter stadium geometry. Microwaves are fed into the billiard by a stripline and form a standing wave pattern within the junction. There are some differences as compared to the billiards discussed in the preceding sections. First, the wave propagation velocity within the junction only amounts to about 3% of the vacuum velocity of light leading to a corresponding reduction in size. Typical side lengths of the junction have been in the range of 0.1 to 0.5 mm. Second, the microwaves are reflected at the open boundaries implying Neumann instead of Dirichlet boundary conditions.

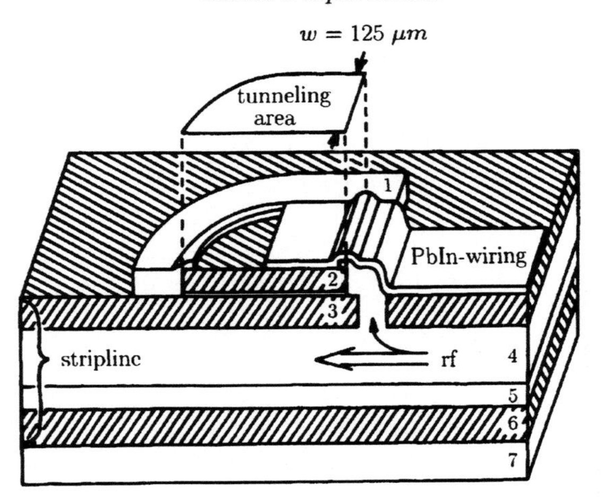

Figure 2.17. Cross-section of an Nb/AlO$_x$/Nb Josephson tunnel junction with quarter stadium geometry. The microwaves are coupled in via a superconducting stripline (1: SiO_2, 275 nm, 2: Nb top electrode, 150 nm, 3: Nb base electrode, 150 nm, 4: SiO_2, 1.7 μm, 5: Nb_2O_5, 45 nm, 6: Nb microstrip ground plate, 7: silicon substrate) [Krü94] (with kind permission from Elsevier Science).

Finally, the Q values of the junctions are only of the order of 100, i.e. they are one to two orders of magnitude smaller than in normally conducting microwave cavities. This allows a broad band excitation of the microwaves, but the resolution of individual resonances is not possible, except for those lying very low.

To study the field distribution within the junction the authors have used the fact that the current through the junction is increased in the presence of microwaves, which is due to a photon-assisted tunnelling of quasiparticles. Scanning the sample by means of a high-resolution electron beam, the junction is now locally heated with a consequent change in the current–voltage characteristics. For spatially varying microwave fields the change in voltage as a function of the position of the electron beam thus maps the field distribution within the junction. In Fig. 2.18 an experimentally obtained pattern is compared to a calculated one using a rectangular billiard, where the microwaves are coupled through a slit from one side. A good correspondence is found. The strong damping is a consequence of the comparatively poor quality of the junction. Similar measurements have been performed for other shapes and coupling geometries. The high experimental effort will probably impede a more extensive application of this technique. Nevertheless, Josephson junctions represent another nice visualization of chaotic wave patterns.

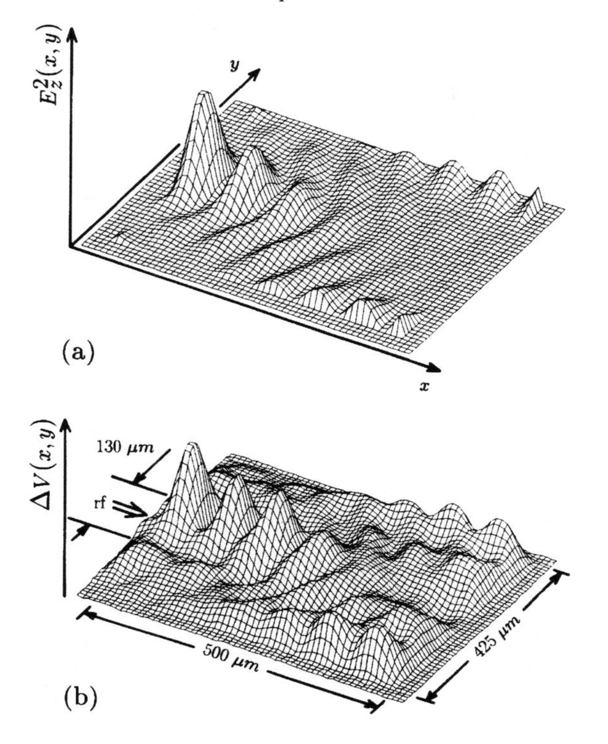

Figure 2.18. Theoretical (a) and experimental (b) field distribution inside a Josephson tunnel junction with rectangular geometry. The microwaves enter through a slit from the left side [Krü94] (with kind permission from Elsevier Science).

2.3 Mesoscopic structures

2.3.1 Antidot lattices

A detailed discussion of the large number of transport studies through mesoscopic structures is beyond the scope of this book. Readers interested in more details should consult the monograph by Datta [Dat95]. There are, however, a number of experiments, performed in billiard-like structures, which will be presented in the following. We start with the experiments by Weiss and coworkers [Wei91, Wei93] in antidot lattices. They have been manufactured from GaAs/Al_xGa_{1-x} As heterostructures with a two-dimensional electron gas layer some 100 Å below the surface. A typical layer sequence of a heterojunction together with its band structure is shown in Fig. 2.19 [Wei95]. In the heterojunction the Si atoms, acting as donors, and the two-dimensional electron gas are spatially separated. This leads to a high electron mobility, which assures

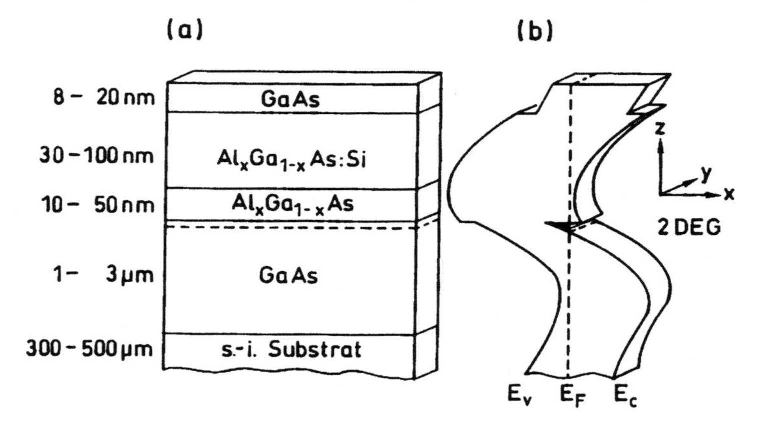

Figure 2.19. Typical GaAs/$Al_xGa_{1-x}As$ heterostructure (a), and corresponding band structure (b). The electrons are confined in the nearly triangular potential at the interface between the GaAs and $Al_xGa_{1-x}As$ layer [Wei95] (with kind permission from Elsevier Science).

that the electron mean free path is large compared to the linear dimensions of the billiard. The antidot structures have been formed by etching a two-dimensional array of nanometre scaled holes through the layers, thus confining the motion of the electrons to the sea between the islands formed by the dots. An antidot lattice may be regarded as a realization of the Lorentz gas. This is the caricature of a real gas, consisting of hard spheres all of which but one are fixed on a regular square or rectangular lattice. The Wigner–Seitz cell of this lattice is the Sinai billiard.

In the experiments the resistance has been studied as a function of the magnetic field. Figure 2.20(a) shows magnetoresistance curves for three different antidot samples. The maxima found in the resistance can be associated with different commensurate cyclotron orbits shown in the insert. The structures can be explained by assuming that at field strengths B, where the commensurability condition holds, the electrons are trapped in the respective orbits and are thereby lost for the transport. Calculations by Fleischmann *et al.* [Fle92] showed that the structures can be quantitatively understood, if the antidot structure is described by a properly adjusted periodic soft potential.

A new phenomenon is found if the probes are cooled down to 0.4 K. Now the mean free path lengths are long enough to allow the observation of quantum interference effects. Figure 2.20(b) shows the magnetoresistance obtained for the sample corresponding to the topmost trace in Fig. 2.20(a). The new oscillations superimposed on the resistance peak are caused by the quantization of the cyclotron orbit. For comparison the magnetoresistance of

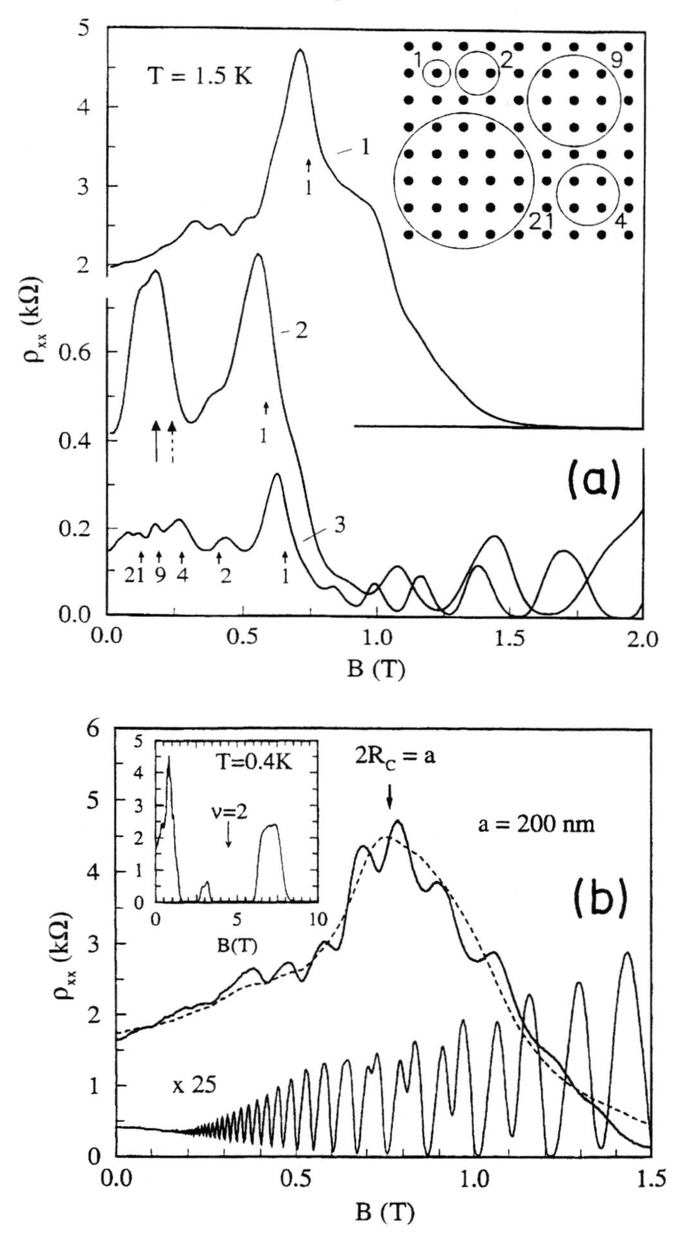

Figure 2.20. (a) Magnetoresistance of three different antidot lattices. The maxima correspond to commensurate cyclotron orbits shown in the insert. (b) Magnetoresistance at 0.4 K (solid line) and 4.7 K (dashed line) for the antidot lattice corresponding to the topmost trace in (a). The additional oscillations at 0.4 K are due to the quantization of the cyclotron orbits. For comparison, in the lower part the magnetoresistance for the unstructured layer is presented, showing the ordinary Shubnikov–de Haas oscillations [Wei93] (Copyright 1993 by the American Physical Society).

the unstructured sample is shown as well. Here the usual Shubnikov–de Haas oscillations are found, caused by the quantization of the ordinary cyclotron orbits which are suppressed in the antidot structures. The authors have calculated the actions for the relevant orbits in the antidot structures by solving the classical equations of motion and have been able to reproduce the oscillations quantitatively. More details on the experiments and their interpretation can be found in Ref. [Wei95].

2.3.2 Quantum dot billiards

Another type of billiard experiments has been initiated by Marcus and coworkers [Mar92], who have studied electronic transport through ballistic microscopic semiconductor structures. These mesoscopic or quantum dot billiards may be considered as tiny relatives of the microwave billiards. There are, however, some differences. First, the mesoscopic billiards are true quantum mechanical systems. To ensure that the de Broglie wavelength of the electron is comparable to the billiard dimensions, its sizes have to be of the order of 1 μm, i.e. microstructuring is necessary. Second, the motion of the electrons can be modified by the presence of magnetic fields due to Lorentz forces leading to a break of time-reversal symmetry. It has already been mentioned that this break has important consequences for the statistical properties of the spectra. With microwaves a break of time-reversal symmetry demands special tricks, as we have seen in Section 2.2.3. Finally, to achieve ballistic electron transport, the measurements have to be performed well below 1 K to suppress phonon-induced scattering.

The billiards were constructed by electron-beam lithography on GaAs/AlGaAs heterostructures similar to those used for the antidot lattices. By applying a voltage to the circular or stadium-shaped gate electrode on the surface the electrons are confined to the region below the electrodes. The inserts of Fig. 2.21 show the geometry of the gates including the two coupling wires used for the voltage supply. In more recent works this technique has also been used to study Sinai and square billiards [Cha95, Cla95]. In all cases the resistance of the sample was measured as a function of the magnetic field applied perpendicularly to the billiard. Figure 2.21 shows the result for a stadium and a circular electrode [Mar92]. We find chaotic conductance fluctuations in the resistance which are similar to those known from the so-called universal conductance fluctuations observed at low temperatures in the magnetoconductance in mesoscopic structures [Dat95]. The fluctuations in the billiards and in the mesoscopic structures are due to the same cause. In both cases the complicated interference patterns are provoked by a coherent super-

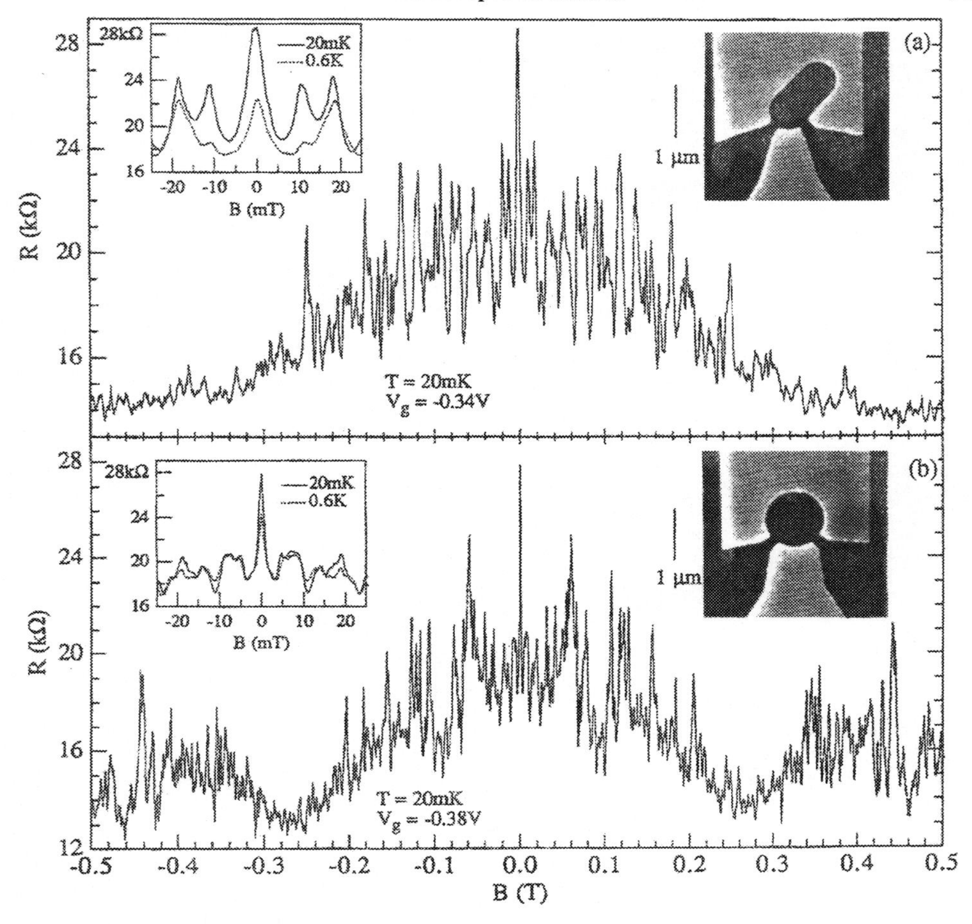

Figure 2.21. Resistance R of a stadium (a) and a circular (b) mesoscopic billiard as a function of the perpendicular magnetic field. The sharp peaks at $B = 0$, shown in greater detail in the inserts, are due to weak localization caused by enhanced coherent backscattering [Mar92] (Copyright 1992 by the American Physical Society).

position of the contributions of all paths connecting the entrance with the exit lead.

The peak in the resistance at $B = 0$ observed in both samples has attracted special attention and has been studied in a more detailed way by several authors [Mar93, Ber94, Cha94]. It is a manifestation of the so-called weak localization phenomenon [Lee85]: for $B = 0$ there is enhanced backscattering due to constructive interference between pairs of paths passing in opposite directions. As soon as a magnetic field is applied, the coherence is destroyed, and the backscattered intensity is reduced by a factor of 2. The complementary effect is seen in the transmission and gives rise to the peak observed in the resistance at

$B = 0$. We shall return to the universal conductance fluctuations and the weak localization in Section 6.3.2.

2.3.3 Quantum well billiards

Quantum wells belong to the semiconductor devices most intensely studied in recent years. Thus it is not astonishing that they have been applied to the study of quantum chaotic questions as well. Figure 2.22 shows a sketch of the heterostructure used in the experiments of Fromhold and coworkers [Fro94, Fro95, Wil96]. The well is formed by a GaAs layer sandwiched on both sides by $Al_{0.4}Ga_{0.6}As$ layers (see Fig. 2.22(a)). The potential shape shown in Fig. 2.22(b) is obtained by applying an external voltage. A two-dimensional electron gas is formed in the triangular region on the emitter side. In addition an external magnetic field of $B = 11.4$ T is applied. The angle between B and the normal to the layer planes could be varied. Some effort is needed to get such extreme inductions. With typical nonsuperconducting laboratory magnets

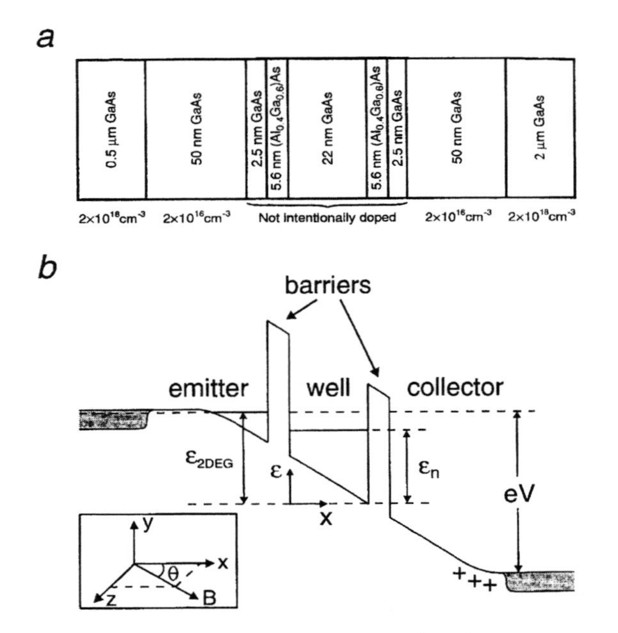

Figure 2.22. Quantum-well heterostructure used for the resonant tunnelling spectroscopy experiments (a), and corresponding band structure (b). A magnetic field B is applied whose angle θ to the normal to the layer planes can be varied. The current I shows maxima dependent upon the applied voltage U, whenever an energy level within the well is in resonance with the lowest Landau level in the emitter [Wil96] (Copyright 1996 by MacMillan Magazines).

the values obtained are only in the 0.1 to 1 T region. The motion of the electrons in the triangular potential is quantized in the direction perpendicular to the planes, whereas the magnetic field causes an additional Landau quantization within the planes. With the high magnetic fields applied and the liquid He temperature used, only the lowest Landau level is occupied.

Within the quantum well the motion of the electrons is described by the Hamiltonian

$$\mathcal{H} = \frac{1}{2m}\left(\boldsymbol{p} - \frac{e}{c}\boldsymbol{A}\right)^2 + e(U_0 - xF), \tag{2.3.1}$$

where we have chosen the x axis perpendicular to the layer planes. m is the reduced mass of the electron, amounting to only 7% of the free electron mass. U_0 is an offset voltage, and F is the electric field strength within the well structure. For a constant magnetic field in the xz plane,

$$\boldsymbol{B} = B(\cos\theta,\ 0,\ \sin\theta), \tag{2.3.2}$$

a suitable choice for the vector potential is

$$\boldsymbol{A} = B(0,\ x\sin\theta - z\cos\theta,\ 0), \tag{2.3.3}$$

whence follows for the Hamiltonian

$$\mathcal{H} = \frac{1}{2m}\left\{p_x^2 + \left[p_y - \frac{eB}{c}(x\sin\theta - z\cos\theta)\right]^2 + p_z^2\right\} + e(U_o - xF). \tag{2.3.4}$$

As $p_y = \hbar k_y$ is a constant of motion this may be written as

$$\mathcal{H} = \frac{1}{2m}(p_x^2 + p_z^2) + V(x, z), \tag{2.3.5}$$

where

$$V(x, z) = \frac{1}{2m}\left[\hbar k_y - \frac{eB}{c}(x\sin\theta - z\cos\theta)\right]^2 + e(U_o - xF) \tag{2.3.6}$$

plays the role of an effective potential [Fro95]. For $\theta = 0°$ the classical motion is regular and can be considered as a superposition of a back and forth movement between the two barriers on the one hand, and a cyclotron motion in the yz plane on the other. We consequently expect a series of bouncing-ball eigenstates for the quantum mechanical spectrum corresponding to standing waves between the barriers. This very phenomenon is demonstrated by the experiment [Fro94]. To study the eigenstates within the barrier, resonant tunnelling spectroscopy has been applied. Using this technique the bias voltage U is changed thus varying continuously the eigenvalue spectrum within the well. Whenever an eigenenergy is in resonance with the energy of the lowest Landau level of the two-dimensional electron gas, a maximum current I through the barrier is observed. Figure 2.23 shows the result. To get rid of a

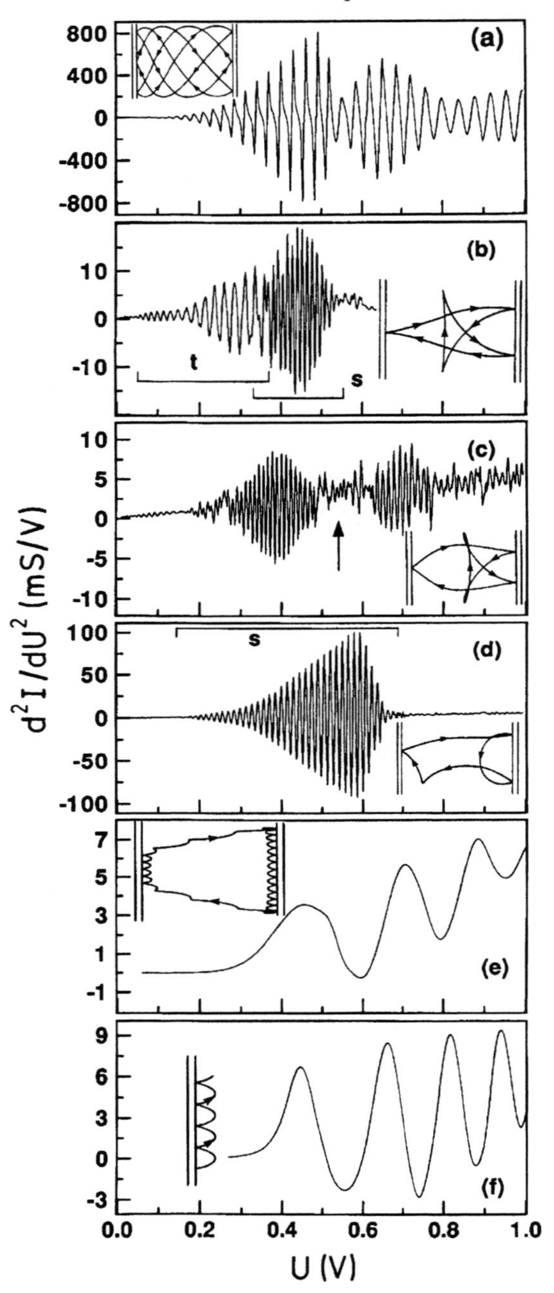

Figure 2.23. Second derivative d^2I/dU^2 of the tunnelling current through a double barrier structure as a function of the applied voltage U with an applied magnetic field of $B = 11.4$ T. The angle between B and the normal to the barrier planes is $\theta = 0°$ (a), 20° (b), 25° (c), 40° (d), 80° (e), 90° (f) [Fro94] (Copyright 1994 by the American Physical Society).

monotonically varying background the second derivative d^2I/dU^2 of the current is plotted. The regular oscillations in the figure directly reflect the bouncing-ball structures of the spectrum.

For angles $\theta > 0°$ the motion becomes chaotic until at $\theta = 90°$ it is regular again. The interpretation of the results requires some knowledge of periodic orbit theory, however, and is postponed to Section 8.1.1.

2.3.4 Quantum corrals

The scanning tunnelling microscope (STM), invented in the early 1980s, allows the mapping of surfaces and adsorbed atoms with a hitherto unprecedented resolution. In the experiments the tunnelling current I between a needle-like tip and a metallic substrate is measured as a function of the applied voltage U, the distance between tip and substrate, and the tip position. For a fixed distance the differential conductance dI/dU directly reproduces the local density of states (LDOS) of the surface. Even the Friedel oscillations, caused by the scattering of the conduction electrons at impurities, have thus been made visible [Cro93b]. In the experiments discussed in this section another feature of the STM is used, namely the possibility to change the position of individual adsorbed atoms with the help of the tip. To achieve this, the tip is placed directly over an adsorbed atom, and is lowered to a position where the attractive force between the tip and the atom is sufficient to keep the atom beneath the tip. Subsequently the tip pulls the atom to the position desired. Finally the tip is withdrawn to a distance where the atom-tip interaction is negligible. Recently this technique has been extended to the construction of two-dimensional billiards of different shapes, called quantum corrals by their designers [Cro93b, Cro95]. In these experiments Fe atoms on a Cu substrate have been used as the wall material. Figure 2.24 shows different stages of assembly of a circular quantum corral. The final circle exhibits a nice standing wave pattern of the conduction electrons resembling closely the pattern we have already found for water surface waves (see Fig. 2.4).

By varying the voltage between top and surface the LDOS can be probed in a range of ± 0.5 eV about the Fermi energy. Figure 2.25 shows the pattern thus observed, and obtained with the tip placed at the centre of the circle. For an ideal circle with radius R the eigenfunctions are given by

$$\psi_{nl}^{(1,2)}(r, \phi) \sim J_l(k_{nl}r) \times \begin{cases} \cos l\phi \\ \sin l\phi \end{cases}, \tag{2.3.7}$$

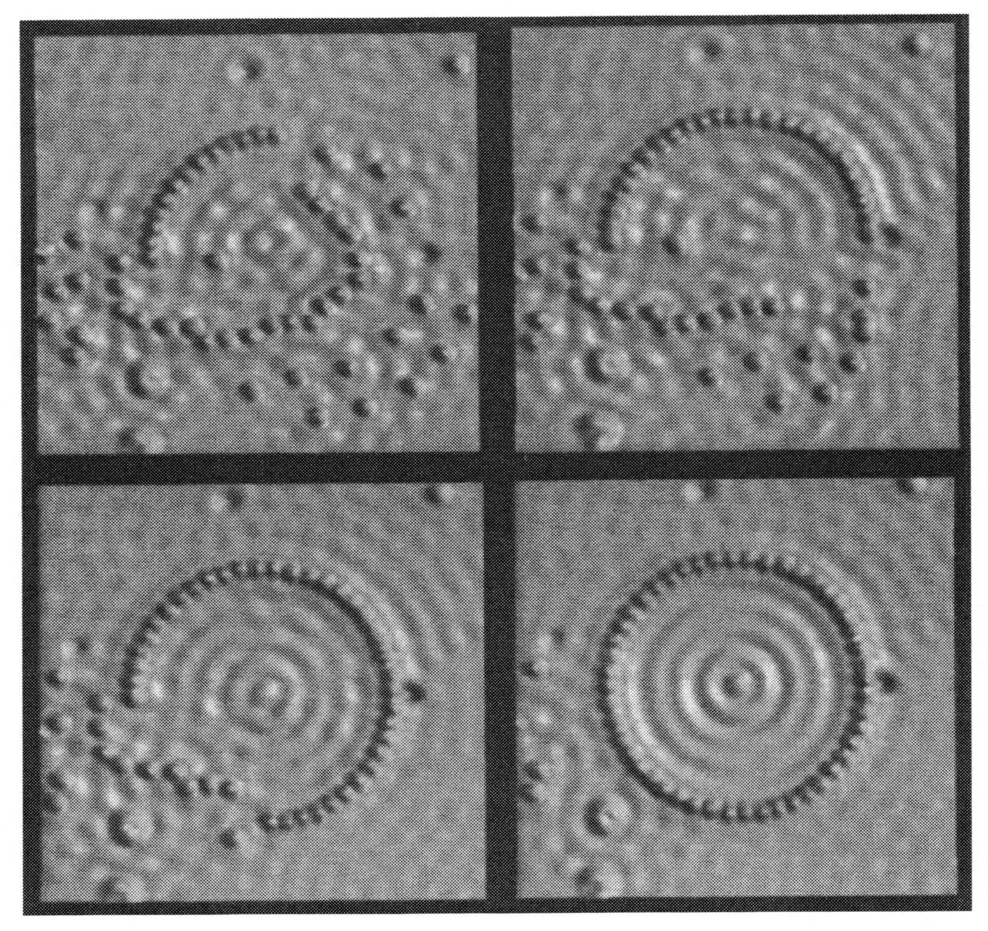

Figure 2.24. Different stages of assembly of a circular quantum corral consisting of a ring of 48 Fe atoms on a Cu(111) surface. Each Fe atom is individually positioned with the tip of a scanning tunnelling microscope [Cro95] (with kind permission from Elsevier Science).

where the corresponding wave numbers k_{nl} are obtained from the boundary condition

$$J_l(k_{nl}R) = 0. \tag{2.3.8}$$

In the centre of the circle only the $l = 0$ eigenfunctions have a nonvanishing amplitude. Hence only these eigenfunctions should contribute to the LDOS. Indeed the peak positions in Fig. 2.25 are in good agreement with the spectrum calculated from Eq. (2.3.7). If the tip is moved slightly from the centre, $l = 1$ contributions become visible, too [Cro93b]. The small discrepancy for the

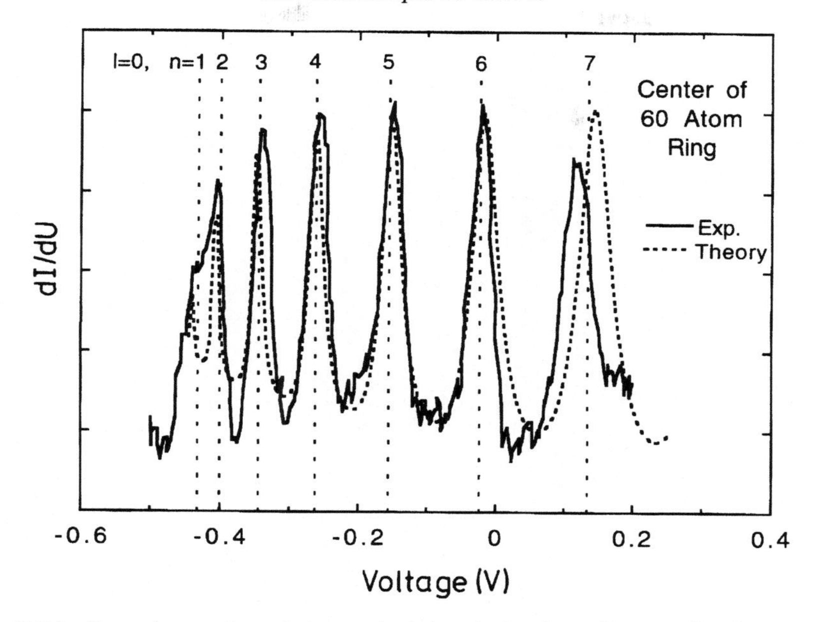

Figure 2.25. Experimental and theoretical local density of states in the centre of a circular quantum dot as a function of the voltage applied. The local density of states is obtained from the differential conductance dI/dU between tunnelling tip and surface. Each maximum corresponds to a Bessel eigenfunction of order zero [Cro93b] (Copyright 1994 by MacMillan Magazines).

($l = 0$, $n = 7$) peak in Fig. 2.25 is caused by the rather strong damping, due to which the boundary condition (2.3.8) becomes somewhat questionable. From the widths of the resonances a mean decay time of 3×10^{-14} s is estimated which exceeds 2×10^{-14} s, needed for an electron to cross the circle, only by a factor of 1.5.

In view of the leaky wall of the quantum corral the particle-in-a-box model is only qualitatively correct. A multiple scattering approach is more realistic. Here the electrons leaving the tip are treated as waves, which are repeatedly scattered by the atoms in the wall until they eventually return to the tip, where they interfere constructively or destructively with the outgoing wave. The technique has been applied to a stadium-shaped quantum corral. Figure 2.26 shows an experimentally obtained LDOS together with a calculation along the sketched lines [Hel94]. To obtain an agreement between experiment and calculation, the authors had to assume that about 25% of the incident amplitude is typically reflected and 25% transmitted, leaving 50% absorbed, presumably into bulk states. Due to this high absorption rate the method cannot be easily applied to quantum chaotical experiments, and

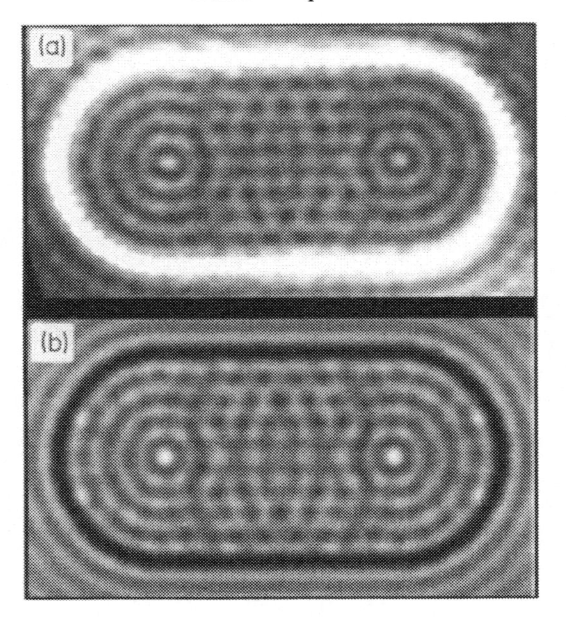

Figure 2.26. Experimental (a) and theoretical (b) local density of states in a stadium-shaped quantum corral. To obtain the theoretical pattern, the authors have used a multi-scattering approach described in the text [Hel94] (Copyright 1994 by MacMillan Magazines).

it is still an open question whether this problem can be solved. But even if this is not the case, quantum corral experiments have a value of their own, as they offer a unique opportunity to make true quantum mechanical wave functions visible to the eye.

3
Random matrices

Random matrix theory was developed in the nineteen fifties and sixties by Wigner, Dyson, Mehta and others. Originally conceived to bring some order into the spectra of complex nuclei, the interest in random matrix theory was renewed enormously when Bohigas, Giannoni and Schmit [Boh84] conjectured that it should be applicable to the spectra of *all* chaotic systems. In the following years overwhelming evidence has been obtained that this conjecture is true. The most important works up to 1965 together with a summarizing text have been reprinted in a book by Porter [Por65] which even up to this day belongs to the standard literature on the topic. The state of the art up to 1980 is compiled in an excellent review article by Brody and others [Bro81]. For the newcomer the most recent survey by Bohigas [Boh89] is recommended.

Two monographs on the subject have to be mentioned. The first one, *Random Matrices*, was written by Mehta [Meh91], one of the pioneers in the field. The first edition appeared in 1967, the second enlarged one in 1991. Probably there is no really important topic on random matrices which cannot be found there. It is, however, a book for specialists. For those who are only interested in the basic principles, it is much too detailed. The other monograph, *Quantum Signatures of Chaos* by Haake [Haa91a], has now, a few years after its first appearance, become the standard introduction into the field. It presents an excellent explanation of the elements of random matrices, and also of Floquet systems and spectral level dynamics under the variation of an external parameter. We shall come back to the last two points in some of the following chapters.

In view of the literature already existing we shall not try to give a comprehensive account of random matrix theory. Instead we shall confine ourselves to the basic concepts of which knowledge is indispensable for beginners. Mathematical derivations will be presented only exemplarily to give an idea of the techniques applied. The main emphasis will be on topics where experimental material is also available, such as the nearest neighbour spacing distribution or the spectral rigidity.

In the last part of this chapter the supersymmetry technique will be introduced. It is a completely new method having come up in recent years to treat ensemble averages. As there is as yet no introductory presentation, a somewhat more detailed account will be given. In view of the rapidly growing number of publications on supersymmetry calculations even a researcher without personal ambitions in this field should get some idea of the background of this method. The section on supersymmetry is not essential for the following text and may be skipped in a first reading. For details see the review articles by Efetov [Efe83] and Verbaarschot, Weidenmüller and Zirnbauer [Ver85a], where the foundations of the supersymmetry technique are laid. Efetov recently published a monograph, *Supersymmetry in Disorder and Chaos* [Efe97].

3.1 Gaussian ensembles

3.1.1 Symmetries

Symmetry is one of the most successful concepts in physics. If all symmetries of a system are known, it is *qualitatively* understood. Only for *quantitative* calculations of the quantum mechanical spectrum are the details of the interactions needed. In classical mechanics symmetries are associated with constants of motion, in quantum mechanics every symmetry gives rise to a new quantum number. The conservative systems, which are invariant with respect to a shift of the time axis, are of special importance. If the Hamilton operator $\mathcal{H}$ does not explicitly depend on time, then the time dependence in the time dependent Schrödinger equation

$$\imath\hbar\frac{\partial\psi}{\partial t} = \mathcal{H}\psi \tag{3.1.1}$$

may be separated with help of the ansatz

$$\psi_n(x, t) = \psi_n(x)\exp\left(\frac{\imath}{\hbar}E_n t\right), \tag{3.1.2}$$

for the wave function, where $\psi_n(x)$ obeys the stationary Schödinger equation

$$\mathcal{H}\psi_n = E_n\psi_n. \tag{3.1.3}$$

For conservative Hamiltonians the energy has thus been found to be a constant of motion, with a corresponding quantum number n labelling the eigenenergies of the system. In completely chaotic systems there are no other constants of motion.

Nonconservative systems frequently found are Floquet systems periodically depending on time. In this case the energies are replaced by the quasienergies. They are defined only up to multiples of h/τ where τ is the period. Floquet systems will be discussed in detail in Chapter 4, but occasionally examples will be given in the following sections.

If there are additional symmetries, the system can be further simplified. To this end we go to the matrix representation of the Schrödinger equation by expanding the eigenfunctions $\psi_n(x)$ into a set of basis functions $\phi_n(x)$,

$$\psi_n(x) = \sum_m a_{nm}\phi_m(x). \tag{3.1.4}$$

The choice of the $\phi_n(x)$ is still arbitrary. They only have to obey the orthogonality relation

$$\langle \phi_n|\phi_m\rangle = \int \phi_n^*(x)\phi_m(x)\,dx = \delta_{nm}, \tag{3.1.5}$$

where we have introduced Dirac's bra-ket notation. For dimension d greater than one the symbol dx has to be interpreted as an abbreviation for $dx_1 \ldots dx_d$. This convention will be adopted throughout this monograph. Inserting expression (3.1.4) into the Schrödinger equation (3.1.3), and using the orthogonality relation, we obtain the matrix representation desired

$$\sum_m H_{nm}a_m = E_n a_n, \tag{3.1.6}$$

where the H_{nm} are the matrix elements of $\mathcal{H}$ in the ϕ_n basis:

$$\begin{aligned} H_{nm} &= \langle \phi_n|\mathcal{H}|\phi_m\rangle \\ &= \int \phi_n^*(x)\mathcal{H}\phi_m(x)\,dx. \end{aligned} \tag{3.1.7}$$

The problem of solving the Schrödinger equation has thus been reduced to the diagonalization of the matrix $H = (H_{nm})$.

Let us now assume that $\mathcal{H}$ is invariant with respect to a symmetry operation, e.g. against a rotation about the z axis,

$$\mathcal{H}(x, y, z) = \mathcal{H}(x', y', z'), \tag{3.1.8}$$

where

$$x' = x\cos\alpha + y\sin\alpha \tag{3.1.9}$$

$$y' = -x\sin\alpha + y\cos\alpha \tag{3.1.10}$$

$$z' = z. \tag{3.1.11}$$

For an infinitesimal rotation Eq. (3.1.8) reads

$$\mathcal{H}(x, y, z) = \mathcal{H}(x + \epsilon y, y - \epsilon x, z)$$
$$= \mathcal{H}(x, y, z) + \epsilon\left(y\frac{\partial}{\partial x} - x\frac{\partial}{\partial y}\right)\mathcal{H}(x, y, z). \tag{3.1.12}$$

The latter relation can be written in terms of the commutator relation

$$[L_z, \mathcal{H}] = 0, \tag{3.1.13}$$

where

$$L_z = \frac{\hbar}{\imath}\left(x\frac{\partial}{\partial y} - y\frac{\partial}{\partial x}\right) \tag{3.1.14}$$

is the z component of the angular momentum. This can be generalized: if $\mathcal{H}$ is invariant with respect to a symmetry relation then there is an associated operator R commuting with $\mathcal{H}$,

$$[R, \mathcal{H}] = 0. \tag{3.1.15}$$

If R commutes with $\mathcal{H}$, then this is the case for $R + R^\dagger$ and $\imath(R - R^\dagger)$ as well. We may hence assume without loss of generality that R is selfadjoint, $R = R^\dagger$. The matrix representation of $\mathcal{H}$ is simplified considerably, if basis functions $\phi_{n,\alpha}$ are used, which are eigenfunctions of R,

$$R\phi_{n,\alpha} = r_n\phi_{n,\alpha}. \tag{3.1.16}$$

Here all r_n are assumed to be different, where the index α labels all eigenfunctions $\phi_{n,\alpha}$ belonging to the same eigenvalue r_n. With the $\phi_{n,\alpha}$ as basis functions the commutator relation (3.1.15) reads

$$0 = \langle\phi_{n,\alpha}|R\mathcal{H} - \mathcal{H}R|\phi_{m,\beta}\rangle$$
$$= (r_n - r_m)\langle\phi_{n,\alpha}|\mathcal{H}|\phi_{m,\beta}\rangle. \tag{3.1.17}$$

As all r_n have been assumed to be different, it follows that

$$\langle\phi_{n,\alpha}|\mathcal{H}|\phi_{m,\beta}\rangle = \delta_{nm}H^{(n)}_{\alpha\beta}, \tag{3.1.18}$$

where

$$H^{(n)}_{\alpha\beta} = \langle\phi_{n,\alpha}|\mathcal{H}|\phi_{n,\beta}\rangle. \tag{3.1.19}$$

The matrix representation of $\mathcal{H}$ has thus been reduced to a block form

$$H = \begin{pmatrix} H^{(1)} & 0 & \cdots \\ 0 & H^{(2)} & \cdots \\ \vdots & \vdots & \ddots \end{pmatrix}, \tag{3.1.20}$$

where $H^{(n)} = (H^{(n)}_{\alpha\beta})$. A large part of the diagonalization process has been achieved automatically by making use of the mere symmetry. If there are further symmetries, the procedure can be repeated, until we end up with a matrix representation for $\mathcal{H}$ that cannot be further reduced.

If we are lucky the resulting matrix is already diagonal. This is the case for the hydrogen atom. Here every state is classified unequivocally by a set of four quantum numbers n, l, m and m_s for energy, angular momentum, and the projection of angular momentum and electron spin onto the z axis, respectively. The number four reflects the degrees of freedom of the electron, three translational ones and one spin degree of freedom. Systems where the number of quantum numbers equals the number of degrees of freedom are called integrable. They are exceptional, though one might get a different impression from beginners' courses on theoretical physics.

An atomic nucleus is a typical example of a nonintegrable system. The nuclear interaction is invariant against rotation about an arbitrary axis and against inversion of the coordinate system. Accordingly the nuclear states are classified by the quantum numbers I^π, m_I, where I is the total angular momentum quantum number, and m_I is the associated magnetic quantum number. π denotes the parity of the state, taking the value '$+$' if it is even, and '$-$' if it is odd under inversion. Other constants of motion do not exist, apart from the total energy. Atomic nuclei are hence described by Hamiltonians allowing a block representation where every block is associated with one I^π label. In the absence of magnetic fields every block enters $(2I+1)$ fold because of the energetic degeneracy of the m_I sublevels.

The energy levels of a large number of nuclei, including their quantum numbers, have been determined in nuclear spectroscopy. For a statistical analysis we must first arrange the levels into subspectra with given angular momentum and parity quantum numbers. Each subspectrum corresponds to the eigenvalues of one of the submatrices entering into the block representation (3.1.20). In a second step the spectra are normalized to an averaged density of states of one, to allow the comparison of different subspectra and the spectra of different nuclei. The distribution $p(s)$ of the spacings s_n between neighbouring eigenvalues E_n and E_{n-1}, which has already been introduced in Chapter 2 is by far the most popular spectral property. As an example we present the results of a famous compilation of energy levels of a large number of different nuclei and symmetry classes by Bohigas and coworkers [Haq82, Boh83]. In the literature the data set is known as the *nuclear data ensemble*. Figure 3.1 shows the histogram of a nearest neighbour spacing distribution obtained by combining the results of different subspectra of some 30 nuclei. Other spectral correlations of the same ensemble of nuclear levels will be presented later. The solid line corresponds to a Wigner distribution

$$p(s) = \frac{\pi}{2} s \exp\left(-\frac{\pi}{4} s^2\right). \tag{3.1.21}$$

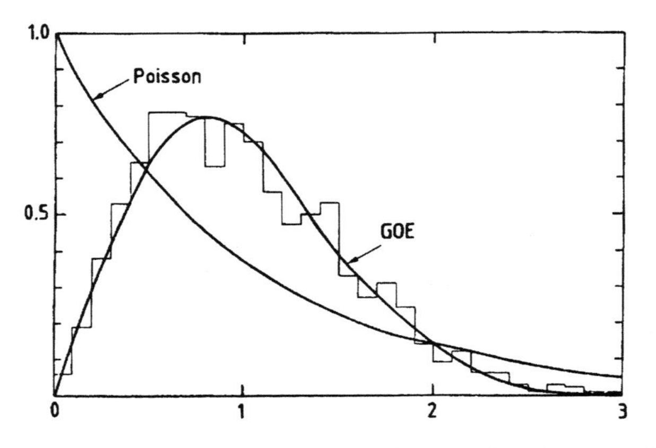

Figure 3.1. Nearest neighbour distance distribution for the so-called nuclear data ensemble. It contains altogether 1726 nuclear energy levels of 36 sequences of 32 different nuclei. The solid line corresponds to the Wigner distribution [Boh83] (with kind permission from Kluwer Academic Publishers).

In contrast to integrable systems the distribution shows a hole at small distances, demonstrating that the eigenvalues tend to repel each other. We have already seen that the same distribution is also well able to describe the spectra of microwave billiards (see Section 2.2.1). The spectra of nuclei and microwave billiards cannot be distinguished, as far as nearest neighbour spacing distributions are concerned! The same is true for a number of other spectral correlations which will be discussed in this chapter. Consequently we cannot hope to learn too much from such an analysis of the spectrum, but it implies the chance to apply similar techniques to a large number of different chaotic systems, including nuclei, microwave billiards, and mesoscopic systems such as quantum dots and quantum wells.

It is essential that only subspectra belonging to the same set of quantum numbers enter into the analysis. For this reason most microwave experiments have been performed in desymmetrized billiards to avoid the complication of a superposition of different spectra of different parity classes. It is instructive to see what happens if two independent spectra are superimposed. Figure 3.2 shows a series of nearest neighbour spacing distributions of acoustic resonances in a quartz block [Ell96]. The experimental technique has been described in Section 2.1.3. In Fig. 3.2(a) the distribution for a brick-shaped block with the dimensions $14 \times 25 \times 40$ mm^3 is shown. We might get the impression that there are three reflection symmetries, but this is not the case. Quartz is an anisotropic crystal with a threefold rotation symmetry about the crystallographic Z axis, the optical axis, and three twofold rotation symmetries about

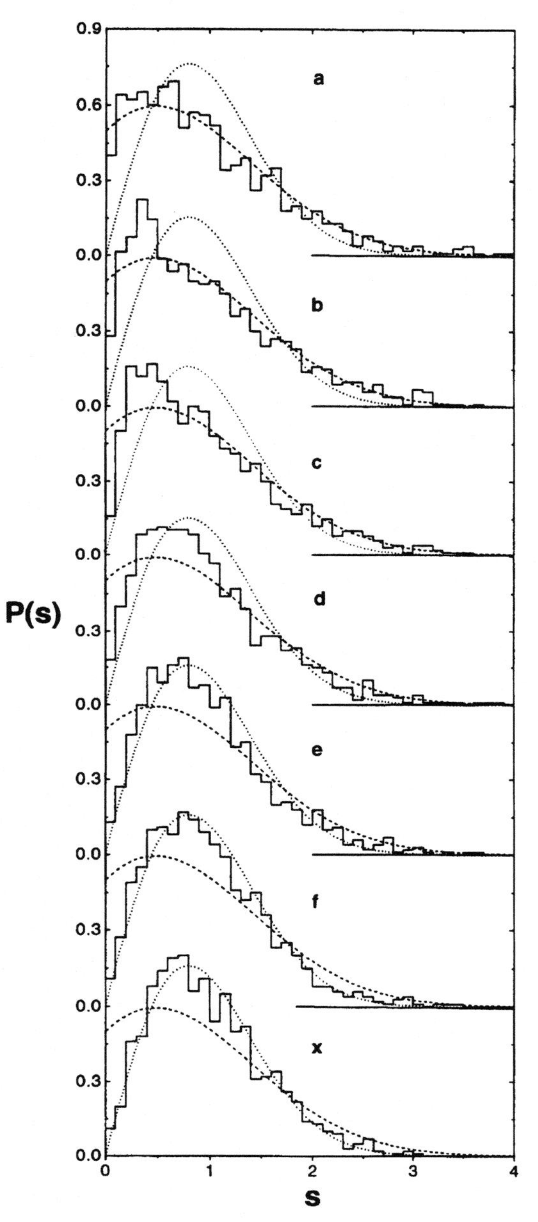

Figure 3.2. Level spacing distribution of the acoustic resonances of a quartz block with the dimensions $14 \times 25 \times 40$ mm^3. An octant of a sphere has been removed from one corner with $r = 0$ mm (a), 0.5 mm (b), 0.8 mm (c), 1.1 mm (d), 1.4 mm (e), 1.7 mm (f), 10 mm (x). The plot shows a gradual change from a distribution corresponding to the uncorrelated superposition of two Wigner distributions (dashed line) to a Wigner distribution (dotted line), which is caused by the break of a rotation symmetry [Ell96] (Copyright 1996 by the American Physical Society).

the crystal's X axes, which are perpendicular to the Z axis. The Z axis has been chosen parallel to the short side of the crystal, and one of the X axes parallel to the long side, leaving only a twofold rotation symmetry about this X axis. Indeed the nearest neighbour distribution obtained experimentally closely follows the distribution expected for a random superposition of two spectra constructed from two Wigner distributions (see appendix A.2 of Ref. [Meh91]). Note that for zero spacing the distribution is unequal to zero in contrast to the Wigner distribution (3.1.21), since eigenvalues belonging to different symmetry classes do not repel each other. In the next step an octant of a sphere has been excised out of one corner with a stepwise increasing radius, thus destroying the remaining symmetry. Figures 3.2(b) to (f) show a continuous change of the distribution, until a Wigner distribution reappears for a radius of 10 mm.

A comparable phenomenology is found in the spectra of nuclei with an equal number of protons and neutrons. Here *isospin*, i.e. the symmetry with respect to an interchange of protons and neutrons, is approximately conserved, as long as the Coulomb interaction is small. Consequently a spectral behaviour corresponding to the superposition of two Wigner distributions is expected. To check this prediction, Mitchell *et al.* analysed the spectrum of ^{26}Al, a nucleus with 13 protons and neutrons [Mit88]. The spectrum showed a behaviour between that of a single Wigner distribution and that of an independent superposition of two Wigner distributions. By a detailed analysis of the data Guhr and Weidenmüller succeeded in extracting the strength of the isospin breaking Coulomb matrix element [Guh90].

Let us now consider the untypical case of a completely integrable system, where the block representation (3.1.20) of $\mathscr{H}$ is already diagonal from symmetry considerations alone. Since any eigenvalue makes up a symmetry class of its own, it is reasonable to assume that the eigenvalues are completely uncorrelated (in Section 7.1.2 this heuristic argument will be founded on more solid grounds). For such a situation the nearest neighbour spacing distribution $p(s)$ is easily calculated. $p(s)\,ds$ is the probability to find an eigenvalue in a distance between s and $s+ds$ from a given eigenvalue but no other one in between. To calculate this probability the interval of length s is subdivided into N subintervals of length s/N. Then we get

$$p(s)\,ds = \lim_{N\to\infty}\left(1-\frac{s}{N}\right)^N ds, \tag{3.1.22}$$

where the first factor on the right hand side gives the probability to find no eigenvalue in any of the subintervals. The second factor gives the probability to find an eigenvalue in a distance between s and $s+ds$. Again a mean level density of one has been assumed. Performing the limit $N \to \infty$ we reach a Poisson nearest neighbour distribution

$$p(s) = \exp(-s). \tag{3.1.23}$$

In the derivation the assumption of uncorrelated eigenvalues has been essential. Only then is it allowed to multiply probabilities as done on the right hand side of Eq. (3.1.22).

Figure 3.3 shows the nearest neighbour spacing distribution for a rectangular billiard studied by Casati *et al.* [Cas85]. The eigenvalues are easily calculated using the relation

$$\begin{aligned} E_n &= \frac{\hbar^2 k_n^2}{2m_e} \\ &= \frac{\hbar^2}{2m_e}\left[\left(\frac{\pi n}{a}\right)^2 + \left(\frac{\pi m}{b}\right)^2\right] \qquad n,\, m = 0,\, 1,\, 2,\, \ldots, \end{aligned} \tag{3.1.24}$$

where a and b are the side lengths of the rectangle, and m_e is the mass of the particle. An irrational side ratio of $\alpha = a/b = \sqrt{\pi/3}$ has been chosen for the calculations. The figure shows that the Poisson distribution holds with high

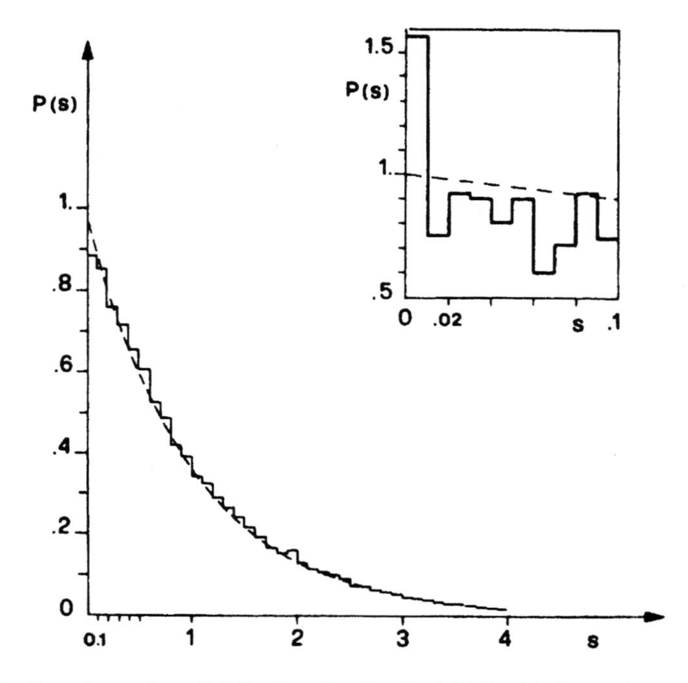

Figure 3.3. Level spacing distribution for the first 100 000 eigenvalues of a rectangular billiard. The dotted line corresponds to a Poisson distribution. For the insert see the text [Cas85] (Copyright 1995 by the American Physical Society).

precision. Thus the eigenvalues of a rectangle behave more or less like a sequence of random numbers, at least as far as nearest neighbour spacings are concerned. For rational values of α number-theoretical degeneracies lead to an extra contribution at $s = 0$ in the level spacing distribution [Con97]. Though in the above rectangle α has been chosen irrational, an excess at very small distances survives nevertheless, as is evident from the insert of Fig. 3.3. This is not surprising, since any irrational number may be approached by rational approximants [Rob98]. A more detailed analysis shows that there are additional significant deviations from random behaviour in the long range correlations of the eigenvalues. We shall come back to this aspect in Section 8.2.1.

3.1.2 Universality classes

We have seen that the matrix representation of a Hamiltonian can be reduced to a block form if symmetries are involved. To avoid the complications induced by uncorrelated superpositions of spectra belonging to different sets of quantum numbers (see Section 3.1.1), it is now assumed that the Hamiltonian entering into the Schrödinger equation

$$\imath\hbar\frac{\partial\psi}{\partial t} = \mathscr{H}\psi \tag{3.1.25}$$

is already reduced to a sub-Hamiltonian corresponding to just one set of quantum numbers. Furthermore the rank of the matrix representing $\mathscr{H}$, which is infinite for all realistic systems, shall be truncated to a finite rank N.

A very important symmetry which has not yet been considered, is time-reversal symmetry. The time-reversal operator T reverses the sign of time,

$$Tf(t) = f(-t). \tag{3.1.26}$$

A change of the sign of time in the time-differential operator $\partial/\partial t$ on the left hand side of the time-dependent Schrödinger equation (3.1.1) can be compensated by replacing $\imath$ by $-\imath$. This can be expressed in terms of the commutator relation

$$\left[CT, \imath\hbar\frac{\partial}{\partial t}\right] = 0, \tag{3.1.27}$$

where C is the conjugate-complex operator, turning any quantity into its complex conjugate,

$$CA = A^*. \tag{3.1.28}$$

The question is, whether $\mathscr{H}$ too commutes with CT. As we have restricted ourselves for the moment to conservative systems, $\mathscr{H}$ commutes with T. Therefore it is sufficient to consider the commutation properties of $\mathscr{H}$ with respect to C. Three situations are possible leading to the concept of the universality classes.

First we consider the Hamiltonian

$$\mathcal{H} = \frac{1}{2m}\left(\boldsymbol{p} - \frac{e}{c}\boldsymbol{A}\right)^2 + V(x) \tag{3.1.29}$$

with a nonvanishing vector potential $\boldsymbol{A}$. As $\boldsymbol{p} = (i/\hbar)\nabla$ is a complex operator, $\mathcal{H}$ does not commute with C. Consequently the time-reversal symmetry is broken for such a system. Classically this is realized for an electron in a magnetic field. Since the Hamiltonian is Hermitian,

$$\mathcal{H} = \mathcal{H}^\dagger, \tag{3.1.30}$$

its matrix elements obey the relation

$$H_{nm} = H^*_{mn}, \tag{3.1.31}$$

where in general $H_{nm} \neq H_{mn}$. In systems with broken time-reversal symmetry it is therefore not possible to represent the Hamiltonians by real matrices. The Hermitian property of H is preserved under the *unitary* transformation

$$H' = UHU^\dagger \tag{3.1.32}$$

where U is unitary, $UU^\dagger = 1$.

For the case that there is a time-reversal symmetry, we have to discriminate two situations. First we assume that there are no interactions with spin $\frac{1}{2}$. This situation holds for the Hamiltonian

$$\mathcal{H} = \frac{p^2}{2m} + V(x). \tag{3.1.33}$$

Since now there is no interaction with an external magnetic field, $\mathcal{H}$ commutes with C. Thus we can find basis functions ϕ_n which are eigenfunctions of C. Because $C^2 = 1$ the only eigenvalues of C are 1 and -1:

$$C\phi_n = \phi_n^* = \pm\phi_n. \tag{3.1.34}$$

The ϕ_n are consequently real or purely imaginary. As the phase of the basis functions may be chosen arbitrarily, ϕ_n can be taken, without loss of generality, as real. The matrix elements of $\mathcal{H}$ are then real too,

$$H_{nm} = H_{mn}. \tag{3.1.35}$$

For systems with time-reversal symmetry and without spin-$\frac{1}{2}$ interactions the Hamiltonians can be represented by real symmetric matrices. This property is preserved under *orthogonal* transformations

$$H' = OHO^T \tag{3.1.36}$$

where O is an orthogonal matrix, $OO^T = 1$. O^T is the transposed matrix of O with the matrix elements $(O^T)_{nm} = O_{mn}$.

Last we consider a system with time-reversal symmetry, and an additional spin-$\frac{1}{2}$ interaction. It may be realized by adding a spin–orbit Hamiltonian

$$\mathscr{H}_{\text{SO}} = A\boldsymbol{L}\boldsymbol{S} \tag{3.1.37}$$

to the Hamiltonian (3.1.33). Here

$$\boldsymbol{L} = \frac{\hbar}{\imath}\boldsymbol{r} \times \boldsymbol{\nabla} \tag{3.1.38}$$

is the the angular momentum operator, and

$$\boldsymbol{S} = \frac{\hbar}{2}\boldsymbol{\sigma} \tag{3.1.39}$$

is the spin operator, where

$$\sigma_x = \begin{pmatrix} \cdot & 1 \\ 1 & \cdot \end{pmatrix}, \quad \sigma_y = \begin{pmatrix} \cdot & -\imath \\ \imath & \cdot \end{pmatrix}, \quad \sigma_z = \begin{pmatrix} 1 & \cdot \\ \cdot & -1 \end{pmatrix} \tag{3.1.40}$$

are the Pauli spin matrices, obeying the commutation rules

$$[\sigma_x, \sigma_y] = 2\imath\sigma_z, \quad [\sigma_y, \sigma_z] = 2\imath\sigma_x, \quad [\sigma_z, \sigma_x] = 2\imath\sigma_y. \tag{3.1.41}$$

As $\mathscr{H}_{\text{SO}}$ contains complex quantities, the total Hamiltonian no longer commutes with C. Alternatively we may express the spin matrices $\boldsymbol{\sigma}$ by the quaternions $\boldsymbol{\tau}$. Both quantities are related to each other via

$$\boldsymbol{\tau} = \imath\boldsymbol{\sigma}, \tag{3.1.42}$$

whence follows for the quaternion matrices

$$\tau_x = \begin{pmatrix} \cdot & \imath \\ \imath & \cdot \end{pmatrix}, \quad \tau_y = \begin{pmatrix} \cdot & 1 \\ -1 & \cdot \end{pmatrix}, \quad \tau_z = \begin{pmatrix} \imath & \cdot \\ \cdot & -\imath \end{pmatrix}. \tag{3.1.43}$$

We immediately verify the relations

$$\tau_x^2 = \tau_y^2 = \tau_y^2 = -1, \tag{3.1.44}$$

and

$$\tau_x\tau_y = -\tau_y\tau_x, \quad \tau_x\tau_z = -\tau_z\tau_x, \quad \tau_y\tau_z = -\tau_z\tau_y. \tag{3.1.45}$$

As the squares of all quaternions are -1, the quaternions may be considered as a generalization of the ordinary complex numbers to complex numbers with three imaginary units. For this reason the quaternions are sometimes denoted by i, j, k. The $\imath$ in the definition (3.1.42) of the quaternions cancels against the $\imath^{-1}$ in the definition (3.1.38) of the angular momentum. This cancellation has motivated the introduction of the quaternions. Therefore the total Hamiltonian of a system with spin-$\frac{1}{2}$ interaction can be written as

$$\begin{aligned} \mathscr{H} &= \mathscr{H}_0 + \boldsymbol{H}\boldsymbol{\tau} \\ &= \mathscr{H}_0 + \mathscr{H}_x\tau_x + \mathscr{H}_y\tau_y + \mathscr{H}_z\tau_z, \end{aligned} \tag{3.1.46}$$

where $\mathscr{H}_0$, $\mathscr{H}_x$, $\mathscr{H}_y$ and $\mathscr{H}_z$ are real operators. τ_x and τ_z, however, are imaginary and thus *anti*commute with C. It is impossible to find representations of the quaternions with real matrices only. But this can be compensated by replacing C by $C\tau_y$. Using the anti-commutation relations (3.1.45), we find that this operator commutes with all the components of $\boldsymbol{\tau}$,

$$[C\tau_y, \tau_x] = [C\tau_y, \tau_y] = [C\tau_y, \tau_z] = 0. \tag{3.1.47}$$

Therefore the total Hamiltonian $\mathscr{H}$ also commutes with $C\tau_y$. If ψ_n is an eigenfunction of $\mathscr{H}$ then $C\tau_y\psi_n$ is another eigenfunction to the same eigenvalue. This is *Kramer's degeneracy*, observed in all spin-$\frac{1}{2}$ systems with time-reversal symmetry. If the eigenfunctions are not yet known then it is useful to choose pairs of spinors ϕ_n and $\overline{\phi}_n = C\tau_y\phi_n$ as basis functions. With these basis functions the matrix elements of $\mathscr{H}$ can be combined into 2×2 matrices,

$$H_{nm} = \begin{pmatrix} \langle\overline{\phi}_n|\mathscr{H}|\overline{\phi}_m\rangle & \langle\overline{\phi}_n|\mathscr{H}|\phi_m\rangle \\ \langle\phi_n|\mathscr{H}|\overline{\phi}_m\rangle & \langle\phi_n|\mathscr{H}|\phi_m\rangle \end{pmatrix}. \tag{3.1.48}$$

The four matrix elements are not independent of each other. As an example the upper left element will be considered:

$$\begin{aligned} \langle\overline{\phi}_n|\mathscr{H}|\overline{\phi}_m\rangle &= \int (C\tau_y\phi_n)^\dagger \mathscr{H}(C\tau_y\phi_m)\, dx \\ &= \int (\phi_n^\dagger\tau_y^\dagger)^* \mathscr{H}(\tau_y\phi_m)^*\, dx \\ &= \left(\int \phi_n^\dagger\tau_y^\dagger\mathscr{H}^*\tau_y\phi_m\, dx\right)^* \\ &= \left(\int \phi_n^\dagger\mathscr{H}\phi_m\, dx\right)^*, \end{aligned} \tag{3.1.49}$$

or

$$\langle\overline{\phi}_n|\mathscr{H}|\overline{\phi}_m\rangle = (\langle\phi_n|\mathscr{H}|\phi_m\rangle)^*. \tag{3.1.50}$$

In the last step in Eq. (3.1.49) we have used (i) that $\mathscr{H}$ commutes with $C\tau_y$, i.e. $H^*\tau_y = \tau_y H$, and (ii) that $\tau^\dagger{}_y\tau_y = 1$. The lower right element of the matrix (3.1.48) is thus the conjugate complex of the upper left element. In the same way we find that the nondiagonal elements are related via

$$\langle\phi_n|\mathscr{H}|\overline{\phi}_m\rangle = -(\langle\overline{\phi}_n|\mathscr{H}|\phi_m\rangle)^*. \tag{3.1.51}$$

H_{nm} can then be written as

$$H_{nm} = \begin{pmatrix} (H_0)_{nm} + \imath(H_z)_{nm} & (H_y)_{nm} + \imath(H_x)_{nm} \\ -(H_y)_{nm} + \imath(H_x)_{nm} & (H_0)_{nm} - \imath(H_z)_{nm} \end{pmatrix}, \tag{3.1.52}$$

or, in shorthand notation,

$$
\begin{aligned}
H_{nm} &= (H_0)_{nm}1 + (H_x)_{nm}\tau_x + (H_y)_{nm}\tau_y + (H_z)_{nm}\tau_z \\
&= (H_0)_{nm}1 + \boldsymbol{H}_{nm}\boldsymbol{\tau},
\end{aligned} \tag{3.1.53}
$$

where $(H_0)_{nm}$ and the components of $\boldsymbol{H}_{nm}$ are real, and where 1 is the 2×2 unit matrix. Matrices with this property are called *quaternion real*. It should be noted that the rank of the matrix representing $\mathcal{H}$ is $2N$, the factor 2 resulting from the two possible spin orientations. It is simply a matter of convenience that we write H as an $N \times N$ matrix, where the matrix elements are 2×2 matrices themselves.

We can again look for the transformations leaving the form (3.1.53) of the matrix elements invariant. These are the *symplectic* transformations,

$$
H' = SHS^R, \tag{3.1.54}
$$

where S is symplectic, $SS^R = 1$. Here S^R is the dual of S defined by

$$
S^R = ZS^T Z^{-1} = -ZS^T Z, \tag{3.1.55}
$$

where Z, looked upon as an $N \times N$ matrix, is diagonal with the matrix elements

$$
Z_{nm} = \delta_{nm}\tau_y. \tag{3.1.56}
$$

The presentation of the universality classes is now complete. We have found that

– Hamiltonians without time-reversal symmetry can be represented by Hermitian matrices being invariant under unitary transformations,
– Hamiltonians with time-reversal symmetry but no spin-$\frac{1}{2}$ interaction can be represented by real matrices being invariant under orthogonal transformations,
– Hamiltonians with time-reversal symmetry plus spin-$\frac{1}{2}$ interactions can be represented by quaternion real matrices being invariant under symplectic transformations.

By far the most experiments have been performed in systems belonging to the orthogonal universality class. For the unitary class only two experiments exist, and there are none for the symplectic class. Therefore mainly the first two cases will be considered in the following.

To illustrate why it is so difficult to realize systems without time-reversal symmetry experimentally, we take as an example the hydrogen atom in an external magnetic field. Its Hamilton operator is given by

$$
\mathcal{H} = \frac{p^2}{2m} - \frac{e^2}{r} - \frac{eB}{2mc}(L_z + gS_z) + \frac{e^2B^2}{8mc^2}(x^2 + y^2). \tag{3.1.57}
$$

It is obtained from Eq. (3.1.29) with a vector potential given by

$$
\boldsymbol{A} = \frac{B}{2}(-y, x, 0). \tag{3.1.58}
$$

The hydrogen atom in a strong magnetic field is the 'guinea pig' of quantum chaos. It will be discussed in detail in Section 8.1.2. Here only the aspect of time-reversal symmetry will be considered. Classically the electron spirals up and down under the combined influence of the external magnetic field and the Coulomb attraction by the nucleus. Obviously time-reversal symmetry is broken by the Lorentz force, and we might think that the system belongs to the unitary class. But the break of time-reversal symmetry can be compensated by a reflection at the xy plane. The system is invariant neither with respect to the time-reversal symmetry T nor to the mirror reflection P. It is, however, invariant with respect to the combined operation PT. It is easy to see that due to this symmetry it is again possible to represent the Hamiltonian by a real symmetric matrix. In principle it should be possible to break the mirror symmetry by applying an inhomogenous magnetic field, but this experiment has not yet been performed. The situation is easier with billiards where it is trivial to avoid mirror symmetries, and indeed the first experiments with broken time-reversal symmetry have been realized in this field [So95, Sto95].

3.1.3 Definition of the Gaussian ensembles

Wigner distributions for nearest neighbour spacings are observed in a large variety of very different systems, ranging from the atomic nucleus to the microwave billiard. This suggests that the details of interaction are not relevant at all. In random matrix theory this fact is accounted for by replacing the Hamiltonian by a matrix whose elements are randomly chosen. This is surely the most radical simplification possible.

We are not completely free in fixing the matrix elements. First of all the universality class restrictions for the matrix elements derived in the last section have to be obeyed. For the orthogonal class which will now be discussed exemplarily, the Hamiltonian can be represented by a real symmetric matrix, leading to $N(N+1)/2$ independent matrix elements. In a fully chaotic system one set of basis functions is a priori suited as well as the other. The correlated probability $p(H_{11}, \ldots, H_{NN})$ for the matrix elements should therefore not depend on the set of basis functions applied. This leads to the invariance property

$$p(H_{11}, \ldots, H_{NN}) = p(H'_{11}, \ldots, H'_{NN}) \tag{3.1.59}$$

H' is obtained from H by an orthogonal transformation $H' = OHO^T$ with $OO^T = 1$. By means of relation (3.1.59) the number of possible forms for $p(H_{11}, \ldots, H_{NN})$ is radically reduced. Functions of the H_{nm} being invariant under orthogonal transformations can depend only on traces of powers of H. This is a consequence of the commutativity property

$$\mathrm{Tr}(OAO^T) = \mathrm{Tr}(AO^TO) = \mathrm{Tr}(A) \tag{3.1.60}$$

of the trace. Thus $p(H_{11}, \ldots, H_{NN})$ can be expressed as

$$p(H_{11}, \ldots, H_{NN}) = f[\mathrm{Tr}(H), \mathrm{Tr}(H^2), \ldots]. \tag{3.1.61}$$

Moreover, we demand that the matrix elements are uncorrelated,

$$p(H_{11}, \ldots, H_{NN}) = p(H_{11})p(H_{12}) \ldots p(H_{NN}). \tag{3.1.62}$$

The only possible functional form for $p(H_{11}, \ldots, H_{NN})$ obeying both Eq. (3.1.61) and Eq. (3.1.62) is given by

$$p(H_{11}, \ldots, H_{NN}) = C \exp[-B\,\mathrm{Tr}(H) - A\,\mathrm{Tr}(H^2)]. \tag{3.1.63}$$

Without loss of generality we may take $B = 0$ as it is always possible to shift the average energy, $1/N\,\mathrm{Tr}(H)$, to zero. The prefactor C is fixed with the help of the normalization condition

$$\int p(H_{11}, \ldots, H_{NN})dH_{11} \ldots dH_{NN} = 1. \tag{3.1.64}$$

For the orthogonal case the nondiagonal matrix elements occur twice in the exponential function each yielding a normalization prefactor $\sqrt{2A/\pi}$, whereas the normalization factor for the diagonal elements is given by $\sqrt{A/\pi}$. We thus end with

$$p(H_{11}, \ldots, H_{NN}) = \left(\frac{A}{\pi}\right)^{N/2} \left(\frac{2A}{\pi}\right)^{N(N-1)/2} \exp\left(-A\sum_{n,m} H_{nm}^2\right). \tag{3.1.65}$$

The constant A may be expressed in terms of the variance either of the diagonal matrix elements

$$\langle H_{nn}^2 \rangle = \sqrt{\frac{A}{\pi}} \int H_{nn}^2 \exp(-AH_{nn}^2)dH_{nn} = \frac{1}{2A}, \tag{3.1.66}$$

or of the nondiagonal elements

$$\langle H_{nm}^2 \rangle = \sqrt{\frac{2A}{\pi}} \int H_{nm}^2 \exp(-2AH_{nm}^2)dH_{nm} = \frac{1}{4A}. \tag{3.1.67}$$

The set of all real random matrices with matrix elements obeying the distribution function (3.1.65) defines the *Gaussian Orthogonal Ensemble* (GOE). In complete analogy we obtain the *Gaussian Unitary Ensemble* (GUE) and the *Gaussian Symplectic Ensemble* (GSE) by the demand that the distribution of the matrix elements is invariant under unitary and symplectic transformations, respectively. For the correlated distribution of matrix elements of the GUE we get

$$p(H_{11}, \ldots, H_{NN}) = \left(\frac{A}{\pi}\right)^{N/2} \left(\frac{2A}{\pi}\right)^{N(N-1)} \times \exp\left\{-A \sum_{n,m} [(H_R)^2_{nm} + (H_I)^2_{nm}]\right\}, \tag{3.1.68}$$

where $(H_R)_{nm}$ and $(H_I)_{nm}$ are the real and imaginary parts of H_{nm}, respectively. The probability distribution for the matrix elements of the GSE is given by

$$p(H_{11}, \ldots, H_{NN}) = \left(\frac{A}{\pi}\right)^{N/2} \left(\frac{2A}{\pi}\right)^{2N(N-1)} \times \exp\left\{-A \sum_{n,m} [(H_0)^2_{nm} + (H_x)^2_{nm} + (H_y)^2_{nm} + (H_z)^2_{nm}]\right\}, \tag{3.1.69}$$

where $(H_0)_{nm}$, $(H_x)_{nm}$, $(H_y)_{nm}$, and $(H_z)_{nm}$ are the quaternionic components of $\boldsymbol{H}_{nm}$ (see Eq. (3.1.53)).

As it is of principal interest, yet another motivation for the distribution of matrix elements of the Gaussian ensembles will be given. It uses an argument from information theory [Bal68]. The information contained in the distribution function $p(H_{11}, \ldots, H_{NN})$ is given by

$$I = -\int p(H_{11}, \ldots, H_{NN}) \mathrm{ld}[p(\mathrm{H}_{11}, \ldots, H_{NN})]\, dH_{11} \ldots dH_{NN}, \tag{3.1.70}$$

where $\mathrm{ld}(x)$ is the logarithm to the base two. To give an idea of the background to this formula to readers not familiar with information theory, let us consider the following example. Take a deck of 32 cards. Ask a friend to select one card. What is now the best strategy to determine the selected card with as few questions as possible? The answer is simple. The first question will consider the colour of the card. If the answer is 'red', the next question will be 'diamond' or 'heart', and so on. After five questions the card will be identified. Several card tricks are based on this principle. Now consider the general case that there are n possible answers each with probability p_i. The total number of possible outcomes after m questions is given by

$$N_m = \frac{m!}{\prod_{i=1}^{n} (p_i m)!}. \tag{3.1.71}$$

To simplify the discussion it has been assumed that all p_i are rational and that m is chosen in such a way that all $p_i m$ are integers. The information contained in a quantity is defined as the number of bits necessary to encode the quantity. For the number of possible outcomes after m questions the information is thus given by $\mathrm{ld}(N_m)$. The average information obtained by *one* question is then given by

$$I = \lim_{m\to\infty} \frac{1}{m} \mathrm{ld}(N_m). \tag{3.1.72}$$

Using Stirling's formula

$$n! \approx \sqrt{2\pi n}\left(\frac{n}{e}\right)^n \tag{3.1.73}$$

we obtain for the information

$$I = -\sum_{i=1}^{n} p_i \mathrm{ld}(p_i). \tag{3.1.74}$$

This expression holds for the case that the number of possible answers is discrete. If one asks for the magnitude of a matrix element, the range of possible answers is continuous, and the sum has to be replaced by an integral.

Let us now ask for the distribution function $p(H_{11}, \ldots, H_{NN})$, normalized to one, which, for a given variance, contains the least information. To this end we have to minimize the expression (3.1.70) where in addition the two conditions

$$\int p(H_{11}, \ldots, H_{NN})\, dH_{11} \ldots dH_{NN} = 1, \tag{3.1.75}$$

and

$$\int \mathrm{Tr}(H^2) p(H_{11}, \ldots, H_{NN})\, dH_{11} \ldots dH_{NN} = \text{const.} \tag{3.1.76}$$

have to be fulfilled. Using the method of the Lagrange multiplier this leads to the functional

$$F(p) = \int p(H_{11}, \ldots, H_{NN})[\ln p(H_{11}, \ldots, H_{NN}) + \lambda_1 + \lambda_2 \,\mathrm{Tr}(H^2)] \times dH_{11} \ldots dH_{NN}. \tag{3.1.77}$$

We have replaced the binary logarithm by the more familiar natural one. This corresponds to the multiplication of $F(p)$ by an irrelevant constant factor. The minimum of $F(p)$ is obtained from the condition that its variation with respect to p is zero,

$$\delta F = \int [1 + \ln p(H_{11}, \ldots, H_{NN}) + \lambda_1 + \lambda_2 \,\mathrm{Tr}(H^2)]\delta p\, dH_{11} \ldots dH_{NN} = 0, \tag{3.1.78}$$

whence follows

$$p(H_{11}, \ldots, H_{NN}) = \exp[-1 - \lambda_1 - \lambda_2 \operatorname{Tr}(H^2)]. \tag{3.1.79}$$

We have again arrived at the distribution function for the matrix elements of the Gaussian ensembles. They can thus be understood alternatively as the ensembles of matrices whose matrix element distribution contains the least information for a given variance.

3.1.4 Correlated eigenenergy distribution

The distribution function of the matrix elements of the Gaussian ensembles derived in the last section does not allow a direct comparison with the experiment, since usually only the eigenenergies are accessible. The correlated distribution function for the eigenenergies is therefore of more practical interest. In the following we shall calculate this quantity for the example of the orthogonal case. Using the fact that every symmetric matrix H can be diagonalized by means of an orthogonal transformation we obtain

$$H = OH_D O^T, \tag{3.1.80}$$

or

$$H_{nm} = \sum_k O_{nk} E_k O_{mk}, \tag{3.1.81}$$

where H_D is the diagonal matrix with the matrix elements

$$(H_D)_{kl} = E_k \delta_{kl}. \tag{3.1.82}$$

An orthogonal matrix O is characterized by $\frac{1}{2}N(N-1)$ independent variables p_α. This is easily seen: a unit vector of length N is determined by $N-1$ variables, a second unit vector perpendicular to the first one by $N-2$ variables, and so on. Then the total number of independent variables is $(N-1)+(N-2)+\cdots+1 = \frac{1}{2}N(N-1)$. In Eq. (3.1.81) the $\frac{1}{2}N(N+1)$ variables H_{nm} on the left hand side are expressed in terms of the N eigenenergies E_n and the $\frac{1}{2}N(N-1)$ variables p_α on the right hand side. To get explicit expressions for E_k and p_α in terms of H_{nm}, we would actually have to diagonalize H, but fortunately this is not needed. The correlated distribution function for the matrix elements in terms of the new variables reads

$$p(H_{11}, \ldots, H_{NN})\, dH_{nm} \sim \exp\left(-A \sum_k E_k^2\right) |J|\, dE_k\, dp_\alpha, \tag{3.1.83}$$

where we have used that

$$\sum_{n,m} H_{nm}^2 = \sum_k E_k^2. \tag{3.1.84}$$

This follows from the independence of the trace of the basis applied. It remains to determine the Jacobi determinant

$$|J| = \left| \frac{\partial(H_{nm})}{\partial(E_k,\ p_\alpha)} \right| \tag{3.1.85}$$

of the transformation. To this end we differentiate Eq. (3.1.80) with respect to E_k and p_α, respectively, and obtain

$$\frac{\partial H}{\partial E_k} = O \frac{\partial H_D}{\partial E_k} O^T, \tag{3.1.86}$$

and

$$\begin{aligned} \frac{\partial H}{\partial p_\alpha} &= \frac{\partial O}{\partial p_\alpha} H_D O^T + O H_D \frac{\partial O^T}{\partial p_\alpha} \\ &= O\big(S_\alpha H_D - H_D S_\alpha\big) O^T, \end{aligned} \tag{3.1.87}$$

where we have introduced the matrix

$$S_\alpha = O^T \frac{\partial O}{\partial p_\alpha} = -\frac{\partial O^T}{\partial p_\alpha} O. \tag{3.1.88}$$

The equality of the last two expressions follows from $O^T O = 1$. From Eqs. (3.1.86) and (3.1.87) we get for the matrix elements

$$\frac{\partial H_{nm}}{\partial E_k} = O_{nk} O_{mk}, \tag{3.1.89}$$

and

$$\begin{aligned} \frac{\partial H_{nm}}{\partial p_\alpha} &= \sum_{i,j} O_{ni} O_{mj} (S_\alpha H_D - H_D S_\alpha)_{ij} \\ &= \sum_{i,j} O_{ni} O_{mj} (S_\alpha)_{ij} (E_j - E_i). \end{aligned} \tag{3.1.90}$$

The matrix elements of the Jacobi determinant can therefore be written as

$$J_{nm,k\alpha} = \sum_{i,j} O_{ni} O_{mj} M_{ij,k\alpha}, \tag{3.1.91}$$

where the $M_{ij,k\alpha}$ are the elements of the matrix

$$M = \begin{pmatrix} \delta_{ij} & 0 \\ 0 & (S_\alpha)_{ij}(E_j - E_i) \end{pmatrix}. \tag{3.1.92}$$

The ordering is such that in the columns the N H_{ii} come first, followed by the $\frac{1}{2}N(N-1)$ H_{ij} with $i > j$. In the rows the N eigenvalues E_k are followed by the $\frac{1}{2}N(N-1) p_\alpha$.

Equation (3.1.91) shows that J can be written as the product of the two $\frac{1}{2}N(N+1) \times \frac{1}{2}N(N+1)$ matrices $\hat{O}$ and M, where $\hat{O}$ is the matrix with the elements $(\hat{O})_{nm,ij} = O_{ni} O_{mj}$. By applying elementary determinant rules we get

$$|J| = |\hat{O}|\cdot|M|$$

$$= |\hat{O}|\cdot|S|\cdot \prod_{i>j}(E_i - E_j). \tag{3.1.93}$$

Here S is the $\frac{1}{2}N(N-1)\times\frac{1}{2}N(N-1)$ matrix with the elements $(S_\alpha)_{ij}$. Neither $\hat{O}$ nor S are known, but this does not matter, since both quantities depend on p_α alone. To obtain the correlated energy distribution function from Eq. (3.1.83), we have to integrate over the p_α and obtain

$$P(E_1, \ldots, E_N) \sim \prod_{n>m}(E_n - E_m)\exp\left(-A\sum_n E_n^2\right) \tag{3.1.94}$$

for the correlated eigenenergy distribution function. The derivation of the corresponding equations for the other ensembles can be found in the book by Mehta [Meh91]. The expressions for all ensembles can be cast into one single formula

$$P(E_1, \ldots, E_N) \sim \prod_{n>m}(E_n - E_m)^\nu \exp\left(-A\sum_n E_n^2\right), \tag{3.1.95}$$

where ν is the universality index. It takes the values 1, 2 and 4 for the GOE, the GUE and the GSE, respectively. For $\nu = 0$ the eigenvalues are uncorrelated. We have already seen in Section 3.1.1 that in this case the nearest neighbour spacings are described by a Poisson distribution. Accordingly the ensemble of matrices with an energy distribution function described by Eq. (3.1.95) with $\nu = 0$ is called the Poisson ensemble.

The correlated energy distribution function contains all relevant information on the Gaussian ensembles. One might therefore believe that with the derivation of an expression for $P(E_1, \ldots, E_N)$ the larger part of the work is already done, but this is not true. Whenever trying to extract explicit expressions, e.g. for the density of states or the nearest neighbour spacing distribution, we are confronted with serious mathematical problems. The major part of Mehta's monograph [Meh91] deals with these questions.

3.1.5 Averaged density of states

A fundamental property of a quantum mechanical system is its density of states

$$\rho(E) = \sum_n \delta(E - E_n). \tag{3.1.96}$$

In the case of the Gaussian ensembles we are not interested in the density of states of an individual random matrix, but in its average $\langle\rho(E)\rangle$ over the complete ensemble. There are several ways to calculate this quantity. We might spontaneously think that the correlated energy distribution function derived in Section 3.1.4 is the best starting point. This approach is, however, a bit more complicated, as we shall see later in Section 3.2.3. Supersymmetry techniques are another alternative. They will be discussed at the end of this chapter, where the calculation of the density of states of the GUE will serve as an illustrative example.

For the beginning a more direct approach is applied. To this end we write the delta function in Eq. (3.1.96) as the limit of a Lorentzian curve with vanishing width,

$$\delta(E) = \lim_{\epsilon \to 0} \frac{\epsilon}{\pi} \frac{1}{E^2 + \epsilon^2}, \tag{3.1.97}$$

resulting in

$$\begin{aligned} \rho(E) &= \lim_{\epsilon \to 0} \frac{\epsilon}{\pi} \sum_n \frac{1}{(E - E_n)^2 + \epsilon^2} \\ &= -\lim_{\epsilon \to 0} \frac{1}{\pi} \operatorname{Im} \left(\sum_n \frac{1}{E - E_n + \imath\epsilon} \right) \end{aligned} \tag{3.1.98}$$

for the density of states. Using

$$\sum_n \frac{1}{E - E_n} = \operatorname{Tr}\left(\frac{1}{E - H} \right), \tag{3.1.99}$$

this can be written as

$$\rho(E) = -\frac{1}{\pi} \operatorname{Im}\left(\operatorname{Tr}\left(\frac{1}{E - H} \right) \right). \tag{3.1.100}$$

In the last equation the limit $\epsilon \to 0$ has no longer been explicitly written to simplify the notation. Whenever we meet an expression of this type in the following, it is understood that there is an infinitesimally small positive imaginary part of E.

The density of states can alternatively be expressed in terms of the quantum mechanical Green function

$$\begin{aligned} G(q_A, q_B, E) &= \sum_n \frac{\psi_n^*(q_A)\psi_n(q_B)}{E - E_n} \\ &= \sum_n \psi_n^*(q_A) \frac{1}{E - H} \psi_n(q_B), \end{aligned} \tag{3.1.101}$$

where $\psi_n(q)$ are the eigenfunctions of H to the eigenvalue E_n (see Chapter 7 of Ref. [Mor53]). Using

$$\sum_n \psi_n^*(q_A)\psi_n(q_B) = \delta(q_A - q_B)$$

$$= \int \delta(q_A - q)\delta(q_B - q)\, dq, \qquad (3.1.102)$$

Eq. (3.1.101) may be written as

$$G(q_A, q_B, E) = \int \delta(q_a - q) \frac{1}{E - H} \delta(q_B - q)\, dq$$

$$= \left\langle q_A \left| \frac{1}{E - H} \right| q_B \right\rangle, \qquad (3.1.103)$$

where we have introduced the delta function $\delta(q_A - q) = |q_A\rangle$ as the eigenfunction of the position operator:

$$q|q_A\rangle = q_A|q_A\rangle. \qquad (3.1.104)$$

The Green function can thus be interpreted as the matrix element of the operator $(E - H)^{-1}$ with the eigenfunctions of the position operator as basis functions. Thus Eq. (3.1.100) for the density of states may be alternatively expressed as

$$\rho(E) = -\frac{1}{\pi} \operatorname{Im}(\operatorname{Tr}(G)). \qquad (3.1.105)$$

Sometimes the Green function entering Eq. (3.1.105) is denoted by G^+ to indicate that E has been assumed to have a small positive imaginary part, whereas the corresponding Green function, where E has a small negative imaginary part, is denoted by G^-. Since only G^+ will be used in the following, this discrimination is not necessary in the present context.

In Eq. (3.1.100) the eigenvalues no longer enter explicitly. Instead the average of the trace of an operator inverse must be calculated

$$S = \left\langle \operatorname{Tr}\left(\frac{1}{E - H}\right) \right\rangle. \qquad (3.1.106)$$

This is done by expanding $(E - H)^{-1}$ into its Taylor series,

$$S = \sum_{n=0}^{\infty} \frac{1}{E^{n+1}} \langle \operatorname{Tr} H^n \rangle. \qquad (3.1.107)$$

The series only converges if $|E|$ exceeds all eigenvalues of H in magnitude. For the calculation of the averaged density of states, on the other hand, we need S in the range of the eigenvalues, where the expansion diverges. In such cases it is advisable to ignore the problem at first. When finally arriving at an analytical expression for S (this will be the case), it will be valid everywhere due to the principle of analytic continuation.

The problem is now reduced to the calculation of the ensemble average of the trace of H^n. We shall perform the average for the example of the GUE. For the other ensembles the calculation proceeds in a similar manner. From the distribution probability $p(H_{11}, \ldots, H_{NN})$ for the GUE (see Eq. (3.1.68)) we obtain

$$\begin{aligned}\langle H_{\alpha\beta} H_{\beta\alpha} \rangle &= \int H_{\alpha\beta} H_{\beta\alpha}\, p(H_{11}, \ldots, H_{NN})\, dH_{11} \ldots dH_{NN} \\ &= \frac{1}{2A}, \end{aligned} \tag{3.1.108}$$

holding both for $\alpha = \beta$ and $\alpha \neq \beta$. The ensemble averages of all other products of two matrix elements vanish,

$$\langle H_{\alpha\beta} H_{\gamma\delta} \rangle = 0 \quad \text{if} \quad (\alpha, \beta) \neq (\delta, \gamma). \tag{3.1.109}$$

From Eq. (3.1.108) we get for the average of the trace of H^2

$$\langle \mathrm{Tr}\, H^2 \rangle = \sum_{\alpha,\beta} \langle |H_{\alpha\beta}|^2 \rangle = \frac{N^2}{2A}. \tag{3.1.110}$$

The averages of the traces of all odd powers of H vanish,

$$\langle \mathrm{Tr} H^{2n+1} \rangle = 0, \tag{3.1.111}$$

as is immediately clear from symmetry considerations. The first nontrivial case is $n = 4$,

$$\langle \mathrm{Tr}\, H^4 \rangle = \left\langle \sum_{\alpha,\beta,\gamma,\delta} H_{\alpha\beta} H_{\beta\gamma} H_{\gamma\delta} H_{\delta\alpha} \right\rangle. \tag{3.1.112}$$

In the ensemble average only terms survive where matrix elements $H_{\alpha\beta}$ and $H_{\beta\alpha}$ occur pairwise. To this end we introduce the bracket notation $\overline{H_{\alpha\beta} H_{\gamma\delta}}$, denoting that only terms with $(\alpha, \beta) = (\delta, \gamma)$ are taken in the sums. Then there are four surviving terms, the first one given by

$$\left\langle \sum_{\alpha,\beta,\gamma,\delta} \overline{H_{\alpha\beta} H_{\beta\gamma}}\,\overline{H_{\gamma\delta} H_{\delta\alpha}} \right\rangle = \left\langle \sum_{\alpha,\beta,\delta} H_{\alpha\beta} H_{\beta\alpha} H_{\alpha\delta} H_{\delta\alpha} \right\rangle = \mathcal{O}(N^3), \tag{3.1.113}$$

where $\mathcal{O}(N^3)$ denotes that the number of terms in the sum is N^3. Two further terms are given by

$$\left\langle \sum_{\alpha,\beta,\gamma,\delta} \overline{H_{\alpha\beta} \underline{H_{\beta\gamma} H}_{\gamma\delta} H_{\delta\alpha}} \right\rangle = \left\langle \sum_{\alpha,\beta,\gamma} H_{\alpha\beta} H_{\beta\gamma} H_{\gamma\beta} H_{\beta\alpha} \right\rangle = \mathcal{O}(N^3), \tag{3.1.114}$$

and

$$\left\langle \sum_{\alpha,\beta,\gamma,\delta} H_{\alpha\beta} H_{\beta\gamma} H_{\gamma\delta} H_{\delta\alpha} \right\rangle = \left\langle \sum_{\alpha,\beta} H_{\alpha\beta} H_{\beta\beta} H_{\beta\alpha} H_{\alpha\alpha} \right\rangle$$

$$= \left\langle \sum_{\alpha} H_{\alpha\alpha}^4 \right\rangle = \mathcal{O}(N). \tag{3.1.115}$$

The case is left where $H_{\alpha\beta}$ and $H_{\beta\alpha}$ occur twice within one term,

$$\left\langle \sum_{\alpha,\beta,\gamma,\delta} H_{\alpha\beta} H_{\beta\gamma} H_{\gamma\delta} H_{\delta\alpha} \right\rangle = \left\langle \sum_{\alpha,\beta} |H_{\alpha\beta}|^4 \right\rangle = \mathcal{O}(N^2). \tag{3.1.116}$$

This shows that sums with interlacing brackets as well as sums containing the same matrix element repeatedly are of lower order in N than the sums with noninterlacing brackets. In the limit of large N therefore only the latter terms have to be considered. This facilitates the calculation considerably. Introducing the abbreviation

$$M_n = \langle \mathrm{Tr}\, H^{2n} \rangle, \tag{3.1.117}$$

Eq. (3.1.107) can be written as

$$S = \sum_{n=0}^{\infty} \frac{1}{E_{2n+1}} M_n. \tag{3.1.118}$$

To derive a recursion formula for the M_n we write

$$M_n = \left\langle \sum_{\alpha} (H^{2n})_{\alpha\alpha} \right\rangle$$

$$= \left\langle \sum_{\alpha} H_{\alpha\beta} (H^{2n-1})_{\beta\alpha} \right\rangle. \tag{3.1.119}$$

In order that a given term survives the averaging, one of the factors of H^{2n-1} must be identical with $H_{\beta\alpha}$. As each of the $(2n-1)$ factors can assume this role we get

$$M_n = \sum_{k=0}^{2n-2} \sum_{\alpha,\beta,\gamma,\delta} \langle H_{\alpha\beta} (H^k)_{\beta\gamma} H_{\gamma\delta} (H^{2n-k-2})_{\delta\alpha} \rangle$$

$$= \sum_{k=0}^{2n-2} \sum_{\alpha,\beta} \langle H_{\alpha\beta} (H^k)_{\beta\beta} H_{\beta\alpha} (H^{2n-k-2})_{\alpha\alpha} \rangle. \tag{3.1.120}$$

Now we use the fact that contributions of the types (3.1.115) and (3.1.116) are negligible in the limit of large N. Then the ensemble average in Eq. (3.1.120) factorizes:

$$M_n = \sum_{k=0}^{2n-2} \sum_{\alpha,\beta} \langle |H_{\alpha\beta}|^2 \rangle \langle (H^k)_{\beta\beta} \rangle \langle (H^{2n-k-2})_{\alpha\alpha} \rangle. \tag{3.1.121}$$

Using Eq. (3.1.108) this may be written as

$$\begin{aligned} M_n &= \frac{1}{2A} \sum_{k=0}^{2n-2} \sum_{\alpha,\beta} \langle (H^k)_{\beta\beta} \rangle \langle (H^{2n-k-2})_{\alpha\alpha} \rangle \\ &= \frac{1}{2A} \sum_{k=0}^{2n-2} \langle \mathrm{Tr}\, H^k \rangle \langle \mathrm{Tr}\, H^{2n-k-2} \rangle, \end{aligned} \tag{3.1.122}$$

or

$$M_n = \frac{1}{2A} \sum_{k=0}^{n-1} M_k M_{n-k-1}, \tag{3.1.123}$$

where in the last step we took into account that only the traces of the even powers of H survive the ensemble averages. By means of the initial condition

$$M_0 = \mathrm{Tr}(1) = N, \tag{3.1.124}$$

Equation (3.1.123) may be used to calculate the M_n recursively. But an explicit knowledge of the M_n is not even needed here. We may instead directly enter the recursion relation (3.1.123) into expression (3.1.118) for S,

$$\begin{aligned} S &= \sum_{n=0}^{\infty} \frac{1}{E^{2n+1}} M_n \\ &= \frac{1}{E} \left(N + \sum_{n=1}^{\infty} \frac{1}{E^{2n}} M_n \right) \\ &= \frac{1}{E} \left(N + \sum_{n=1}^{\infty} \frac{1}{E^{2n}} \frac{1}{2A} \sum_{k=0}^{n-1} M_k M_{n-k-1} \right). \end{aligned} \tag{3.1.125}$$

By changing the order of summation and subsequently shifting the summation index we get

$$
\begin{aligned}
S &= \frac{1}{E}\left(N + \frac{1}{2A}\sum_{k=0}^{\infty}\sum_{n=k+1}^{\infty}\frac{1}{E^{2n}}M_k M_{n-k-1}\right) \\
&= \frac{1}{E}\left(N + \frac{1}{2A}\sum_{k=0}^{\infty}\sum_{n=0}^{\infty}\frac{1}{E^{2(n+k+1)}}M_k M_n\right) \\
&= \frac{1}{E}\left(N + \frac{1}{2A}\sum_{k=0}^{\infty}\frac{1}{E^{2k+1}}M_k\sum_{n=0}^{\infty}\frac{1}{E^{2n+1}}M_n\right) \\
&= \frac{1}{E}\left(N + \frac{1}{2A}S^2\right).
\end{aligned}
\tag{3.1.126}
$$

We have ended with a quadratic equation for S which is easily solved:

$$
S = \left\langle \mathrm{Tr}\left(\frac{1}{E-H}\right)\right\rangle = AE\left(1 - \sqrt{1 - \frac{2N}{AE^2}}\right). \tag{3.1.127}
$$

For the sign of the square root the negative sign has been taken, as $S \to 0$ for $E \to \infty$ (see Eq. (3.1.118)).

Using Eq. (3.1.100) the averaged density of states is now immediately obtained:

$$
\langle \rho(E) \rangle = \begin{cases} \dfrac{A}{\pi}\sqrt{\dfrac{2N}{A} - E^2} & |E| < \dfrac{\sqrt{2N}}{A} \\ 0 & |E| > \dfrac{\sqrt{2N}}{A} \end{cases}. \tag{3.1.128}
$$

In the limit $E \to 0$ the ensemble averaged density of states becomes constant, $\langle \rho(E) \rangle = \sqrt{2NA}/\pi$. It is a common practice to normalize this quantity to one by taking

$$
A = \frac{\pi^2}{2N}, \tag{3.1.129}
$$

with the result

$$
\langle \rho(E) \rangle = \begin{cases} \sqrt{1 - \left(\dfrac{\pi E}{2N}\right)^2} & |E| < \dfrac{2N}{\pi} \\ 0 & |E| > \dfrac{2N}{\pi} \end{cases}. \tag{3.1.130}
$$

We have thus derived Wigner's *semicircle law*. The averaged density of states for the GUE is described by a semicircle, if the energy is measured in units of $2N/\pi$. The same semicircle law is also observed in the other Gaussian

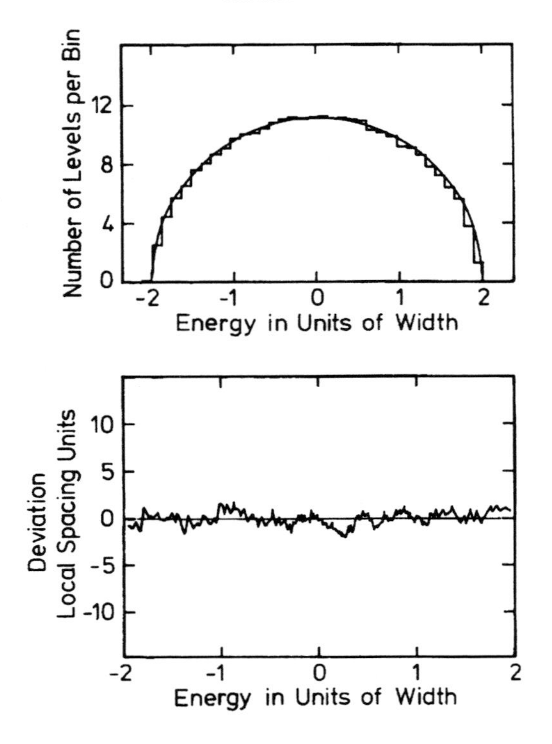

Figure 3.4. Averaged density of states for an ensemble of 50 matrices of the rank of 294 taken from the GOE (top). The bottom part of the figure shows the deviation of the density of states from the semicircle law [Bro81] (Copyright 1981 by the American Physical Society).

ensembles. Figure 3.4 shows a numerical example obtained by superimposing the spectra of 50 GOE random matrices.

For energies $|E| \ll N$ the density of states becomes constant. All future discussions will be restricted to this range. It does not make much sense to go towards the limits of the semicircle. These limits are an artifact of the finite rank of the random matrices. As the Hamiltonians of real systems are represented by matrices of infinite rank, the semicircle law can no longer be a realistic description of a real chaotic system if $|E|$ is of the order of N.

3.2 Spectral correlations

3.2.1 Nearest neighbour distance distribution

There is probably no spectral correlation studied more intensely than the nearest neighbour spacing distribution. It can be calculated from the correlated eigenenergy distribution functions of the Gaussian ensembles derived in

Section 3.1.4, but unfortunately this calculation is rather elaborate and goes beyond the scope of an introductory presentation.

The calculation is performed instead for the special case of the Gaussian ensembles of 2×2 matrices. Here the nearest neighbour spacing distribution $p(s)$ is easily obtained from the correlated energy distribution function $P(E_1, E_2)$ (see Eq. (3.1.95)) by

$$\begin{aligned} p(s) &= \int_{-\infty}^{\infty} dE_1 \int_{-\infty}^{\infty} dE_2 P(E_1, E_2)\delta(s - |E_1 - E_2|) \\ &= C\int_{-\infty}^{\infty} dE_1 \int_{-\infty}^{\infty} dE_2 |E_1 - E_2|^{\nu} \exp\left(-A\sum_n E_n^2\right) \\ &\quad \times \delta(s - |E_1 - E_2|). \end{aligned} \tag{3.2.1}$$

The constants A and C are fixed by the two normalization conditions

$$\int_0^{\infty} p(s)\, ds = 1, \tag{3.2.2}$$

$$\int_0^{\infty} s p(s)\, ds = 1. \tag{3.2.3}$$

The total probability is normalized to one by the first condition, the mean level spacing by the second. The integrations are elementary and yield

$$p(s) = \begin{cases} \dfrac{\pi}{2} s \exp\left(-\dfrac{\pi}{4} s^2\right) & \nu = 1 \text{ (GOE)} \\[2ex] \dfrac{32}{\pi^2} s^2 \exp\left(-\dfrac{4}{\pi} s^2\right) & \nu = 2 \text{ (GUE)}\,. \\[2ex] \dfrac{2^{18}}{3^6 \pi^3} s^4 \exp\left(-\dfrac{64}{9\pi} s^2\right) & \nu = 4 \text{ (GSE)} \end{cases} \tag{3.2.4}$$

These are the Wigner distributions we have already met for the GOE case. Though the distributions have been calculated for Gaussian ensembles of 2×2 matrices only, they are nevertheless able to describe the spectra of random matrices of arbitrary rank with high precision.

As an illustrative example we take the spectra of a spin-$\frac{1}{2}$ particle in a three-dimensional anharmonic oscillator potential with the Hamiltonian [Cau89]

$$\begin{aligned} \mathscr{H} = \frac{1}{2}\left(p_x^2 + p_y^2 + p_z^2\right) &+ x^4 + \frac{1}{2} y^4 + \frac{1}{10} z^4 \\ &+ 12x^2y^2 + 14x^2z^2 + 16y^2z^2 \\ &+ r^2 z(ax + by) + cr\boldsymbol{LS} \end{aligned} \tag{3.2.5}$$

where $r = (x^2 + y^2 + z^2)^{1/2}$. The universal behaviour of the system depends on the values of the parameters a and b. For $a = b = 0$ there are three reflection symmetries with respect to the xy, xz, and yz planes. In this case it is possible to find a real matrix representation of $\mathscr{H}$. Thus the system belongs to the GOE class, despite the presence of the spin-orbit interaction. If either a or b differ from zero, one of the reflection symmetries is destroyed, promoting the system to the GUE class. If both a and b are different from zero, all reflection symmetries are destroyed, and due to the presence of the spin-orbit term we now have a GSE system. Figure 3.5 shows the level spacing distributions obtained for the three classes. The solid curves represent the Wigner distributions (3.2.4). In all three cases perfect agreement is found. The three ensembles mainly differ in their small distance level repulsion behaviour according to $p(s) \sim s^\nu$. The different repulsion exponents $\nu = 1, 2, 4$ are clearly seen in the figure.

In Fig. 3.6(a) the differences between the exact distance distributions and their approximations (3.2.4) are plotted for the three ensembles (see Section 4.4 of Ref. [Haa91a]). The largest deviation is found for the GOE but even here it never exceeds two per cent. For the GUE and the GSE it is smaller than one per cent. Therefore a really excellent database is needed to make visible the difference. Figure 3.6(b) shows an example taken from the class of the kicked tops. Altogether some 10^7 data points entered into the histogram. The kicked tops belong to the Floquet systems. Here the energies are replaced by the quasi-energies, as has already been mentioned at the beginning of this chapter. The figure shows the distribution of the quasi-energy distances. In Section 4.1.2 we shall learn that the level spacing distributions of conservative and Floquet systems are identical in the limit of large N.

We shall now discuss a number of experimental verifications of nearest neighbour spacing distributions. It has already been shown in Section 3.1.1 that the nearest neighbour spacings for the nuclear spectra compiled by Bohigas and coworkers [Boh83] could be perfectly described by a Wigner distribution. Bohigas, Giannoni and Schmit subsequently studied the spectrum of a desymmetrized Sinai billiard [Boh84]. Again a Wigner nearest neighbour distribution was found (see Fig. 3.7(a)). A corresponding result for a microwave stadium billiard has already been shown in Section 2.2.1. We should be aware of the fact that nuclei and billiards differ in one important aspect. The nonintegrability of the nuclei is due to the existence of a large number of protons and neutrons with many degrees of freedom. In the billiard there are only two degrees of freedom, and the nonintegrability is a consequence of the boundary conditions. Nevertheless the distance distribution is identical in both cases. In view of these findings Bohigas, Giannoni and Schmit formulated the following

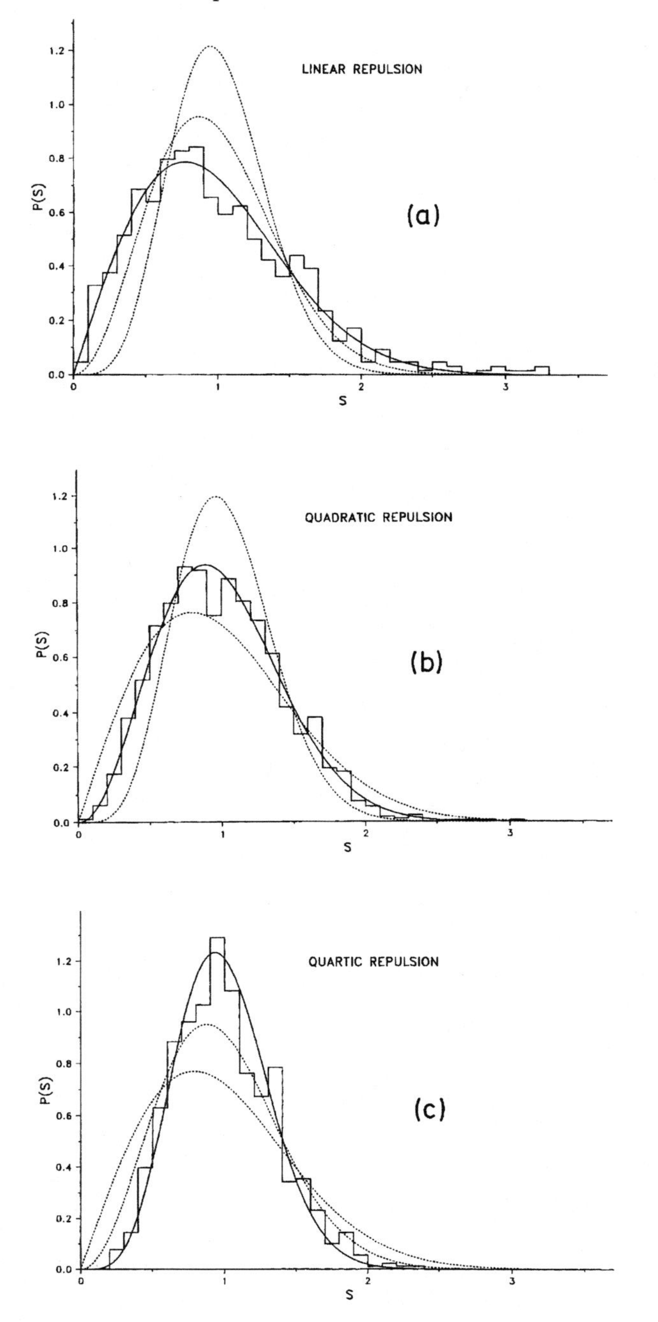

Figure 3.5. Level spacing distribution of a three-dimensional anharmonic oscillator (see Eq. (3.2.5)) with $a = 0$, $b = 0$ (a), $a = 0$, $b \neq 0$ (b), $a \neq 0$, $b \neq 0$ (c). The solid lines correspond to the GOE, GUE, and GSE predictions, respectively [Cau89] (with kind permission from Elsevier Science).

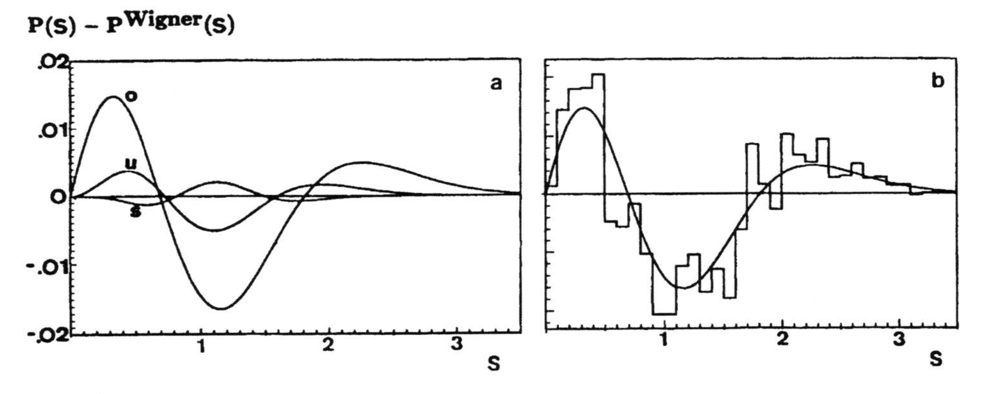

Figure 3.6. (a) Deviations of the exact level spacing distributions from the respective Wigner distributions for the three Gaussian ensembles (see Eq. (3.2.4)). (b) Numerical verification for the GOE case, taken from a Floquet spectrum of a kicked top with time-reversal symmetry [Haa91a] (Copyright 1991 by Springer Verlag).

conjecture which will be cited in full as it initiated a complete programme of quantum chaos research:

Spectra of time-reversal-invariant systems whose classical analogs are K systems show the same fluctuation properties as predicted by GOE (alternative stronger conjectures that cannot be excluded would apply to less chaotic systems, provided that they are ergodic). If the conjecture happens to be true, it will then have been established the *universality of the laws of level fluctuations* in quantal spectra already found in nuclei and to a lesser extent in atoms. Then, they should also be found in other quantal systems, such as molecules, hadrons, etc.

In K systems all parts of the classical phase space show chaotic dynamics [Ott93]. We shall see later that for systems with a mixed phase space, where some parts behave regularly and other parts chaotically, the nearest neighbour spacing distribution has to be modified. Today there is overwhelming evidence that the Bohigas–Giannoni–Schmit (BGS) conjecture is true. Figure 3.7 presents a collection of results from a number of very different systems. In addition to the already discussed example of the Sinai billiard, the nearest neighbour spacing distributions for a hydrogen atom in a strong magnetic field, the excitation spectrum of an NO_2 molecule, the vibration spectrum of a Sinai-shaped quartz block, the microwave spectrum of a three-dimensional chaotic cavity, and the vibration spectrum of a quarter-stadium-shaped plate are shown. In all these cases excellent agreement with the Wigner distribution is found. This is the more remarkable as only the first three examples represent true quantum mechanical systems. For the acoustic and the three-dimensional microwave measurements there is no longer a correspondence with quantum

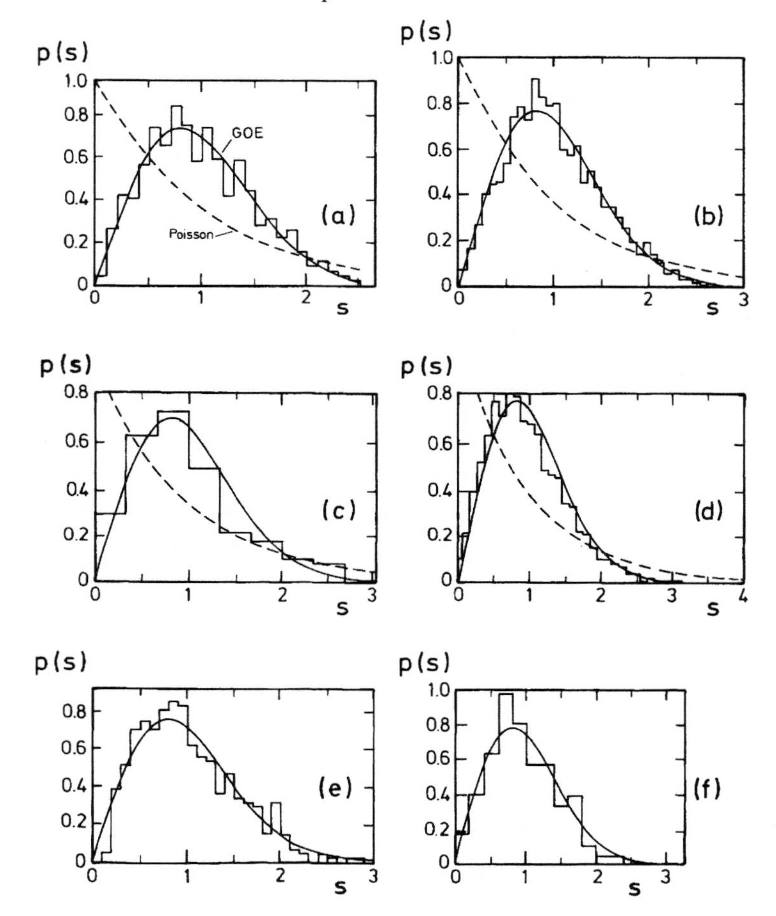

Figure 3.7. Level spacing distribution for a Sinai billiard [Boh84] (a), a hydrogen atom in a strong magnetic field [Hön89] (b), the excitation spectrum of an NO_2 molecule [Zim88] (c), the acoustic resonance spectrum of a Sinai-shaped quartz block [Oxb95] (d), the microwave spectrum of a three-dimensional chaotic cavity [Deu95] (e), and the vibration spectrum of a quarter-stadium shaped plate [Leg92] (f). In all cases a Wigner distribution is found though only in the first three cases are the spectra quantum mechanical in origin (Copyright 1984–95 by the American Physical Society).

mechanics, although the wave propagation is still described by the Helmholtz equation. In the last example even this is not the case, as it is the square of the Laplace operator which enters into the wave equation of vibrating plates (see Section 2.1.1). Obviously the BGS conjecture is not restricted to quantum mechanical systems, but remains valid in a much more general context. Recently two more or less equivalent proofs of the BGS conjecture were claimed [And96, Muz95]. They raised complicated issues in ergodic theory and are still discussed [Ley97, Zir99]. For the present state of the art Section 5 of the recently published review article by Guhr *et al.* [Guh98] is recommended.

One may ask whether the BGS conjecture originally formulated only for the GOE, also holds for the two other universality classes. According to numerical studies there is little doubt that this is the case. The example given above has already shown that for the three-dimensional anharmonic oscillator the nearest neighbour spacing distributions can be described by generalized Wigner distributions for all universality classes. In the experimental field the situation is less favourable. At present there is no example for the GSE case, for the GUE there are only two experimental realizations, both of them in microwave billiards [So95, Sto95]. In microwave billiards the time-reversal symmetry can be broken by introducing ferrites into the resonators, as has been explained in Section 2.2.3. In one of the two experiments mentioned the reflection properties of one wall were changed by means of a ferrite coating [So95], in the second example the unidirectional properties of a microwave isolator were used [Sto95]. The second system is not conservative because of the presence of an absorbing channel. Its discussion will therefore be postponed to Section 6.1.3. If the microwaves are reflected by a ferrite plane they acquire a phase shift depending on the direction of propagation. The ferrite thus has the same effect on the microwaves as a magnetic field has on the electrons in a mesoscopic billiard. It should therefore induce a transition from the GOE to the GUE regime by merely switching on and off the magnetization of the ferrite. The result is shown in Fig. 3.8. To reduce the statistical fluctuations, the authors have plotted the integrated spacing distribution

$$I(s) = \int_0^s p(s)\, ds. \tag{3.2.6}$$

Due to the integration the repulsion exponent increases by one. To make the repulsion exponents clearly visible, $I(s)$ is plotted versus s in a double logarithmic scale. Both for the GOE and the GUE case an excellent agreement with the theoretical curves is found. At least in this one experiment the BGS conjecture has proved to be true for the GUE case as well.

3.2.2 *From the integrable to the nonintegrable regime*

Most real systems are neither integrable nor completely chaotic, but have a mixed classical phase space. In these systems the nearest neighbour spacing distribution is somewhere between Poisson and Wigner-like behaviour. There

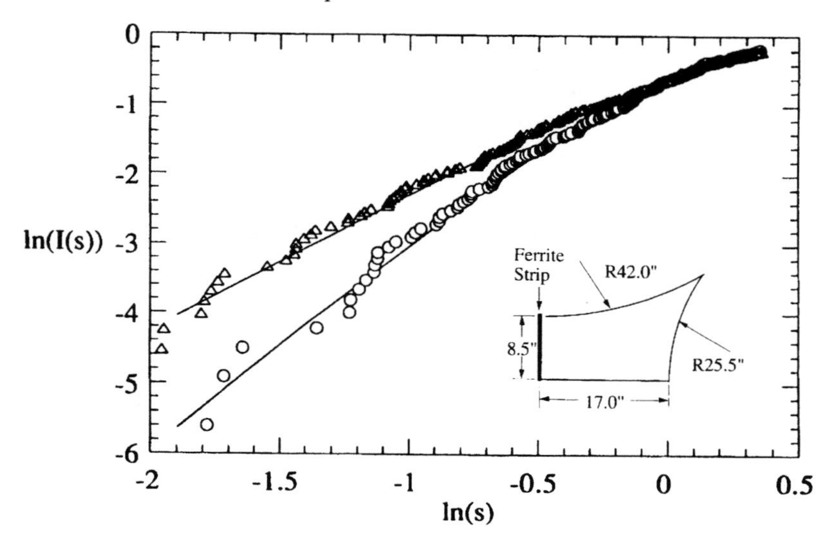

Figure 3.8. Integrated level spacing distribution in a microwave billiard, where time-reversal symmetry is broken by coating one billiard wall with a ferrite strip. The insert shows the geometry of the billiard. Circles and triangles correspond to data with and without magnetization of the ferrite, respectively. Theoretical curves for the GOE and the GUE are superimposed [So95] (Copyright 1995 by the American Physical Society).

are different approaches to treat such a situation, which will be discussed in the following.

In the purely phenomenological Brody approach [Bro73, Bro81] the repulsion exponent is varied to obtain a gradual change from the integrable to the nonintegrable region. Assuming a probability density

$$r(s) = As^{\nu}, \tag{3.2.7}$$

to find an eigenvalue in an interval of length s, the nearest neighbour spacing distribution $p(s)$ is then obtained from the equation

$$p(s)\,ds = \left(\int_s^\infty p(s)\,ds\right) r(s)\,ds. \tag{3.2.8}$$

The equation expresses the probability $p(s)\,ds$ to find the eigenvalue next to a given eigenvalue in a distance between s and $s + ds$ as the product of (i) the probability $\int_s^\infty p(s)\,ds$ to find the next eigenvalue in a distance larger than s, and (ii) the probability $r(s)\,ds$ to find an eigenvalue in the interval $\langle s, s + ds\rangle$. From Eq. (3.2.8) $p(s)$ is obtained by integration as

$$p(s) = Cr(s)\exp\left(-\int_0^s r(s)\,ds\right). \tag{3.2.9}$$

Inserting expression (3.2.7) for $r(s)$ and fixing the constants A and C by the normalization conditions

$$\int_0^\infty p(s)\, ds = 1, \tag{3.2.10}$$

$$\int_0^\infty sp(s)\, ds = 1, \tag{3.2.11}$$

we obtain the Brody distribution

$$p(s) = (\nu + 1)a_\nu s^\nu \exp\left(-a_\nu s^{\nu+1}\right) \tag{3.2.12}$$

where

$$a_\nu = \left[\Gamma\left(\frac{\nu+2}{\nu+1}\right)\right]^{\nu+1}. \tag{3.2.13}$$

Here

$$\Gamma(\nu) = \int_0^\infty x^{\nu+1} e^{-x}\, dx \tag{3.2.14}$$

is Euler's gamma function. ν is called the Brody parameter. For $\nu = 0$ the Brody distribution reduces to the Poisson distribution, and for $\nu = 1$ the Wigner distribution for the GOE is obtained. The Brody distribution thus interpolates between the integrable and the nonintegrable regime. Unfortunately the derivation contains a flaw. The product ansatz on the right hand side of Eq. (3.2.8) is only justified if the probabilities are uncorrelated, i.e. for $\nu = 0$. Here (and only here!) the derivation is correct and yields an independent proof of the fact that for uncorrelated eigenvalues a Poisson nearest neighbour spacing is expected. For all other cases the derivation can be at most qualitatively correct. For the GUE and the GSE not even the Gaussian decrease of $p(s)$ at large distances, which is expected in all Gaussian ensembles, is reproduced correctly. The Brody distribution nevertheless enjoys considerable popularity. But it is perhaps not unfair to say that the Brody parameter ν, though it is doubtlessly well suited to characterize a given distribution, has no deeper physical meaning.

An approach by Izrailev, originally formulated to describe eigenvalue distance distributions in systems showing quantum localization [Izr88, Izr90] is on more solid ground. In a disordered system quantum mechanical wave functions tend to localize due to destructive interference of the matter waves scattered at the imperfections. Experimental demonstrations of these facts with water waves and microwaves have already been presented in Sections 2.1.2 and 2.2.2, respectively. Localization reduces the repulsion between neighbouring eigenvalues, as there is no cause for a mutual repulsion, if the respective wave functions have no or only little overlap. By this heuristic argumentation Izrailev has been led to the following ansatz for the nearest neighbour spacing distribution:

$$p(s) = As^{\nu} \exp\left[-\frac{\pi^2}{16}\nu s^2 - \left(C - \frac{\nu}{2}\right)\frac{\pi}{2}s\right], \tag{3.2.15}$$

where the constants A and C can again be fixed by means of the normalization conditions (3.2.10) and (3.2.11). The Izrailev parameter ν measures the degree of localization of the wave function. Details can be found in Refs. [Izr88, Izr90]. For $\nu = 0$ we once more obtain the Poisson distribution. For $\nu = 1, 2$ the Izrailev distribution coincides with the GOE- and GUE-Wigner distributions (3.2.4) on the 5% level.

Though the Izrailev distribution was originally conceived for localization problems, it has been used in other contexts too, because it represents a very convenient interpolation formula between the different ensembles. In one application of this type Silberbauer *et al.* [Sil95] studied the motion of an electron in a two-dimensional periodic potential with a perpendicularly applied magnetic field. The spin-orbit interaction has been neglected in the calculation. The studied Hamiltonian is

$$\mathscr{H} = \frac{1}{2m_e}\left(\boldsymbol{p} - \frac{e}{c}\boldsymbol{A}\right)^2 + V_0 \cdot \left(\cos\frac{\pi x}{d}\sin\frac{\pi y}{d}\right)^n. \tag{3.2.16}$$

By means of the parameter n the steepness of the potential can be varied. In the calculations a value of $n = 8$ has been used. A convenient choice for the vector potential is again

$$\boldsymbol{A} = \frac{B}{2}(-y, x, 0). \tag{3.2.17}$$

In the absence of a magnetic field the Hamiltonian commutes with the translational operators T_x, T_y defined by

$$T_x f(x, y) = f(x + d, y), \tag{3.2.18}$$

$$T_y f(x, y) = f(x, y + d). \tag{3.2.19}$$

From the Bloch theorem of solid state physics we know that because of these translational invariances the wave functions can be classified by k vectors, which may be restricted to the first Brillouin zone defined by $|k_x| \leq \pi/d$, $|k_y| \leq \pi/d$. In the presence of a magnetic field the translational symmetry is destroyed by the vector potential term. But now $\mathscr{H}$ commutes with the modified translational operators $T_x \exp(-\imath\pi\alpha y/d)$ and $T_y \exp(\imath\pi\alpha x/d)$, where $\alpha = eBd^2/hc$ counts the number of magnetic flux quanta per unit cell. It is easy to show (see Ref. [Sil95] and the references cited therein) that the wave function can again be classified by generalized wave vectors, now called magnetic wave vectors θ, if α is an integer or a rational number p/q. In the latter case, which has been studied by the authors, the unit cell of the reciprocal lattice is smaller by a factor of q than it is in the absence of the magnetic field.

The authors have calculated the spectra of $\mathscr{H}$ for different values of the magnetic wave vector θ. B has been fixed to a value corresponding to a rational number of flux quanta per unit cell. The nearest neighbour spacing distributions obtained have been parameterized by means of the Izrailev distribution (3.2.15). In Fig. 3.9 the Izrailev parameters obtained are plotted in a grey scale ranging from 0 to 2. In the insert prominent lines and points of the Brillouin zone are labelled by Greek letters, as is usual in solid state physics. To understand the figure we have to consider the symmetries at the different points of the Brillouin zone. At a point K off any symmetry line GUE behaviour is expected because of the broken time-reversal symmetry. In this case Izrailev parameters close to 2 are found. Along the symmetry lines Σ, Δ and Z the broken time-reversal symmetry is compensated by an additional mirror reflec-

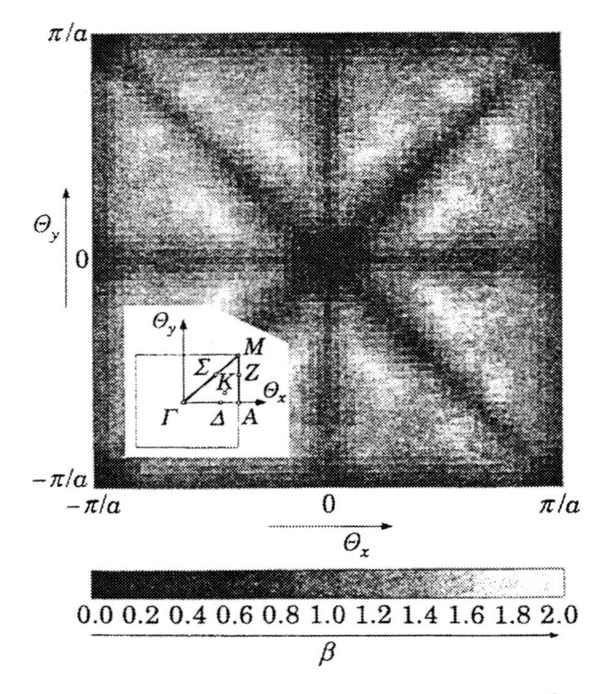

Figure 3.9. Grey-scale plot of the Izrailev parameters, obtained from the subspectra of an electron in a two-dimensional periodic potential and a perpendicularly applied magnetic field. Every pixel belongs to a different magnetic wave vector θ of the first magnetic Brillouin zone [Sil95].

tion. Consequently we find a GOE behaviour. The situation is analogous to the hydrogen atom in a strong magnetic field (see Section 3.1.2). At the points Γ, M and A the symmetry is highest, leading to an independent superposition of spectra of up to four different symmetry classes and yielding a nearest neighbour spacing distribution close to Poisson behaviour. Here the Izrailev parameters have indeed been found to be about zero. Thus we have the extraordinary situation that within one system three of the four universality classes as well as transitions between them are realized simultaneously.

A modified Izrailev distribution has been used by Casati *et al.* [Cas91] in the interpretation of nearest neighbour energy spacings of band random matrices. Their studies have been motivated by the fact that in Hamilton operators of practical interest all nonzero matrix elements are typically close to the diagonal, the far-off diagonal elements being zero. Therefore band matrices should be better realizations of real systems than random matrices with nonzero matrix elements everywhere. But the authors have found that relatively narrow band matrices have the same statistical properties as full random matrices.

Another approach to interpolate between the different ensembles has been proposed by Lenz and Haake [Len91]. They studied matrices of the form

$$H = \frac{1}{\sqrt{1+\lambda^2}}(H_0 + \lambda V), \tag{3.2.20}$$

where H_0 belongs to one universality class and V to another. Obviously H undergoes a continuous transition between the two classes, if λ is varied between 0 and ∞. There is no restriction in the universality classes of H_0 and V, but we shall discuss here only that case where H_0 and V are taken from the Poissonian and from the Gaussian orthogonal ensemble, respectively. If the discussion is moreover restricted to 2×2 matrices, the nearest neighbour spacing distribution can be calculated analytically, though the result is already quite complicated:

$$p(s) = \frac{u(\lambda)^2}{\lambda} s \exp\left(-\frac{u(\lambda)^2}{4\lambda^2}s^2\right) \int_0^\infty e^{-(x^2+2x\lambda)} I_0\left(\frac{u(\lambda)}{\lambda}xs\right) dx. \tag{3.2.21}$$

Here $I_0(x)$ is a modified Bessel function. $u(\lambda)$ can be expressed in terms of Kummer's function $U(a, b, x)$ as

$$u(\lambda) = \sqrt{\pi}U\left(-\frac{1}{2}, 0, \lambda^2\right). \tag{3.2.22}$$

Detailed information on both functions can be found in monographs on the mathematical functions of physics such as the books by Magnus, Oberhettinger and Soni [Mag66] or by Abramowitz and Stegun [Abr70]. In the limits $\lambda \to 0$

and $\lambda \to \infty$ the Poisson and the Wigner distribution are obtained again for the GOE, although this cannot be easily recognized in the equation.

The Lenz–Haake distribution has been used to analyze the nearest neighbour spacing distributions in a rectangular microwave billiard [Haa91b]. Though the ideal rectangle is integrable and obeys a Poisson nearest neighbour distribution, the system has become pseudointegrable by the presence of the antenna. The influence of the antenna is small in the low frequency range, but becomes more and more pronounced with increasing frequency. This is demonstrated in Fig. 3.10 where nearest neighbour spacing distributions for a rectangular microwave billiard are shown for three different frequency ranges. The solid lines are fits with the Lenz–Haake distribution. The obtained Lenz–Haake parameter increases almost linearly with the frequency thus demonstrating the increasing influence of the antenna at higher frequencies. The situation is completely analogous to the vibrating rectangular glass plates fixed in the centre, which have already been discussed in Section 2.1.1.

We shall not enter into a detailed discussion of pseudointegrable systems, but a few remarks are appropriate. The term was introduced by Richens and Berry [Ric81] who studied the motion of a particle in a polygonal billiard with angles being rational multiples of π. At a polygon vertex an incoming bundle of parallel trajectories is split. As a consequence the motion in phase space takes place no longer on a torus, but on a multi-handled sphere, the number of handles depending on the vertex angles. The system is therefore not integrable. But it is not fully chaotic either, as the motion is still on a compact two-dimensional subspace of the four-dimensional phase space. Polygonal billiards are discussed in Ref. [Shi95]. Another possibility to make the motion in an integrable billiard pseudointegrable is the introduction of a point-like scatterer [Šeb90]. In the example of a rectangular microwave billiard discussed above the antenna has acted exactly as such a point-like scatterer. More details can be found in Refs. [Alb91, Shi94].

In all the distributions interpolating between integrable and nonintegrable ensembles discussed up to now the parameters describing the different distributions have been just fit parameters, and there has not been any prescription to link them with system properties. The situation is different for the Berry–Robnik distribution [Ber84] where the interpolation parameter can be determined directly from properties of the classical phase space. This will be illustrated with Pascal's limaçon, better known as the Robnik billiard in the quantum chaos community. It is obtained by a conformal mapping of the unit circle in the complex plane by means of the function

$$w = z + \lambda z^2, \quad |z| \leqslant 1. \tag{3.2.23}$$

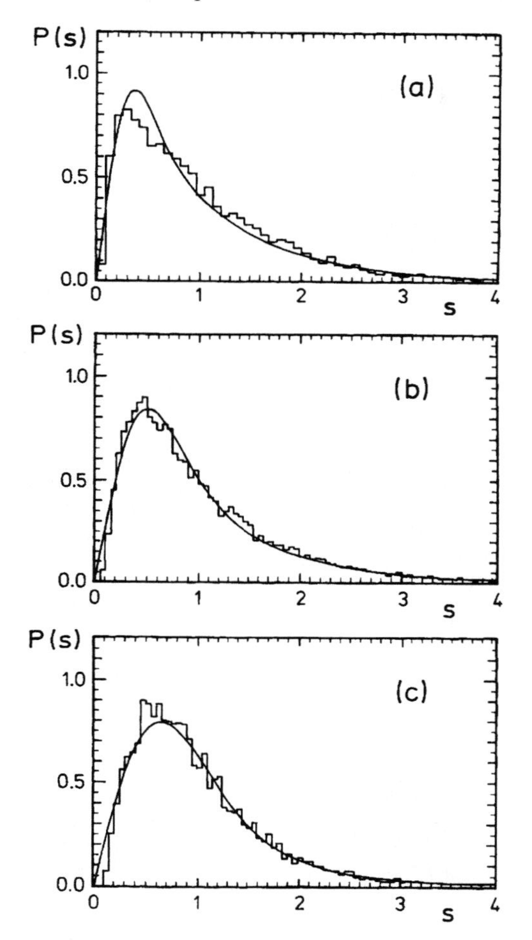

Figure 3.10. Level spacing distributions collected from 69 rectangular microwave billiards of different sizes for three frequency ranges 5 to 10 GHz (a), 10 to 15 GHz (b), 15 to 18 GHz (c). The solid lines are fits with the Lenz–Haake distribution with λ = 0.19, 0.34, and 0.58, respectively [Haa91b] (Copyright 1991 by the American Physical Society).

The parameter λ can vary between 0 and $\frac{1}{2}$. The two limiting cases correspond to the unit circle and the cardioïd billiard, respectively. Figure 3.11 shows the billiard for different λ values, each with a typically classical trajectory [Rob83]. Whereas for $\lambda = 0.05$ the situation is still similar to that of the circle, for $\lambda = 0.375$ the trajectory is completely chaotic. In the right column the corresponding Poincaré sections are shown. With increasing λ we observe a gradual transition of the classical dynamics from a regular to a completely chaotic behaviour.

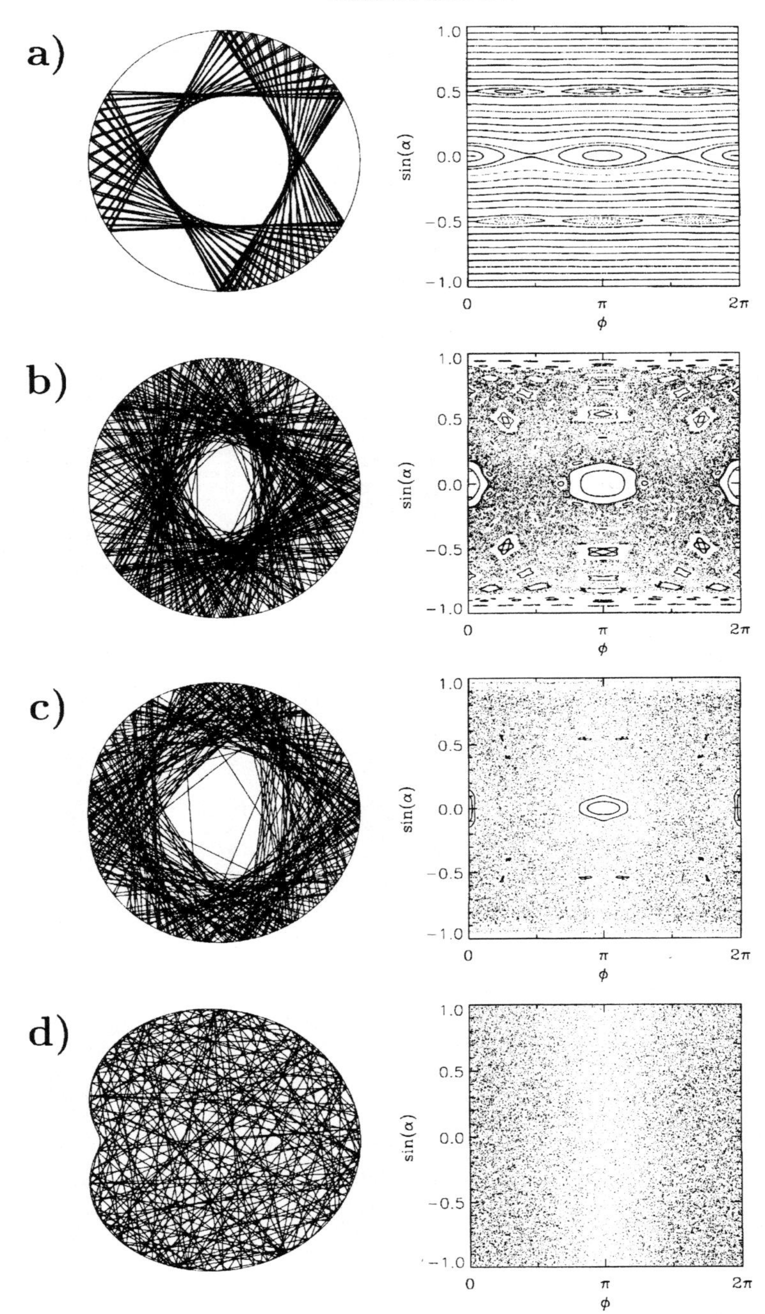

Figure 3.11. Robnik billiard for $\lambda = 0.05$ (a), 0.15 (b), 0.2 (c), 0.375 (d). The right column shows the corresponding Poincaré sections. The abscissa corresponds to the polar angle determining the position of the collision point on the boundary, the ordinate corresponds to the sine of the reflection angle.

Berry and Robnik postulated that for a mixed phase space the contributions from the regular and the chaotic regions to the spectrum should superimpose uncorrelated [Ber84]. We should be aware that this can be true at most approximately, as the uncertainty relation does not really allow a strict separation of the different phase space regions.

For the following discussion it will be assumed that the phase space contains just one regular and one chaotic region. The densities of states ρ_1 and ρ_2 of the two parts of the spectrum directly reflect the corresponding portions of phase space volume. This follows from the fact that every quantum mechanical state occupies a phase state volume of $\hbar^d$ where d is the dimension of the underlying space. It is therefore no longer possible to normalize the density of the subspectra. Only the total level density $\rho_1 + \rho_2$ can be normalized to one. We then obtain

$$p_1(s) = \rho_1 \exp(-\rho_1 s) \tag{3.2.24}$$

for the level spacing distribution of the regular part, and

$$p_2(s) = \frac{\pi}{2} \rho_2^2 s \exp\left(-\frac{\pi}{4} \rho_2^2 s^2 \right) \tag{3.2.25}$$

for that of the chaotic part of the spectrum.

For the calculation of the level spacing distribution of the superimposed spectra we essentially follow the derivation given in Appendix A.2 of Ref. [Meh91]. We begin with the probability $E(s)$ that an arbitrary interval of length s is free of eigenvalues. The connection between $E(s)$ and the level spacing distribution $p(s)$, in which we are really interested, is obtained as follows. Let $q(s)ds$ be the probability that the eigenvalue next to a given one is found in the distance between s and $s + ds$. It can be expressed as the probability that an interval of length s is empty of eigenvalues, minus the probability that an interval of length $s + ds$ is empty, i.e.

$$q(s)\, ds = E(s) - E(s + ds) = -E'(s)\, ds. \tag{3.2.26}$$

On the other hand $q(s)$ may be written as a product of the density of states ρ times the probability that the next eigenvalue has a distance larger than s to a given eigenvalue, i.e.

$$q(s) = \rho \int_s^\infty p(s)\, ds. \tag{3.2.27}$$

Combining Eqs. (3.2.26) and (3.2.27) we obtain the relation

$$p(s) = \frac{1}{\rho} \frac{d^2 E}{ds^2}. \tag{3.2.28}$$

The $E_1(s)$ and $E_2(s)$ for the two subspectra are obtained from Eqs. (3.2.24) and (3.2.25) by twofold integration, resulting in

$$E_1(s) = \exp(-\rho_1 s) \tag{3.2.29}$$

for the integrable, and

$$E_2(s) = \mathrm{erfc}\left(\frac{\sqrt{\pi}}{2}\rho_2 s\right) \tag{3.2.30}$$

for the chaotic part. Here

$$\mathrm{erfc}(x) = \frac{2}{\sqrt{\pi}}\int_x^\infty e^{-t^2}\,dt \tag{3.2.31}$$

is the complement of the error function. If the two subspectra are uncorrelated, the probability to find an interval of length s free of eigenvalues in the superimposed spectrum is the product of the probabilities to find an empty interval in both subspectra, i.e.

$$E(s) = E_1(s)E_2(s). \tag{3.2.32}$$

This multiplicative relation for the $E_i(s)$ has motivated switching from the $p_i(s)$ to the $E_i(s)$. Inserting the expressions (3.2.29) and (3.2.30) for the $E_i(s)$ and differentiating twice we get the Berry–Robnik distribution

$$p(s) = \left[\rho_1^2\,\mathrm{erfc}\left(\frac{\sqrt{\pi}}{2}\rho_2 s\right) + \left(2\rho_1\rho_2 + \frac{\pi}{2}\rho_2^3 s\right)\exp\left(-\frac{\pi}{4}\rho_2^2 s^2\right)\right]\exp(-\rho_1 s). \tag{3.2.33}$$

In the two limiting cases $\rho_1 = 1$, $\rho_2 = 0$ and $\rho_1 = 0$, $\rho_2 = 1$ the Poisson and the Wigner distributions for the GOE, respectively, are obtained once more. The same technique can be used to calculate the level spacing distribution for the uncorrelated superposition of the two GOE spectra met in Section 3.1.1. As an illustration, Fig. 3.12 shows the cumulative level spacing distribution

$$I(s) = \int_0^s p(s)\,ds \tag{3.2.34}$$

for the Robnik billiard for the same set of λ values as in Fig. 3.11 [Pro93]. The solid line corresponds to the best fit with the Berry–Robnik distribution. The figure clearly shows the gradual transition from a Poisson-like to a Wigner-like behaviour. The fitted Berry–Robnik parameter should coincide with the value obtained from the phase space portions of the integrable and the nonintegrable parts, but only qualitative agreement has been found. In a more recent work on a periodically kicked top the authors could verify the expectations quantitatively [Pro94]. It has, however, been necessary to extend the calculations up to the 8000th eigenvalue to reach the Berry–Robnik regime. Before reaching that limit, the level spacing distribution could be described much better by a Brody distribution, though the latter lacks a sound theoretical foundation. An explanation for this fact is still missing.

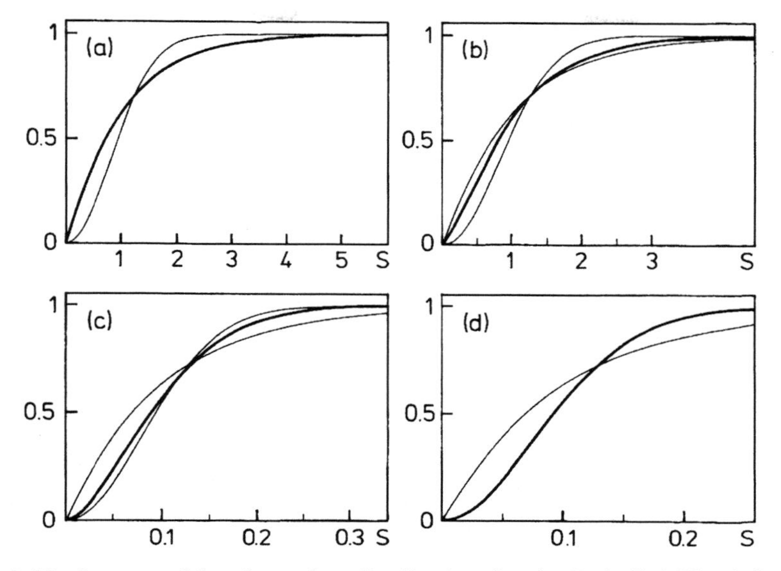

Figure 3.12. Integrated level spacing distribution for the Robnik billiard for $\lambda = 0.05$ (a), 0.15 (b), 0.2 (c), 0.375 (d), showing a gradual transition from a Poisson-like towards a Wigner-like distribution [Pro93].

3.2.3 *n*-point correlation function

A number of spectral correlations to be discussed later depend on the so-called *two-point correlation function*. This is the probability density to find two eigenenergies E_1 and E_2 at two given energies, irrespective of the position of all other eigenenergies. The more general n-point correlation function is defined by

$$R_n(E_1, \ldots, E_n) = \frac{N!}{(N-n)!} \int P_n(E_1, \ldots, E_N)\, dE_{n+1} \ldots dE_N. \qquad (3.2.35)$$

The prefactor takes into account all combinative possibilities to select n eigenvalues out of a total number of N.

The calculation of $R_n(E_1, \ldots, E_n)$ by successive integration will be demonstrated for the GUE, applying the method given in Mehta's monograph [Meh91]. The calculations are always easiest for the GUE. For the GOE, which is more interesting from the practical point of view, the calculation is considerably more complicated.

The procedure is based on the idea of expressing the correlated eigenenergy distribution function

$$P(E_1, \ldots, E_N) \sim \prod_{n>m} (E_n - E_m)^2 \exp\left(-A \sum_n E_n^2\right) \tag{3.2.36}$$

(see Eq. (3.1.95)) as a determinant. Again the density of states for small energies is normalized to one by putting $A = \pi^2/2N$ (see Eq. (3.1.129)). Furthermore we introduce the rescaled energies

$$x = \sqrt{A}E = \frac{\pi}{\sqrt{2N}} E, \tag{3.2.37}$$

to eliminate the factor A in the exponential in Eq. (3.2.36). In terms of the rescaled energies the correlated energy distribution function reads

$$P(x_1, \ldots, x_N) \sim \prod_{n>m} (x_n - x_m)^2 \exp\left(-\sum_n x_n^2\right). \tag{3.2.38}$$

In a first step we express the factor $\prod_{n>m}(x_n - x_m)$ in terms of Vandermonde's determinant

$$\prod_{n>m} (x_n - x_m) = \begin{vmatrix} 1 & \cdots & 1 \\ x_1 & \cdots & x_N \\ \vdots & & \vdots \\ x_1^{N-1} & \cdots & x_N^{N-1} \end{vmatrix}. \tag{3.2.39}$$

By a suitable combination of rows this can be alternatively written as

$$\prod_{n>m} (x_n - x_m) \sim \begin{vmatrix} H_0(x_1) & \cdots & H_0(x_N) \\ \vdots & \ddots & \vdots \\ H_{N-1}(x_1) & \cdots & H_{N-1}(x_N) \end{vmatrix}, \tag{3.2.40}$$

where $H_n(x)$ is an arbitrary polynomial of degree n. Here we take the Hermitian polynomials

$$H_n(x) = \exp(x^2)\left(-\frac{d}{dx}\right)^n \exp(-x^2). \tag{3.2.41}$$

The motivation for this choice will soon become clear. Using again elementary determinant rules, we write in the next step

$$\prod_{n>m} (x_n - x_m) \exp\left(-\frac{1}{2}\sum_n x_n^2\right) \sim \begin{vmatrix} \phi_0(x_1) & \cdots & \phi_0(x_N) \\ \vdots & \ddots & \vdots \\ \phi_{N-1}(x_1) & \cdots & \phi_{N-1}(x_N) \end{vmatrix}, \tag{3.2.42}$$

where the $\phi_n(x)$ are the harmonic oscillator eigenfunctions

$$\phi_n(x) = \frac{1}{(2^n n! \sqrt{\pi})^{1/2}} H_n(x) \exp\left(-\frac{x^2}{2}\right). \tag{3.2.43}$$

The $\phi_n(x)$ obey the orthogonality relation

$$\int_{-\infty}^{\infty} \phi_n(x)\phi_m(x)\,dx = \delta_{nm}. \tag{3.2.44}$$

Collecting the results we see that the correlated eigenenergy distribution function can be written as

$$P(x_1, \ldots, x_N) \sim |M^T M|, \tag{3.2.45}$$

where M is the matrix with the elements

$$M_{n\alpha} = \phi_{n-1}(x_\alpha). \tag{3.2.46}$$

The matrix elements of $M^T M$ are given by

$$(M^T M)_{\alpha\beta} = K_N(x_\alpha, x_\beta) = \sum_{k=0}^{N-1} \phi_k(x_\alpha)\phi_k(x_\beta). \tag{3.2.47}$$

Anticipating the result that the proportionality constant in Eq. (3.2.45) is given by $(N!)^{-1}$, we get the following compact expression for the correlated eigenenergy distribution function:

$$P(x_1, \ldots, x_N) = \frac{1}{N!}|K_N(x_\alpha, x_\beta)|. \tag{3.2.48}$$

Now the integration over the x_n can easily be performed by means of the following:

Theorem: If M_N is a matrix of rank N, whose elements $M_{\alpha\beta}$ obey the following conditions,

(a) $M_{\alpha\beta} = f(x_\alpha, x_\beta)$,
(b) $\int f(x, x)\,dx = c$,
(c) $\int f(x, y)f(y, z)\,dy = f(x, z)$,

then the relation

$$\int |M_N|\,dx_N = (c + 1 - N)|M_{N-1}| \tag{3.2.49}$$

holds, where M_{N-1} is the matrix obtained by removing the last row and the last column of M_N.

For the proof the determinant is developed along the last row or column, and the integral over x_N is performed. Using conditions (b) and (c) the statement results immediately.

It is easy to see that the matrix (3.2.47) with the elements $K_N(x_\alpha, x_\beta)$ obeys the conditions of the theorem. Condition (c) in particular holds because of the orthogonality relation (3.2.44). The constant c is given by $\int K_N(x, x)\,dx = N$. Retrospectively we see that all the manipulations performed above have been aimed at obtaining an expression for the correlated energy distribution function for which the theorem (3.2.49) is applicable. After these preparative steps the n-point correlation function can immediately be written down:

$$R_n(x_1, \ldots, x_n) = |K_N(x_\alpha, x_\beta)|_{\alpha,\beta=1,\ldots,n}. \tag{3.2.50}$$

For the one-point correlation function we get in particular

$$\begin{aligned} R_1(x) &= K_N(x, x) \\ &= \sum_{k=0}^{N-1} |\phi_k(x)|^2, \end{aligned} \tag{3.2.51}$$

and for the two-point correlation function

$$R_2(x, y) = K_N(x, x)K_N(y, y) - |K_N(x, y)|^2. \tag{3.2.52}$$

These two expressions will be discussed in a bit more detail. The one-point correlation function gives the probability density of finding an eigenvalue at point x. It is thus identical with the ensemble averaged density of states $\langle \rho_N(x) \rangle$. Integrating expression (3.2.51) over x, we obtain N. The choice of the prefactor in Eq. (3.2.48) is thus posteriorly justified.

With Eq. (3.2.50) we have obtained a closed expression for the n-point correlation function, which holds for arbitrary values of n and N. Therefore the present derivation is by far superior to that in Section 3.1.5 yielding only an expression for $\langle \rho(E) \rangle$ in the limit of large N. In the limit $N \to \infty$ Eq. (3.2.51) must of course reduce to the old result. It is quite instructive to see how this works. $\phi_n(x)$ is the solution of the Schrödinger equation

$$\phi_n''(x) = -(2E_n - x^2)\phi_n(x) \tag{3.2.53}$$

for the harmonic oscillator, where $E_n = n + \frac{1}{2}$. For $|x| < (2E_n)^{1/2}$, i.e. in the classically allowed region, the solution shows an oscillatory behaviour, whereas outside a rapid decay is observed. Furthermore we know that wave functions with even and odd parity alternate, starting with a wave function with even parity. This suggests the following ansatz,

$$\phi_n(x) = a_n(x) \cos\left(\int_0^x \sqrt{2E_n - x^2}\, dx - \frac{n\pi}{2} \right), \tag{3.2.54}$$

for $\phi_n(x)$. Here $a_n(x)$ is an amplitude function varying only slowly with x in the classically allowed region and vanishing outside. Entering this ansatz into the Schrödinger equation (3.2.53), and neglecting the second derivative of $a_n(x)$, we obtain

$$a_n(x) = \sqrt{\frac{2}{\pi}} \left(2n + 1 - x^2\right)^{-1/4} \tag{3.2.55}$$

in the classically allowed, and $a_n(x) = 0$ in the classically forbidden region. In principle we are still free to choose the sign, but with the positive sign for $a_n(x)$

we are in accordance with the usual sign convention for the Hermitian polynomials (3.2.41). From Eqs. (3.2.54) and (3.2.55) we obtain for the quantum mechanical probability density in the classically allowed region

$$|\phi_n(x)|^2 = \frac{1}{\pi}\frac{1}{\sqrt{2n+1-x^2}}\left[1+(-1)^n \cos\left(\int_0^x 2\sqrt{2n+1-x^2}\,dx\right)\right]. \quad (3.2.56)$$

Apart from the oscillatory term this is exactly the classical probability density for a harmonic oscillator with the energy $E_n = n + \frac{1}{2}$. This is a manifestation of Bohr's correspondence principle. Figure 3.13 shows the quantum mechanical and the classical probability densities for different oscillator eigenfunctions. We can see that the approximation (3.2.56) is already good for quantum numbers $n \geqslant 10$. Readers familiar with quantum mechanics will have noticed that we have applied the so-called WKB approximation to obtain an asymptotic expression for the oscillator eigenfunctions. This technique will become the main key to the development of semiclassical quantum mechanics in Chapter 7.

Now expression (3.2.56) is inserted into the sum in Eq. (3.2.51). In the limit of large N the oscillatory terms are washed out, and the sum can be replaced by an integral, whence follows after performing the integration

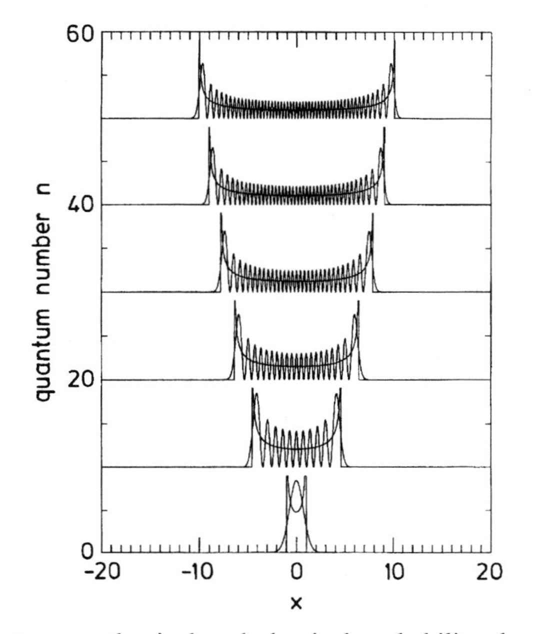

Figure 3.13. Quantum mechanical and classical probability density of the harmonic oscillator for the quantum numbers $n = 0, 10, 20, 30, 40, 50$.

$$R_1(x) = \begin{cases} \frac{1}{\pi}\sqrt{2N+1-x^2} & |x| < \sqrt{2N+1} \\ 0 & |x| > \sqrt{2N+1} \end{cases}. \quad (3.2.57)$$

Resubstituting x by E (see Eq. (3.2.37)), we end again with the semicircle law (3.1.130), the only difference being that in Eq. (3.2.57) $2N$ is replaced by $2N + 1$.

We now turn to the two-point correlation function (3.2.52). Substituting here, too, the x, y by the energies and performing the limit $N \to \infty$, we get in the asymptotic range for the two-point correlation function

$$R_2(E_1 - E_2) = \lim_{N\to\infty} \left\{ \frac{\pi^2}{2N} \left[K_N\left(\frac{\pi}{\sqrt{2N}}E_1, \frac{\pi}{\sqrt{2N}}E_1\right) K_N\left(\frac{\pi}{\sqrt{2N}}E_2, \frac{\pi}{\sqrt{2N}}E_2\right) - K_N\left(\frac{\pi}{\sqrt{2N}}E_1, \frac{\pi}{\sqrt{2N}}E_2\right)^2 \right] \right\}. \quad (3.2.58)$$

In the notation on the left hand side we have anticipated that the two-point correlation function depends only on the difference of the two energies.

The first expression on the right hand side yields $\langle\rho(E_1)\rangle\langle\rho(E_2)\rangle$. If both energies are far from the limits of the semicircle, this quantity can be replaced by one. The second quantity can be simplified by means of the following relation for the oscillator eigenfunctions,

$$K_N(x, y) = \sum_{n=0}^{N-1} \phi_n(x)\phi_n(y) = \sqrt{\frac{N}{2}} \frac{\phi_N(x)\phi_{N-1}(y) - \phi_{N-1}(x)\phi_N(y)}{x - y}, \quad (3.2.59)$$

holding for $x \neq y$ (see textbooks on the functions of mathematical physics [Mag66, Abr70], and Appendix A.10 of Ref. [Meh91]). Inserting the approximate expression (3.2.54) for $\phi_n(x)$ we obtain in the limit of small x, y (corresponding to the limit of large N)

$$K_N(x, y) = \frac{1}{\pi} \frac{\sin\sqrt{2N}(x - y)}{x - y}. \quad (3.2.60)$$

Inserting this into Eq. (3.8.58) we finally end with the simple expression

$$R_2(E) = 1 - \left(\frac{\sin \pi E}{\pi E}\right)^2 \quad (3.2.61)$$

for the two-point correlation function of the GUE. An equivalent quantity often used is its complement

$$Y_2(E) = 1 - R_2(E) = \left(\frac{\sin \pi E}{\pi E}\right)^2, \tag{3.2.62}$$

the so-called *two-level cluster function*. For the GOE the corresponding calculation can be found in Section 6.3 of Ref. [Meh91]. It is technically much more complicated but essentially proceeds along the same paths. The result for the two-level cluster function is

$$Y_2(E) = \left(\frac{\sin \pi E}{\pi E}\right)^2 + \left[\frac{\pi}{2}\,\mathrm{sgn}(E) - \mathrm{Si}(\pi E)\right]\left[\frac{\cos \pi E}{\pi E} - \frac{\sin \pi E}{(\pi E)^2}\right], \tag{3.2.63}$$

where $\mathrm{sgn}(E)$ is the signum function,

$$\mathrm{sgn}(E) = \begin{cases} 1 & E>0 \\ 0 & E=0\,, \\ -1 & E<0 \end{cases} \tag{3.2.64}$$

and

$$\mathrm{Si}(x) = \int_0^x \frac{\sin t}{t}\, dt \tag{3.2.65}$$

is the integral sine function. Figure 3.14 shows $R_2(E)$ for the GOE and the GUE. At large distances where the correlation between the eigenvalues disappears, $R_2(E)$ approaches one in both cases. At small distances the repulsion between the eigenvalues is responsible for the hole at $E = 0$. The different behaviour at small distances

$$R_2(E) \sim \begin{cases} E & (GOE) \\ E^2 & (GUE) \end{cases} \tag{3.2.66}$$

again reflects the different repulsion exponent already found in the nearest neighbour spacing distribution.

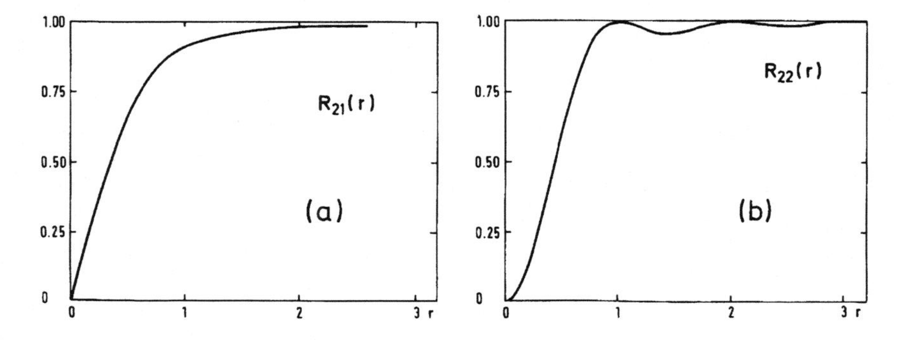

Figure 3.14. Two-level correlation functions $R_2(r)$ for the GOE (a) and the GUE (b) [Meh91].

3.2.4 Σ^2 and Δ_3 statistics

It might be expected that in the study of the spectra of chaotic systems the main interest focusses on the n-level correlation functions. They can be easily extracted from experimental or numerical spectra, and compact analytical expressions are available. We have already seen, however, that the nearest neighbour spacing distribution is much more popular. It unfortunately depends in a complicated way on *all* n-level correlation functions. But two other quantities, which are often used, can be easily derived from the two-level correlation function.

Let us consider the number $n(E, L)$ of eigenvalues in an interval of length L, centred at the energy E. It is obtained from the density of states as

$$n(E, L) = \int_{E-L/2}^{E+L/2} \rho(E)\, dE. \tag{3.2.67}$$

The ensemble average of $n(E, L)$ is independent of E and is given by

$$\langle n(E, L)\rangle = L. \tag{3.2.68}$$

As before it has been assumed that the mean level density is normalized to one. We now proceed to the variance of the number of eigenvalues in this range,

$$\begin{aligned} \Sigma^2(L) &= \langle (n(E, L) - L)^2\rangle \\ &= \langle n(E, L)^2\rangle - L^2, \end{aligned} \tag{3.2.69}$$

the so-called *number variance*. It has already been anticipated in the notation that $\Sigma^2(L)$ will show up to be independent of E. For the quadratically averaged $n(E, L)$ we have

$$\langle n(E, L)^2\rangle = \int_{E-L/2}^{E+L/2} dE_1 \int_{E-L/2}^{E+L/2} dE_2 \langle \rho(E_1)\rho(E_2)\rangle. \tag{3.2.70}$$

For $E_1 \neq E_2$ the average in the integrand can be interpreted as the ensemble averaged probability to find eigenvalues at the energies E_1 and E_2, irrespective of the positions of all other eigenvalues. This is just the two-level correlation function:

$$\langle \rho(E_1)\rho(E_2)\rangle = R_2(E_1 - E_2), \quad \text{if } E_1 \neq E_2. \tag{3.2.71}$$

The case $E_1 = E_2$ has to be treated separately. Here we start directly from the definition of the density of states:

$$\begin{aligned}\langle \rho(E)^2 \rangle &= \left\langle \sum_{n,m} \delta(E - E_n)\delta(E - E_m) \right\rangle \\ &= \left\langle \sum_{n,m} \delta(E - E_n)\delta(E_n - E_m) \right\rangle \\ &= \delta(0) \left\langle \sum_{n} \delta(E - E_n) \right\rangle \\ &= \delta(0)\langle \rho(0) \rangle \\ &= \delta(0). \end{aligned} \tag{3.2.72}$$

The expressions (3.2.71) and (3.2.72) can be combined into one single formula

$$\begin{aligned}\langle \rho(E_1)\rho(E_2) \rangle &= R_2(E_1 - E_2) + \delta(E_1 - E_2) \\ &= 1 + \delta(E_1 - E_2) - Y_2(E_1 - E_2). \end{aligned} \tag{3.2.73}$$

Entering this expression into Eq. (3.2.70) we get

$$\Sigma^2(L) = L - 2\int_0^L (L - E) Y_2(E)\, dE. \tag{3.2.74}$$

For integrable systems the two-level cluster function $Y_2(E)$ vanishes, and we have $\Sigma^2(L) = L$. It may be easily verified that this very behaviour is expected if the eigenvalues are uncorrelated. For the Gaussian ensembles, however, the level repulsion tends to equalize the distance between neighbouring eigenvalues. Consequently the number variance increases only logarithmically for large L,

$$\Sigma^2(L) = \frac{2}{\nu\pi^2} \ln(L) + a_\nu + \mathcal{O}(L^{-1}). \tag{3.2.75}$$

The constant a_ν differs for the three ensembles. The detailed formulas can be found in Appendix A.38 of Ref. [Meh91]). Figure 3.15(a) shows the number variance for the same nuclear data ensemble we have already met in the context of the level spacing distribution [Boh83]. A complete agreement with the GOE curve is found. In Fig. 3.15(b) $\Sigma^2(L)$ is shown for the spectrum of a hydrogen atom in a strong magnetic field [Hön89]. Again an overall agreement with the GOE distribution is found, but there are systematic deviations. We shall see later that the universal behaviour, which is described excellently by random matrix theory, breaks down if the energy ranges considered are of the order of $\hbar/T$ or larger, where T is the period time of a particle on the shortest classical periodic orbit of the system. The deviations seen in Fig. 3.15(b) are a first indication of the influence of these *individual* system properties, which of

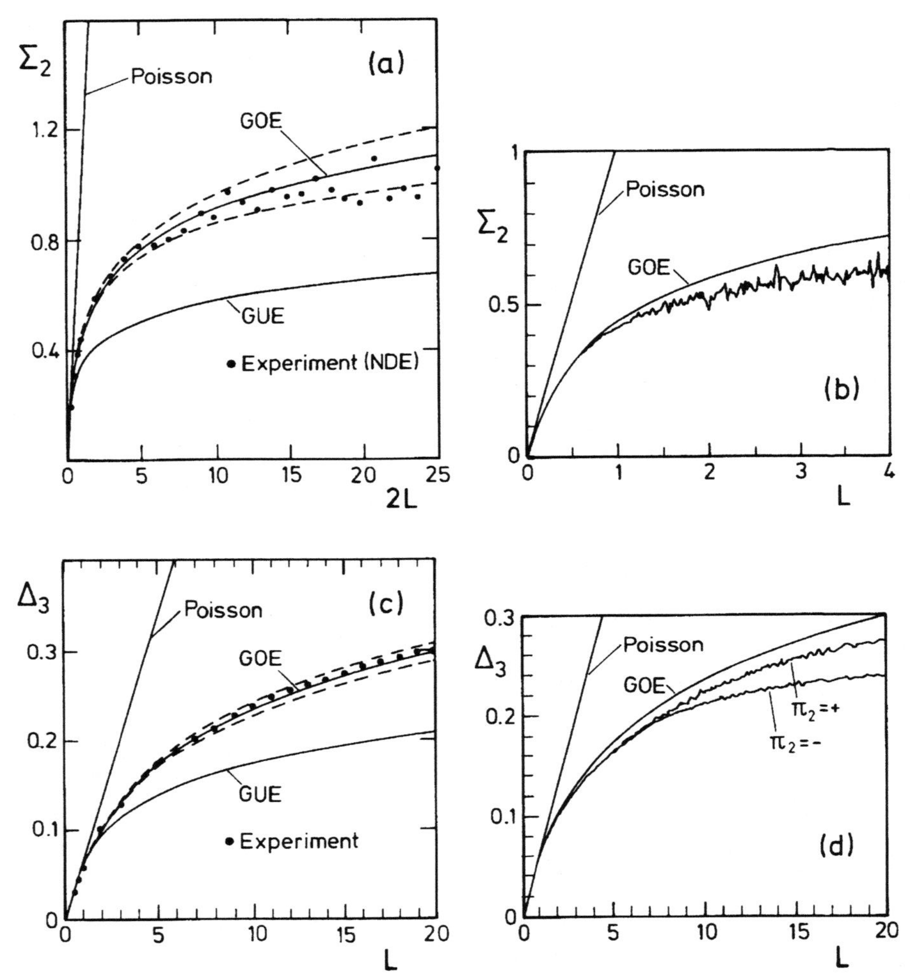

Figure 3.15. Number variance $\Sigma^2(L)$ for the nuclear data ensemble [Boh83] (a) and the hydrogen atom in a strong magnetic field [Hön89] (b), and the spectral rigidity $\Delta_3(L)$ for these two systems (c, d) (Copyright 1983+89 by the American Physical Society).

course cannot be modeled appropriately by the *universal* approach represented by random matrix theory.

The other frequently used quantity based on the two-level correlation function is the *spectral rigidity*, for historical reasons also called Δ_3 *statistics*. It is defined in a somewhat peculiar way. Consider the integrated density of states

$$n(E) = \int_0^E \rho(E)\, dE. \tag{3.2.76}$$

As $\rho(E)$ is a sum over delta functions with a mean spacing of one, $n(E)$ is a step function with a mean slope of one. Now a straight line is fitted to $n(E)$ in the interval $\langle E - \frac{L}{2}, E + \frac{L}{2} \rangle$. The ensemble average of the minimum of the χ^2 function obtained in the fit, i.e.

$$\Delta_3(L) = \left\langle \min_{a,b} \int_{E-L/2}^{E+L/2} [n(E) - a - bE]^2 \, dE \right\rangle, \tag{3.2.77}$$

is the spectral rigidity. The determination of a and b is a standard least squares fit problem and is left as an exercise. Again the result is independent of E. The resulting expression for $\Delta_3(L)$ can be written in terms of the two-level cluster function as

$$\Delta_3(L) = \frac{L}{15} - \frac{1}{15L^4} \int_0^L (L - E)^3 (2L^2 - 9LE - 3E^2) Y_2(E) \, dE. \tag{3.2.78}$$

For integrable systems we therefore have $\Delta_3(L) = L/15$. Alternatively we can express $\Delta_3(L)$ in terms of the number variance,

$$\Delta_3(L) = \frac{2}{L^4} \int_0^L (L^3 - 2L^2 E + E^3) \Sigma^2(E) \, dE. \tag{3.2.79}$$

The asymptotic behaviour of $\Delta_3(L)$ is similar to that of $\Sigma^2(L)$ and is given by

$$\Delta_3(L) = \frac{1}{\nu \pi^2} \ln(L) + b_\nu + \mathcal{O}(L^{-1}) \tag{3.2.80}$$

(see Appendix A.40 of Ref. [Meh91]). Equation (3.2.79) shows that $\Delta_3(L)$ may be looked upon as a smoothed version of $\Sigma^2(L)$. Probably that is why the spectral rigidity is much more popular than the more or less equivalent number variance. In the lower part of Fig. 3.15 the spectral rigidity for the nuclear data ensemble and the hydrogen atom in a strong magnetic field are shown, using the same data which have already entered into the calculation of the number variance. A comparison of the figures directly demonstrates the smoothing effect mentioned. We could have also added here the Δ_3 statistics for all the spectra whose level spacing distributions have been presented in Fig. 3.7. In all these cases, including the spectra of non-quantum mechanical origin, complete agreement with the Δ_3 distribution expected for the GOE has been found, demonstrating again that the applicability of random matrix theory is not restricted to quantum mechanical systems.

Both Σ^2 and Δ_3 statistics test only the two-point correlation function. There are two further quantities involving three- and four-point correlations as well. These are the *skewness*

$$\gamma_1 = \frac{M_3}{(M_2)^{3/2}}, \tag{3.2.81}$$

and the *excess*

$$\gamma_2 = \frac{M_4}{M_2^2} - 3. \tag{3.2.82}$$

Here M_k is the kth moment of the distribution of the fluctuations of $n(E, L)$ about its average value L,

$$M_k = \langle [n(E, L) - L]^k \rangle. \tag{3.2.83}$$

Calculating the product and proceeding as above we see that M_k involves all l-point correlation functions with $l \leqslant k$. The skewness γ_1 is just a normalized third moment. As for symmetric distributions all odd moments vanish, γ_1 measures essentially the asymmetry of the distribution. The excess γ_2 measures

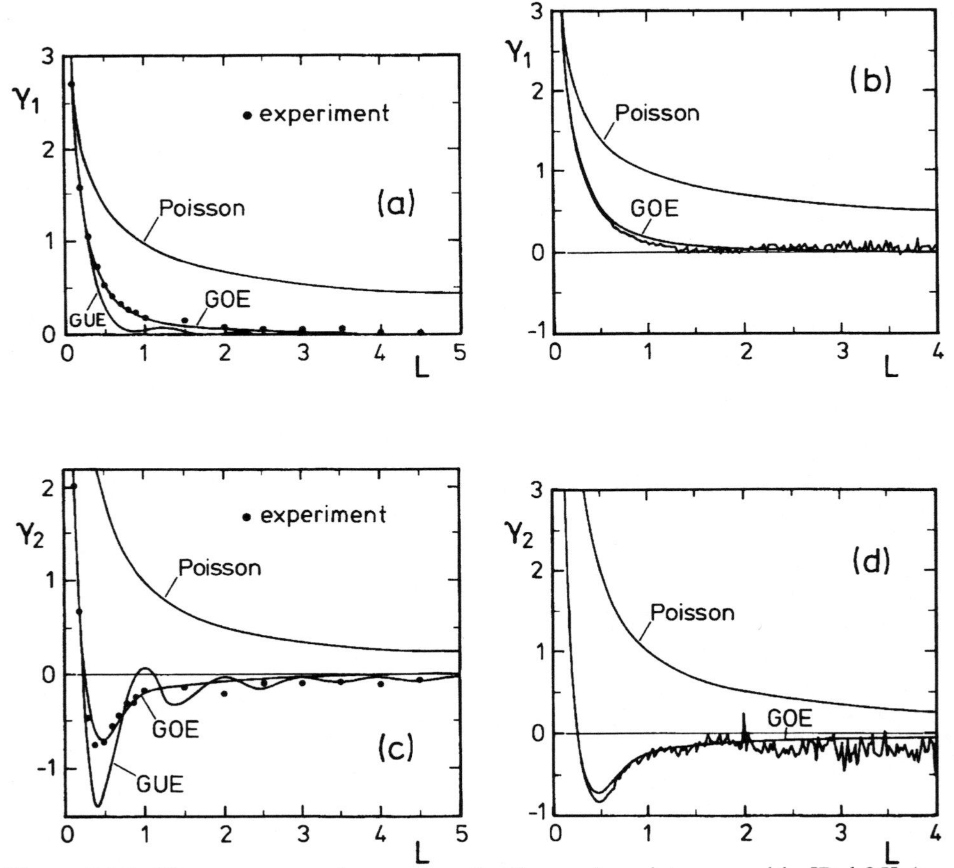

Figure 3.16. Skewness γ_1 and excess γ_2 for the nuclear data ensemble [Boh85] (a, c) and the hydrogen atom in a strong magnetic field [Hön89] (b, d) (Copyright 1985+89 by the American Physical Society).

the deviation of the distribution from a Gaussian one, as here the fourth moment amounts to just three times the square of the second one. Detailed expressions for γ_1 and γ_2 can be found in Section 16.7 of Ref. [Meh91]. Figure 3.16 shows skewness and excess for the same data sets as used above for number variance and spectral rigidity. Needless to say that again the random matrix predictions are fulfilled with high accuracy.

3.2.5 Spectral form factor

A quantity closely related to the two-point correlation function $R_2(E)$ and the two-level cluster function $Y_2(E)$ is the *spectral autocorrelation function*

$$C(E) = \left\langle \rho\left(\overline{E} + \frac{E}{2}\right)\rho\left(\overline{E} - \frac{E}{2}\right)\right\rangle - 1, \tag{3.2.84}$$

where the brackets denote an averaging over $\overline{E}$. The spectral autocorrelation function as well as its Fourier transform

$$K(t) = \int_{-\infty}^{\infty} C(E)\exp\left(-\frac{\imath}{\hbar}Et\right)dE, \tag{3.2.85}$$

the *spectral form factor*, are often used in the analysis of experimental spectra. We shall find a number of experimental examples in Chapters 6 and 8. The spectral autocorrelation function may alternatively be expressed in terms of the autocorrelation function of the fluctuating part of the density of states as

$$\begin{aligned} C(E) &= \left\langle \Delta\rho\left(\overline{E} + \frac{E}{2}\right)\Delta\rho\left(\overline{E} - \frac{E}{2}\right)\right\rangle \\ &= \delta(E) - Y_2(E), \end{aligned} \tag{3.2.86}$$

where $\Delta\rho(E) = \rho(E) - 1$. In the second equality Eq. (3.2.73) has been used. If there are no correlations between the eigenvalues, the two-level cluster function is zero, and the spectral autocorrelation function reduces to a delta function. The delta shape reflects the fact that the spectrum itself is a sum over delta functions. In real spectra the lines are always broadened, resulting in a corresponding broadening of the delta peak in $C(E)$. Therefore the spectral autocorrelation function is especially well suited to studying spectral line shapes in real systems. We shall come back to this aspect in Section 6.3.1.

For the spectral form factor we obtain

$$K(t) = 1 - b_2(t), \tag{3.2.87}$$

where

$$b_2(t) = \int_{-\infty}^{\infty} Y_2(E)\exp\left(-\frac{\imath}{\hbar}Et\right)dE \tag{3.2.88}$$

is the Fourier transform of the two-point cluster function.

All spectral correlations depending on the two-point correlation function, such as number variance or spectral rigidity, may be alternatively expressed in terms of the spectral form factor. This is illustrated here for the spectral rigidity. The calculation for the number variance proceeds in a completely analogous manner. From the inverse Fourier transform of Eq. (3.2.88) we obtain

$$Y_2(E) = \delta(E) - \frac{1}{\pi\hbar}\int_0^\infty K(t)\cos\left(\frac{E}{\hbar}t\right)dt, \tag{3.2.89}$$

where we used that $K(t)$ is symmetric in t. Entering this expression into Eq. (3.2.78) for the spectral rigidity, we have

$$\Delta_3(L) = \frac{1}{\pi\hbar}\int_0^\infty dt\, K(t)\frac{1}{15L^4}\int_0^L (L-E)^3(2L^3 - 9LE - 3E^2)\cos\left(\frac{E}{\hbar}t\right)dE. \tag{3.2.90}$$

The E integration is straightforward, though somewhat tiresome, and yields

$$\Delta_3(L) = \frac{\hbar}{\pi}\int_0^\infty \frac{dt}{t^2}K(t)G\left(\frac{Lt}{2\hbar}\right), \tag{3.2.91}$$

where we have introduced

$$G(x) = 1 - \left(\frac{\sin x}{x}\right)^2 - 3\left[\left(\frac{\sin x}{x}\right)'\right]^2. \tag{3.2.92}$$

The function $G(x)$ will become very important for the semiclassical theory of spectral rigidity to be discussed in Section 8.2.1.

Equation (3.2.87) shows that the spectral form factor takes the constant value one for integrable systems. For nonintegrable systems the term $-b_2(t)$ leads to a reduction of $K(t)$ at small t values, which is known in the literature as the *correlation hole*.

The spectral form factor can be expressed alternatively in terms of the Fourier transform of the spectrum. Therefore it is directly accessible to experiments using Fourier transform spectroscopy. For purposes of demonstration we go back to the definition (3.2.86) of the spectral autocorrelation function and start with the calculation of the average over $\overline{E}$. To this end we have to introduce a window function $w(E)$, normalized to one, by which the spectrum is truncated to a finite range containing say N eigenvalues. The exact shape of $w(E)$ is of no importance. A convenient choice is a box-like window

$$w(E) = \begin{cases} \frac{1}{N} & E < N \\ 0 & E > N \end{cases} . \tag{3.2.93}$$

By means of the window function the energy average in Eq. (3.2.86) may be written as

$$C(E) = \int_{-\infty}^{\infty} \Delta\rho_w\left(\overline{E} + \frac{E}{2}\right)\Delta\rho_w\left(\overline{E} - \frac{E}{2}\right) d\overline{E}, \tag{3.2.94}$$

where

$$\Delta\rho_w(E) = w(E)\Delta\rho(E) = w(E)[\rho(E) - 1] \tag{3.2.95}$$

is the fluctuating part of the density of states, weighted by the window function. From Eq. (3.2.94) the spectral form factor is obtained by Fourier transformation as

$$K(t) = |\Delta\hat{\rho}_w(t)|^2, \tag{3.2.96}$$

where

$$\Delta\hat{\rho}_w(t) = \int_{-\infty}^{\infty} w(E)\Delta\rho(E)\exp\left(-\frac{\imath}{\hbar}Et\right) dE \tag{3.2.97}$$

is the Fourier transform of the spectral density weighted by the window function. In the derivation of Eq. (3.2.96) the Fourier convolution theorem has been used.

If the Fourier transform of the spectrum is available, $K(t)$ can be obtained directly without resolving the individual eigenvalues. Thus Levandier *et al.* [Lev86] and Pique *et al.* [Piq87] have been able to detect the correlation hole in the spectra of molecules. It could also be observed in the nuclear data ensemble [Lom94], and in the spectra of microwave billiards [Kud94, Alt97].

Figure 3.17 shows an experimental example, taken from the spectrum of a microwave billiard in the shape of a symmetry-reduced hyperbola billiard [Alt97]. The solid line shows the experimental spectral form factor, the dashed line has been calculated from Eq. (3.2.87) assuming GOE statistics.

It has been stated above that experimental results are not yet available for the symplectic case. Though this is true, with a trick we may nevertheless study GSE distributions even in spectra obeying GOE statistics! This is made possible by using the remarkable fact that the sequence of levels obtained by picking out every second level of a GOE spectrum obeys GSE statistics. For a discussion of these properties see Chapter 10.7 of Mehta's monograph [Meh91]. Thus Alt *et al.* [Alt97] were able to determine the spectral form factor for the GSE case, too, by taking only every second eigenvalue of the spectrum of the same hyperbola billiard as described above. The lower part of

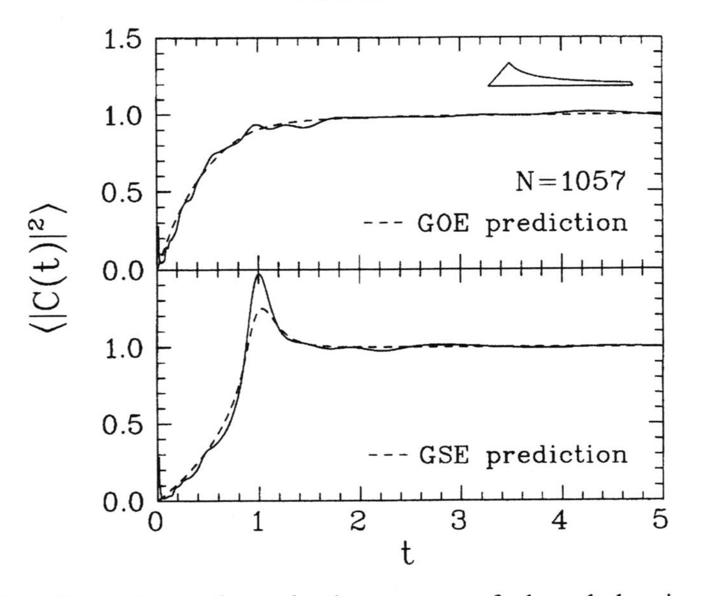

Figure 3.17. Spectral form factor for the spectrum of a hyperbola microwave billiard. In the upper part the complete spectrum has been used; in the lower part only every second level has been selected. The solid lines correspond to the experimental results, the dashed lines have been calculated from Eq. (3.2.87), assuming GOE and GSE statistics, respectively [Alt97] (Copyright 1997 by the American Physical Society).

Fig. 3.17 shows the result. Again a good agreement between experiment and theory is found.

3.3 Supersymmetry method

3.3.1 Replica trick

The recently developed supersymmetry techniques open up a completely new approach to calculating ensemble averages in Gaussian ensembles. Unfortunately the method implies rather complicated integrals for comparatively simple questions. But the underlying ideas are easy to grasp and will be presented here. As an illustrative example, the density of states of the GUE will again be calculated. This is the simplest of all possible problems, but it contains all the main ingredients used in the technique. It is therefore especially suited to making a first acquaintance with the method.

We start with the expression for the density of states already derived in Section 3.1.5,

$$\rho(E) = -\frac{1}{\pi} \operatorname{Im}[\operatorname{Tr}(G)]$$
$$= -\frac{1}{\pi} \operatorname{Im}\left[\operatorname{Tr}\left(\frac{1}{E-H}\right)\right]. \tag{3.3.1}$$

It should be remembered that here E has been assumed to have an infinitely small imaginary part to shift the poles of G off the real axis. In the next step we express the trace of G as a determinant. In the basis system of eigenfunctions of H we have

$$\operatorname{Tr}(G) = \sum_n \frac{1}{E - E_n}$$
$$= \frac{d}{dE} \sum_n \ln(E - E_n)$$
$$= \frac{d}{dE} \ln \prod_n (E - E_n)$$
$$= \frac{d}{dE} \ln|E - H|$$
$$= \frac{1}{|E-H|} \frac{d|E-H|}{dE}. \tag{3.3.2}$$

The last two relations, though derived in the basis system of eigenfunctions of H, hold in any basis system. This is a consequence of the invariance of the determinant with respect to basis transformations. The next step is motivated by the need to find an expression for the determinant which is especially suited to performing the ensemble average. We start with the Gaussian integral

$$\int_{-\infty}^{\infty} dx \int_{-\infty}^{\infty} dy\, e^{\imath(E-E_n)(x^2+y^2)} = \frac{\pi\imath}{E - E_n}. \tag{3.3.3}$$

The integral is well defined since we have assumed that E has a small positive imaginary part. The notation can be made a bit more concise by introducing the complex variables

$$z = x + \imath y, \qquad z^* = x - \imath y. \tag{3.3.4}$$

In terms of these variables Eq. (3.3.3) reads

$$\int dz^* \int dz\, e^{\imath(E-E_n)z^* z} = \frac{2\pi\imath}{E - E_n}, \tag{3.3.5}$$

where the additional factor 2 on the right hand side results from the functional determinant $|\partial(z, z^*)/\partial(x, y)|$. Taking on both sides the product over n we have

$$\int dz^* dz \exp\left[\imath \sum_n (E - E_n) z_n^* z_n\right] = \frac{(2\pi\imath)^N}{|E - H|}, \tag{3.3.6}$$

where now $dz^* dz$ is an abbreviation for $dz_1^* dz_1 \ldots dz_N^* dz_N$. The expression on the right hand side of the equation already contains the determinant desired. The expression on the left hand side, however, is not yet suitable for the ensemble average, since it still depends explicitly on the eigenenergies. But this can easily be changed. With

$$\delta_{nm} E_n = \sum_{\alpha,\beta} H_{\alpha\beta} r_{\alpha n}^* r_{\beta m}, \tag{3.3.7}$$

where the $r_{\alpha n}$ are the elements of the unitary matrix R diagonalizing H, we have

$$\begin{aligned} \sum_n (E - E_n) z_n^* z_n &= \sum_{n,m} (E - E_n) \delta_{nm} z_n^* z_m \\ &= \sum_{n,m,\alpha,\beta} (E\delta_{\alpha\beta} - H_{\alpha\beta}) r_{\alpha n}^* r_{\beta m} z_n^* z_m \\ &= \sum_{\alpha,\beta} (E\delta_{\alpha\beta} - H_{\alpha\beta}) w_\alpha^* w_\beta, \end{aligned} \tag{3.3.8}$$

where

$$w_\alpha^* = \sum_n r_{\alpha n}^* z_n^*, \qquad w_\beta = \sum_m r_{\beta m} z_m. \tag{3.3.9}$$

Now the w_α^*, w_β are introduced as new integration variables. For the functional determinant of the substitution we get, in a straightforward manner,

$$\left|\frac{\partial(w_\alpha^*, w_\beta)}{\partial(z_n^*, z_m)}\right| = \left|\begin{matrix} r_{\alpha n}^* & \cdot \\ \cdot & r_{\beta m} \end{matrix}\right| = |R^\dagger||R| = 1, \tag{3.3.10}$$

whence follows

$$I = \int dw^* \, dw \prod_{\alpha,\beta} e^{\imath(E\delta_{\alpha\beta} - H_{\alpha\beta}) w_\alpha^* w_\beta} = \frac{(2\pi\imath)^N}{|E - H|}. \tag{3.3.11}$$

This equation is one of the two foundations of the supersymmetry technique. It has a property which is decisive for the following calculations: the $H_{\alpha\beta}$ dependences factorize in the integrand on the left hand side. The ensemble average of I can therefore be performed trivially. But this is not exactly wanted in the present case. Differentiating Eq. (3.3.11) on both sides with respect to E, we obtain

$$\frac{\partial I}{\partial E} = -\frac{(2\pi\imath)^N}{|E-H|^2}\frac{\partial |E-H|}{\partial E}$$

$$= -\frac{(2\pi\imath)^N}{|E-H|}\mathrm{Tr}(G), \tag{3.3.12}$$

where expression (3.3.2) has been inserted for $\mathrm{Tr}(G)$. Taking the ensemble average we do not get $\langle \mathrm{Tr}(G)\rangle$, but $\langle |E-H|^{-1}\mathrm{Tr}(G)\rangle$. If we do not succeed in eliminating the unwanted factor $|E-H|^{-1}$ nothing is gained.

One way to tackle the problem is the *replica trick*. It is only sketched here as it has been more or less superseded by the much more elegant and mathematically better founded supersymmetry technique to be described in the next sections. In the replica trick the ensemble average of the nth power of I is calculated,

$$I^n = \int dw^{(1)*}dw^{(1)}\dots dw^{(n)*}dw^{(n)}\prod_{\alpha,\beta}\exp\left[\imath(E\delta_{\alpha\beta} - H_{\alpha\beta})\sum_k w_\alpha^{(k)*}w_\beta^{(k)}\right], \tag{3.3.13}$$

where the upper index (k) runs over the n different 'replicas'. The calculation of the ensemble average of I^n is as simple as that of I, since the $H_{\alpha\beta}$ dependences still factorize. For the derivative of I^n we now have

$$\frac{\partial I^n}{\partial E} = -\frac{n(2\pi\imath)^{nN}}{|E-H|^n}\mathrm{Tr}(G). \tag{3.3.14}$$

In this way the ensemble average of $|E-H|^{-n}\,\mathrm{Tr}(G)$ can be calculated for any natural number n. In the last step the limit $n \to 0$ is performed, leading to the desired ensemble average $\langle \mathrm{Tr}(G)\rangle$!

Every mathematician will shudder at this procedure. Expression (3.3.14), calculated for natural numbers n only, is inadmissibly extrapolated to zero. The principle of analytic continuation is not applicable here since it needs the function values on a finite compact region, not for isolated points. Nevertheless the replica trick has yielded a number of reasonable results, including the density of states for the Gaussian ensembles [Edw76]. But there are also examples where the unjustified extrapolation leads to wrong results (see Ref. [Ver85b]).

In another context the replica trick is probably known to all readers. It is used to calculate the Gaussian integral

$$E = \int_{-\infty}^{\infty} e^{-x^2}\,dx. \tag{3.3.15}$$

The integral can be calculated by taking the square and performing the double integration in polar coordinates,

$$E^2 = \int_{-\infty}^{\infty} dx \int_{-\infty}^{\infty} dy\, e^{-(x^2+y^2)}$$

$$= \int_0^{\infty} r\, dr \int_0^{2\pi} d\phi\, e^{-r^2}. \tag{3.3.16}$$

This yields the well-known result $E = \sqrt{\pi}$. The technique is easily extended to n replicas,

$$E^n = \int_0^{\infty} r^{n-1} e^{-r^2} \int_{\Omega_n} d\Omega_n, \tag{3.3.17}$$

where $d\Omega_n$ and Ω_n are surface element and surface of the n dimensional unit sphere, respectively. This allows a simple calculation of the surface:

$$\Omega_n = \frac{2\pi^{n/2}}{\Gamma\left(\frac{n}{2}\right)}. \tag{3.3.18}$$

3.3.2 Anticommuting variables

The mathematical problems with the replica trick can be very elegantly overcome with the introduction of *anticommuting* or *Grassmann* variables, as was noticed for the first time by Efetov [Efe83]. Originally invented to calculate ensemble averages in disordered systems, Verbaarschot *et al.* [Ver85a] extended the technique to the study of scattering matrices in nuclear reactions. The following presentation is based on these two review articles.

A set of variables χ_i ($i = 1, \ldots, N$) is called antisymmetric if the χ_i obey the relation

$$\chi_i \chi_j + \chi_j \chi_i = 0 \tag{3.3.19}$$

for all i, j. As a consequence the square of an anticommuting variable vanishes,

$$\chi_i^2 = 0. \tag{3.3.20}$$

Therefore any function of an anticommuting variable can be at most linear. Integrals over anticommuting variables are defined in a purely formal manner as

$$\int d\chi = 0, \quad \int \chi\, d\chi = \frac{1}{\sqrt{2\pi}}. \tag{3.3.21}$$

The factor $(2\pi)^{-1/2}$ in the second equation is arbitrary but will prove suitable in the following. It was introduced by Verbaarschot *et al.* [Ver85a], but was missing in Efetov's original article [Efe83]. Differentials anticommute too, hence

$$\iint \chi_1 \chi_2 \, d\chi_1 \, d\chi_2 = -\iint \chi_1 \, d\chi_1 \chi_2 \, d\chi_2 = -\frac{1}{2\pi}. \qquad (3.3.22)$$

We are now going to calculate the integral

$$J = \int d\chi^* d\chi \exp\left[\imath \sum_{\alpha,\beta} (E\delta_{\alpha\beta} - H_{\alpha\beta}) \chi_\alpha^* \chi_\beta \right], \qquad (3.3.23)$$

where $d\chi^* d\chi$ stands for $d\chi_1^* d\chi_1 \dots d\chi_N^* d\chi_N$. Note that this is exactly the integral we have calculated in Section 3.3.1, but then for ordinary commuting variables (see Eq. (3.3.11)). The χ_i^* anticommute with each other as well as with all χ_i. It is customary to call the χ_i^* the 'complex conjugate' of the χ_i, but in fact the χ_i^* are just another set of anticommuting variables.

For the integration we first diagonalize H. To this end we introduce new anticommuting variables η_n^*, η_m by

$$\chi_\alpha^* = \sum_n r_{\alpha n}^* \eta_n^*, \quad \chi_\beta = \sum_m r_{\beta m} \eta_m, \qquad (3.3.24)$$

where the $r_{\alpha n}$ are the matrix elements of the unitary matrix R diagonalizing H (see Eq. (3.3.7)). In terms of the new variables η_n^*, η_m the argument of the exponential function in Eq. (3.3.23) reads

$$\sum_{\alpha,\beta} (E\delta_{\alpha\beta} - H_{\alpha\beta}) \chi_\alpha^* \chi_\beta = \sum_{\alpha,\beta,n,m} (E\delta_{\alpha\beta} - H_{\alpha\beta}) r_{\alpha n}^* r_{\beta m} \eta_n^* \eta_m$$

$$= \sum_n (E - E_n) \eta_n^* \eta_n. \qquad (3.3.25)$$

With anticommuting variables the substitution of variables works exactly as with commuting variables, bearing in mind that the differentials anticommute, too. We first get from Eq. (3.3.24)

$$d\chi_1 \dots d\chi_N = \sum_{n_1, \dots, n_N} r_{1n_1} \dots r_{Nn_N} d\eta_{n_1} \dots d\eta_{n_N}. \qquad (3.3.26)$$

Using the anticommutative properties of the differentials, the product of the differentials on the right hand side can be written as

$$d\eta_{n_1} \dots d\eta_{n_N} = \epsilon_{n_1 \dots n_N} d\eta_1 \dots d\eta_N, \qquad (3.3.27)$$

where $\epsilon_{n_1 \dots n_N}$ is zero, if two arguments are equal; $+1$, if $(n_1, \dots, n_N)$ is an even; and -1, if it is an odd permutation of $(1, \dots, N)$. Inserting expression (3.3.27) into Eq. (3.3.26) we obtain

$$d\chi_1 \dots d\chi_N = |R| d\eta_1 \dots d\eta_N. \qquad (3.3.28)$$

In the same way we get

$$d\chi_1^* \dots d\chi_N^* = |R^\dagger| d\eta_1^* \dots d\eta_N^*, \qquad (3.3.29)$$

whence follows

$$d\chi_1^* d\chi_1 \ldots d\chi_N^* d\chi_N = d\eta_1^* d\eta_1 \cdots d\eta_N^* d\eta_N. \tag{3.3.30}$$

The functional determinant of transformation (3.3.24) is one, exactly as for the commuting variables. Inserting expressions (3.3.25) and (3.3.30) into Eq. (3.3.23) we get

$$J = \int d\eta^* d\eta \exp\left[\imath \sum_n (E - E_n)\eta_n^* \eta_n\right]$$
$$= \prod_n \left[\int d\eta_n^* d\eta_n \, e^{\imath(E-E)\eta_n^* \eta_n}\right]. \tag{3.3.31}$$

The integral in brackets is easily calculated using the fact that all higher powers of the antisymmetric variables vanish. We obtain

$$\int d\eta_n^* \, d\eta_n \, e^{\imath(E-E_n)\eta_n^* \eta_n} = \int d\eta_n^* \, d\eta_n [1 + \imath(E - E_n)\eta_n^* \eta_n]$$
$$= \imath(E - E_n) \int d\eta_n^* \, d\eta_n \eta_n^* \eta_n$$
$$= \frac{(E - E_n)}{2\pi\imath}, \tag{3.3.32}$$

where we have used relations (3.3.21) and (3.3.22) for antisymmetric integrals. Collecting the results we end with

$$J = \int d\chi^* d\chi \prod_{\alpha,\beta} e^{\imath(E\delta_{\alpha\beta} - H_{\alpha\beta})\chi_\alpha^* \chi_\beta} = \frac{|E - H|}{(2\pi\imath)^N}. \tag{3.3.33}$$

This is exactly the reciprocal of the corresponding result (3.3.11) for commuting variables. We have arrived at the central theorem of supersymmetry,

$$IJ = \int d[w] \prod_{\alpha,\beta} e^{\imath(E\delta_{\alpha\beta} - H_{\alpha\beta})(w_\alpha^* w_\beta + \chi_\alpha^* \chi_\beta)} = 1, \tag{3.3.34}$$

where we have introduced the abbreviation

$$d[w] = dw_1^* \, dw_1 \ldots dw_N^* \, dw_N \, d\chi_1^* \, d\chi_1 \ldots d\chi_N^* \, d\chi_N. \tag{3.3.35}$$

The commuting and anticommuting variables enter symmetrically into Eq. (3.3.34), which has given the name to the technique. For the trace of the Green function we obtain, by combining Eqs. (3.3.12) and (3.3.33),

$$\mathrm{Tr}(G) = -\frac{\partial I}{\partial E} J$$
$$= -\imath \int d[w] \sum_\alpha w_\alpha^* w_\alpha \prod_{\alpha,\beta} e^{\imath Z_{\alpha\beta}(E\delta_{\alpha\beta} - H_{\alpha\beta})}, \tag{3.3.36}$$

with

$$Z_{\alpha\beta} = w_\alpha^* w_\beta + \chi_\alpha^* \chi_\beta. \tag{3.3.37}$$

Due to the introduction of the anticommuting variables the determinant factor has disappeared without the need to apply such questionable techniques as the replica trick. As the $H_{\alpha\beta}$ dependences still factorize on the right hand side of Eq. (3.3.36), the ensemble average can be directly performed:

$$\langle \mathrm{Tr}(G) \rangle = -\imath \int d[w] \sum_\alpha w_\alpha^* w_\alpha \prod_\alpha e^{\imath E Z_{\alpha\alpha}} \left\langle \prod_{\alpha,\beta} e^{-\imath Z_{\alpha\beta} H_{\alpha\beta}} \right\rangle. \tag{3.3.38}$$

For the calculation of the average we decompose $H_{\alpha\beta}$ into its real and imaginary parts, $H_{\alpha\beta} = (H_R)_{\alpha\beta} + \imath (H_I)_{\alpha\beta}$, and obtain

$$\begin{aligned} \left\langle \prod_{\alpha,\beta} e^{-\imath Z_{\alpha\beta} H_{\alpha\beta}} \right\rangle = & \prod_\alpha \langle e^{-\imath Z_{\alpha\alpha} (H_R)_{\alpha\alpha}} \rangle \\ & \times \prod_{\alpha > \beta} \langle e^{-\imath (Z_{\alpha\beta} + Z_{\beta\alpha})(H_R)_{\alpha\beta}} \rangle \\ & \times \prod_{\alpha > \beta} \langle e^{(Z_{\alpha\beta} - Z_{\beta\alpha})(H_I)_{\alpha\beta}} \rangle. \end{aligned} \tag{3.3.39}$$

By means of the matrix element probability distribution function for the GUE (see Eq. (3.1.68)) we get

$$\langle e^{-\imath Z_{\alpha\alpha} (H_R)_{\alpha\alpha}} \rangle = \exp\left[-\frac{(Z_{\alpha\alpha})^2}{4A} \right], \tag{3.3.40}$$

$$\langle e^{-\imath (Z_{\alpha\beta} + Z_{\beta\alpha})(H_R)_{\alpha\beta}} \rangle = \exp\left[-\frac{(Z_{\alpha\beta} + Z_{\beta\alpha})^2}{8A} \right], \tag{3.3.41}$$

$$\langle e^{(Z_{\alpha\beta} - Z_{\beta\alpha})(H_I)_{\alpha\beta}} \rangle = \exp\left[\frac{(Z_{\alpha\beta} - Z_{\beta\alpha})^2}{8A} \right]. \tag{3.3.42}$$

Entering these expressions into Eq. (3.3.39) we obtain

$$\left\langle \prod_{\alpha,\beta} e^{-\imath Z_{\alpha\beta} H_{\alpha\beta}} \right\rangle = \exp\left(-\frac{1}{4A} \sum_{\alpha,\beta} Z_{\alpha\beta} Z_{\beta\alpha} \right). \tag{3.3.43}$$

Such a simple expression is only obtained for the GUE. For the GOE the last factor on the right hand side of Eq. (3.3.39) is absent, and the resultant expression for the ensemble average is somewhat more complicated. But it is still sufficiently simple to allow a similar treatment as for the GUE. For the reader not yet familiar with the supersymmetry technique the corresponding calculation for the GOE is an excellent exercise.

By inserting expression (3.3.37) for $Z_{\alpha\beta}$ into Eq. (3.3.43) the sum in the exponential function can be written as

$$\sum_{\alpha,\beta} Z_{\alpha\beta} Z_{\beta\alpha} = S^2 + 2\sigma^*\sigma - T^2, \tag{3.3.44}$$

where

$$S = \sum_\alpha w_\alpha^* w_\alpha, \quad T = \sum_\alpha \chi_\alpha^* \chi_\alpha,$$
$$\sigma = \sum_\alpha w_\alpha^* \chi_\alpha, \quad \sigma^* = \sum_\alpha w_\alpha \chi_\alpha^*. \tag{3.3.45}$$

Using the relation

$$\sum_\alpha Z_{\alpha\alpha} = S + T \tag{3.3.46}$$

following immediately from the definitions, we can write the ensemble average (3.3.38) as

$$\langle \mathrm{Tr}(G) \rangle = -\imath \int d[w] S \exp\left[\imath E(S+T) - \frac{1}{4A}(S^2 + 2\sigma^*\sigma - T^2)\right]. \tag{3.3.47}$$

With the introduction of the anticommuting variables the step normally posing the most problems, namely the calculation of the ensemble average, has become completely trivial. What is left now are the integrations over $2N$ commuting and $2N$ anticommuting variables. This is not a simple task, since the integral on the right hand side of Eq. (3.3.47) contains the variables up to the fourth power in the argument of the exponential function. Two further techniques have still to be developed until the integrations can be performed. This will be done in the next two sections.

3.3.3 Hubbard–Stratonovitch transformation

The supersymmetric notation can formally be simplified considerably using the *graded matrices*, also called *super matrices*. A graded matrix is given by

$$F = \begin{pmatrix} a & \beta \\ \gamma & d \end{pmatrix}, \tag{3.3.48}$$

where a, d are matrices whose matrix elements are ordinary commuting variables, whereas β, γ are matrices with anticommuting variables as matrix elements. Graded matrices have a number of somewhat unusual properties. The transpose of F is defined as

$$F^T = \begin{pmatrix} a^T & -\gamma^T \\ \beta^T & d^T \end{pmatrix}, \tag{3.3.49}$$

differing from the ordinary transpose in the minus sign in the upper right corner. This minus sign is mandatory if the ordinary product relation for transposed matrices,

$$(F_1 F_2)^T = F_2^T F_1^T, \tag{3.3.50}$$

is to also hold for graded matrices. The trace of a graded matrix, too, is not defined in the usual manner:

$$\mathrm{Trg}(F) = \mathrm{Tr}(a) - \mathrm{Tr}(d). \tag{3.3.51}$$

Here 'Trg' stands for '*graded trace*'. This notation was introduced by Verbaarschot *et al.* [Ver85a]. Efetov [Efe83] uses the symbol 'STr' for '*super trace*'. His super trace is the negative of the graded trace defined by Eq. (3.3.51). The minus sign with the second term on the right hand side of Eq. (3.3.51) is again unavoidable, if the commutative relation

$$\mathrm{Trg}(F_1 F_2) = \mathrm{Trg}(F_2 F_1) \tag{3.3.52}$$

is to also hold for graded traces.

With these definitions Eq. (3.3.44) can be written compactly as

$$\sum_{\alpha,\beta} Z_{\alpha\beta} Z_{\beta\alpha} = \mathrm{Trg}(Z^2), \tag{3.3.53}$$

where Z is the graded matrix given by

$$Z = \begin{pmatrix} S & \sigma^* \\ \sigma & T \end{pmatrix}. \tag{3.3.54}$$

Using this notation, Eq. (3.3.47) reads

$$\langle \mathrm{Tr}(G) \rangle = -\imath \int d[w] S \exp\left[\imath E(S+T) - \frac{1}{4A} \mathrm{Trg} Z^2\right]. \tag{3.3.55}$$

In the next step the fourth powers of the integration variables, which enter into the argument of the exponential function, have to be removed. To this end we start with the identity

$$\int d[x] \exp\left(-\frac{A}{4} \mathrm{Trg}\, X^2\right) = 1, \tag{3.3.56}$$

where X is the graded matrix given by

$$X = \begin{pmatrix} x & \xi^* \\ \xi & \imath y \end{pmatrix}, \tag{3.3.57}$$

and where $d[x]$ is an abbreviation for $dx\, dy\, d\xi^*\, d\xi$.

The integrals over the ordinary variables run from $-\infty$ to $+\infty$. The identity is proved by inserting $\mathrm{Trg}(X^2) = x^2 + 2\xi^* \xi + y^2$ and performing the integrations. Note that it has been necessary to introduce an imaginary element $\imath y$ in the lower right corner of X to make the integrals well-defined. Now the variables of integration are shifted by substituting X by $X - \imath Z/A$. It is easily shown that such a shift is allowed both for the commuting and the anticommuting components of X. The result is the *Hubbard–Stratonovich transformation*

$$\exp\left[-\frac{1}{4A}\mathrm{Trg}(Z^2)\right] = \int d[x]\exp\left[-\frac{A}{4}\mathrm{Trg}(X^2) + \frac{\imath}{2}\mathrm{Trg}(XZ)\right]. \tag{3.3.58}$$

It may be considered as a generalization of the well-known relation (1.22) for Gaussian integrals to commuting *and* anticommuting variables. Inserting expression (3.3.58) into Eq. (3.3.55) we get

$$\begin{aligned}\langle \mathrm{Tr}(G)\rangle &= -\imath\int d[w]\,d[x]S\exp\left[\imath E(S+T) - \frac{A}{4}\mathrm{Trg}(X^2) + \frac{\imath}{2}\mathrm{Trg}(XZ)\right] \\ &= -\imath\int d[w]\,d[x]S\exp\Big[\imath E(S+T) \\ &\quad - \frac{A}{4}(x^2 + 2\xi^*\xi + y^2) + \frac{\imath}{2}(xS + \xi^*\sigma - \xi\sigma^* - \imath yT)\Big].\end{aligned} \tag{3.3.59}$$

By means of the Hubbard–Stratonovich transformation we have obtained an expression containing the integration variables in the exponential function only up to the second order. Now it is no longer a problem to perform the integrations over the w variables. But first the prefactor S in front of the exponential function is removed by means of a partial integration over x:

$$\begin{aligned}\int dxS\exp\left(-\frac{A}{4}x^2 + \frac{\imath}{2}xS\right) &= -2\imath\int dx\exp\left(-\frac{A}{4}x^2\right)\frac{d}{dx}\exp\left(\frac{\imath}{2}xS\right) \\ &= 2\imath\int dx\left[\frac{d}{dx}\exp\left(-\frac{A}{4}x^2\right)\right]\exp\left(\frac{\imath}{2}xS\right) \\ &= -\imath A\int dx\,x\exp\left(-\frac{A}{4}x^2 + \frac{\imath}{2}xS\right).\end{aligned} \tag{3.3.60}$$

Inserting this into Eq. (3.3.59) and re-substituting the expressions (3.3.45) for S, T, σ, σ^* we have achieved a complete factorization of the integration over the w and the χ variables:

$$\langle \mathrm{Tr}(G)\rangle = -A\int d[x]\,x\exp\left[-\frac{A}{4}\,\mathrm{Trg}(X^2)\right]\left[F(x, y, \xi, \xi^*)\right]^N, \tag{3.3.61}$$

where

$$F(x, y, \xi, \xi^*) = \int dw^* \, dw \, d\chi^* d\chi \exp\left\{ \imath \left[\left(E + \frac{x}{2} \right) w^* w + \left(E - \frac{\imath y}{2} \right) \chi^* \chi + \frac{1}{2} \xi^* w^* \chi - \frac{1}{2} \xi w \chi^* \right] \right\}. \quad (3.3.62)$$

The integrations in Eq. (3.3.62) are easily performed by removing the terms linear in w and w^* by a shift of the integration variables. The result is

$$F(x, y, \xi, \xi^*) = \frac{1}{2E + x} \left(2E - \imath y + \frac{\xi^* \xi}{2E + x} \right). \quad (3.3.63)$$

Let us stop for a moment and see what has been achieved. By means of the anticommuting variables the ensemble average of the N^2 different matrix elements has become trivial, at the cost, however, of the introduction of $4N$ commuting and anticommuting new variables. Applying a Hubbard–Stratonovich transformation these $4N$ integrations could be carried out. To achieve this aim, only 4 new variables had to be introduced. This is a dramatic reduction in complexity. All these steps are routinely performed in the supersymmetry technique. The real problems only arise afterwards. For the calculation of the density of states for the GOE we end with integrations over 4 commuting and 4 anticommuting variables. Reference [Ver85a] treats a problem leading to a total of 32 integrations. In such cases a lot of additional skill is necessary to reduce the problem further and to end, if we are lucky, with an analytical expression.

In the present case things are easier. The integrations of the anticommuting variables ξ and ξ^* are trivial and yield

$$\langle \mathrm{Tr}(G) \rangle = -\frac{A}{2\pi} \int dx \, dy \, x \exp\left[-\frac{A}{4}(x^2 + y^2) \right] \times \frac{1}{(2E + x)^N} \left[\frac{A}{2}(2E - \imath y)^N - N \frac{(2E - \imath y)^{N-1}}{2E + x} \right]. \quad (3.3.64)$$

Substituting x by $2x(N/A)^{1/2}$, and y by $2y(N/A)^{1/2}$, Eq. (3.3.64) reads

$$\langle \mathrm{Tr}(G) \rangle = -\frac{A^{1/2} N^{3/2}}{\pi} \int dx \, dy \, x \, e^{-N(x^2+y^2)} \times \left(\frac{\epsilon - \imath y}{\epsilon + x} \right)^N \left(2 - \frac{1}{(\epsilon + x)(\epsilon - \imath y)} \right), \quad (3.3.65)$$

where

$$\epsilon = \sqrt{\frac{A}{N}} E. \quad (3.3.66)$$

Both remaining integrations can be carried out in a closed form. The y integration leads to an integral representation of the Hermitian polynomials (see Chapter 5 of Ref. [Mag66]). Proceeding further in this direction, we finally end with the exact expression for the averaged density of states already derived in Section 3.2.3. In the next section another method will be applied. Though it only yields an asymptotic expression, valid in the limit of large N, it has the advantage that it can be generalized to a large class of different problems.

3.3.4 Saddle point integration

The two integrals in Eq. (3.3.65) are both of the type

$$I_N = \int_a^b dz\, f(z) e^{-Ng(z)}. \tag{3.3.67}$$

In the limit of large N such integrals allow an asymptotic solution which will be derived in the following. To avoid unnecessary complications, it is assumed that $g(z)$ has exactly one minimum in the interval $\langle a, b\rangle$ at a point z_s, and that $f(z)$ is nonzero at z_s. Inserting the Taylor expansions of $f(z)$ and $g(z)$ at $z = z_s$ into the integral (3.3.67) we get

$$I_N = \int_a^b dz[f(z_s) + (z - z_s) f'(z_s) + \cdots] \times \exp\left\{-N\left[g(z_s) + \frac{(z - z_s)^2}{2} g''(z_s) + \cdots\right]\right\}. \tag{3.3.68}$$

With the substitution $x = N^{1/2}(z - z_s)$ we obtain

$$I_N = \int_{-\sqrt{N}(z_s - a)}^{\sqrt{N}(b - z_s)} dx \left[f(z_s) + \frac{x}{\sqrt{N}} f'(z_s) + \cdots\right] \times \exp\left[-Ng(z_s) - \frac{g''(z_s)}{2} x^2 - \cdots\right]. \tag{3.3.69}$$

In the limit $N \to \infty$ only the leading term in the Taylor series for $f(z)$ and the first two terms in the series for $g(z)$ survive. Then the limits of the integral can be extended from $-\infty$ to $+\infty$. The integral is now easily performed with the result

$$I_N = \left[\frac{2\pi}{Ng''(z_s)}\right]^{1/2} f(z_s) e^{-Ng(z_s)}. \tag{3.3.70}$$

This is the saddle point integration. If $g(z)$ has more than one minimum within the limits of integration, the different contributions have to be added. The method can be applied to calculate the integral

$$N! = \int_0^\infty dx\, x^N e^{-x}. \tag{3.3.71}$$

The integrand is not yet of the form needed for the application of the method, but after the substitution $x = Nz$ all conditions are fulfilled. Here the saddle point integration leads to the *Stirling formula*

$$N! \simeq \sqrt{2\pi N}\left(\frac{N}{e}\right)^N. \qquad (3.3.72)$$

Readers not familiar with the method should verify Eq. (3.3.72) as an exercise.

The method is now applied to the x integration in Eq. (3.3.65):

$$I_x = \int_{-\infty}^{\infty} dx\, x \frac{\exp(-Nx^2)}{(\epsilon + x)^N}\left[2 - \frac{1}{(\epsilon + x)(\epsilon - \imath y)}\right]. \qquad (3.3.73)$$

Comparison with Eq. (3.3.67) shows that $g(x)$ is given by

$$g(x) = x^2 + \ln(\epsilon + x). \qquad (3.3.74)$$

The saddle point condition $g'(x_s) = 0$ yields the two solutions

$$x_s^{\pm} = -\frac{\epsilon}{2} \pm \sqrt{\frac{\epsilon^2}{4} - \frac{1}{2}}. \qquad (3.3.75)$$

For the second derivative of $g(x)$ at the saddle point we get

$$g''(x_s) = 2 - \frac{1}{(\epsilon + x_s)^2}. \qquad (3.3.76)$$

To illustrate the topography of the integrand, Fig. 3.18 shows a contour plot of

$$F(x) = |e^{-g(x)}| = \left|\frac{e^{-x^2}}{x + \epsilon}\right|. \qquad (3.3.77)$$

Two situations have to be discriminated. For $\epsilon^2 > 2$ both saddle points x_s^+ and x_s^- are on the real axis. This situation is depicted in Fig. 3.18(a). For $\epsilon^2 < 2$ the two saddle points are complex conjugates of each other (see Fig. 3.18(b)). For the integration a path is chosen starting from $-\infty$, contouring the singularity at $x = -\epsilon$ *above* the real axis (here it is used that E has been assumed to have a small positive imaginary part!), passing the saddle point at $x = x_s^+$ and finally extending to $+\infty$. In both cases the other saddle point at $x = x_s^-$ is not touched. The term 'saddle point integration' is immediately evident from the figure. Applying Eq. (3.3.70) we obtain for the integral (3.3.73)

$$I_x = \left[\frac{2\pi}{Ng''(x_s^+)}\right]^{1/2} x_s^+ \frac{\exp(-Nx_s^{+2})}{(\epsilon + x_s^+)^N}\left[2 - \frac{1}{(\epsilon + x_s^+)(\epsilon - \imath y)}\right]. \qquad (3.3.78)$$

Inserting this result into Eq. (3.3.65) we get

$$\langle \mathrm{Tr}(G)\rangle = -\frac{A^{1/2}N^{3/2}}{\pi}\left[\frac{2\pi}{Ng''(x_s^+)}\right]^{1/2} x_s^+ \frac{\exp(-Nx_s^{+2})}{(\epsilon + x_s^+)^N} I_y, \qquad (3.3.79)$$

where

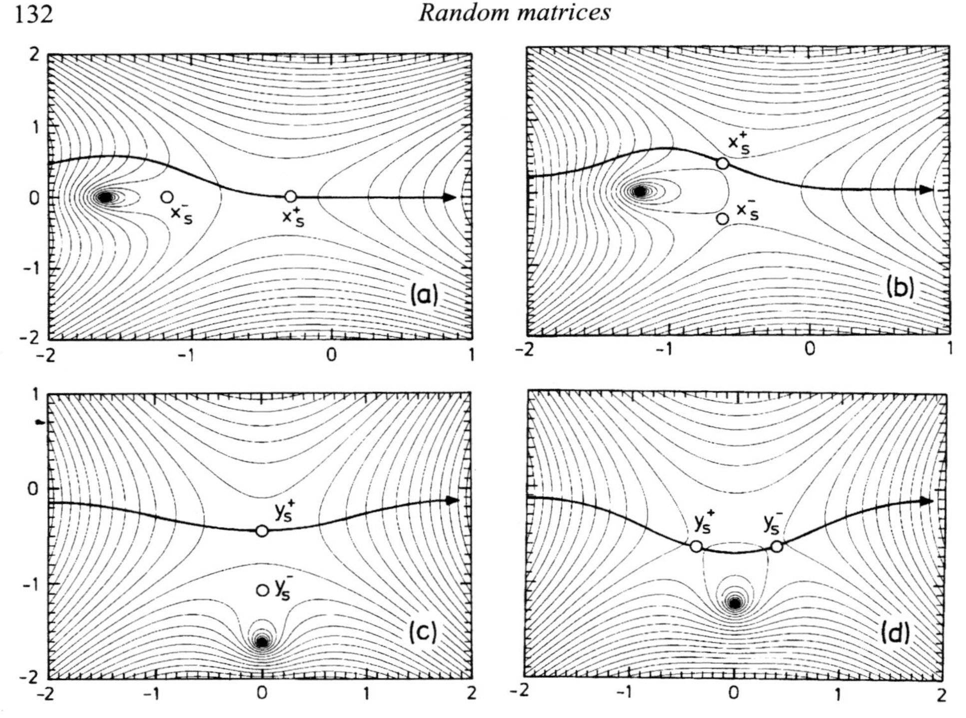

Figure 3.18. Contour plot of the function $F(x)$ (see Eq. (3.3.77)) and path of integration for the two cases $\epsilon^2 > 2$ (a) and $\epsilon^2 < 2$ (b). The bottom part shows the corresponding contour plots of $\overline{F}(y)$ (see Eq. (3.3.84)) for the two cases $\epsilon^2 > 2$ (c) and $\epsilon^2 < 2$ (d).

$$I_y = \int_{-\infty}^{\infty} dy(\epsilon - \imath y)^N \exp(-Ny^2)\left[2 - \frac{1}{(\epsilon + x_s^+)(\epsilon - \imath y)}\right]. \tag{3.3.80}$$

The remaining y integration is performed in a completely analogous manner. Now we have

$$\overline{g}(y) = y^2 - \ln(\epsilon - \imath y), \tag{3.3.81}$$

whence follows for the two saddle points

$$y_s^{\pm} = \imath\left(-\frac{\epsilon}{2} \pm \sqrt{\frac{\epsilon^2}{4} - \frac{1}{2}}\right) = \imath x_s^{\pm}. \tag{3.3.82}$$

For the second derivative of $\overline{g}(y)$ at the saddle points we have

$$\overline{g}''(y_s) = 2 - \frac{1}{(\epsilon - \imath y_s)^2} = 2 - \frac{1}{(\epsilon + x_s)^2} = g''(x_s). \tag{3.3.83}$$

In the bottom part of Fig. 3.18 the contour plots of

$$\overline{F}(y) = |e^{-\overline{g}(y)}| = |(\epsilon - \imath y)e^{-y^2}| \tag{3.3.84}$$

are shown, again for the two cases $\epsilon^2 > 2$ and $\epsilon^2 < 2$. In the first case the integration path again crosses only the saddle point at $y = y_s^+$, but in the second case it touches both saddle points. For the case $\epsilon^2 > 2$ we get

$$I_y = \left[\frac{2\pi}{Ng''(y_s^+)}\right]^{1/2} (\epsilon - \imath y_s^+)^N \exp\left(-Ny_s^{+2}\right)\left[2 - \frac{1}{(\epsilon + x_s^+)(\epsilon - \imath y_s^+)}\right]$$

$$= \left[\frac{2\pi}{Ng''(x_s^+)}\right]^{1/2} (\epsilon + x_s^+)^N \exp\left(Nx_s^{+2}\right)\left[2 - \frac{1}{(\epsilon + x_s^+)^2}\right]. \qquad (3.3.85)$$

For $\epsilon^2 < 2$ on the right hand side a second corresponding expression has to be added, where y_s^+ is replaced by y_s^-. If the resulting expression for I_y is inserted into Eq. (3.3.79) we see, however, that this second contribution vanishes for large N values. In this limit we therefore obtain for both cases, $\epsilon^2 > 2$ and $\epsilon^2 < 2$,

$$\begin{aligned} \langle \mathrm{Tr}(G) \rangle &= -2\sqrt{AN} x_s^+ \\ &= \sqrt{AN}\left(\epsilon - \sqrt{\epsilon^2 - 2}\right) \\ &= AE\left(1 - \sqrt{1 - \frac{2N}{AE^2}}\right), \end{aligned} \qquad (3.3.86)$$

where we have inserted expression (3.3.66) for ϵ. This is identical with the expression (3.1.127) for $\langle \mathrm{Tr}(G) \rangle$, which has been derived in a completely different manner in Section 3.1.5.

Comparing the two different derivations, it must be admitted that the more or less direct approach of Section 3.1.5 is considerably shorter and possibly easier for readers not familiar with the supersymmetry technique. But such a comparison is unfair. The calculation of the density of states for the GUE has been taken as a mere example to illustrate the potential of the method. The technique can be extended to all problems demanding ensemble averages of quantities which can be expressed in terms of Green functions. The first steps up to the Hubbard–Stratonovich transformation are always identical. From the mathematical point of view the problem can already be considered as solved after this step since the ensemble average desired is expressed in terms of an integral over a finite number of variables. The essential point is that the rank N of the original matrix no longer enters into the number of integrations, but has become a mere parameter in the integrand.

This cannot content the physicist, who actually has to solve the integrals. Here the individual problems begin. From this point of view the problem studied was a rather simple one. Only two integrations had to be performed, which was an easy task to do using the saddle point integration.

There is an increasing number of examples where problems otherwise inaccessible can be solved by means of the supersymmetry technique. A recent example, impressively demonstrating the state of the art, can be found in Ref. [Plu95], where electronic transport through mesoscopic structures is studied (see Section 6.3.2). More examples will be given in the following chapters.

4

Floquet and tight-binding systems

In their studies of the kicked rotator Casati *et al.* [Cas79] discovered that the quantization of classically chaotic systems can lead to the suppression of chaos. This is one of the most spectacular results in quantum chaos research. The classical dynamics of the kicked rotator is described by the famous standard map, and it has been known for some time that for sufficiently large kick strengths the classical system shows diffusive behaviour. But for its quantum mechanical counterpart the diffusion is stopped after some time by a process called *dynamical localization*. The dynamical localization of electrons has also been observed experimentally in the microwave ionization of highly excited hydrogen atoms [Gal88, Bay89]. For atoms the effect has also been demonstrated recently [Moo94].

All the examples mentioned belong to the Floquet systems, i.e. periodic time dependences are involved. Soon after the discovery of dynamical localization Fishman *et al.* [Fis82] found that there is a close relation between Floquet systems and the Anderson model, which has been used for a long time to study the influence of disorder on electronic wave functions in crystalline lattices. From this equivalence it became clear that dynamical localization is just a special case of the well-known Anderson localization.

For this reason Floquet systems and the Anderson model are discussed here together. For the Anderson model, however, only those aspects will be considered which are of relevance for dynamical localization. A detailed discussion of the Anderson model as a whole is not intended, as there are already a large number of review articles written by the pioneers in this field. The chapter ends with a discussion of the band model for electrons on a square lattice with a perpendicularly applied magnetic field. It gives rise to the famous *Hofstadter butterfly* [Hof76], one of the rare cases where the difference between rational and irrational numbers becomes manifest in physical systems. We shall see that these commensurability effects can have surprising consequences for spectral correlations such as the level spacing distribution or the spectral rigidity.

4.1 Hamiltonians with periodic time dependences

4.1.1 Floquet operator

Most spectroscopic techniques are based on the application of oscillating electric or magnetic fields. Examples are nuclear magnetic resonance, or electron spin resonance, and laser spectroscopy. All these methods, which, from the technical point of view, are very different, share the property that the underlying Hamiltonian depends periodically on time,

$$\mathcal{H}(t) = \mathcal{H}_0 + V(t). \tag{4.1.1}$$

Here $\mathcal{H}_0$ is the Hamiltonian of the atom or nucleus, and $V(t)$ describes the coupling with the oscillating field, obeying the relation $V(t + \tau) = V(t)$. As $\mathcal{H}$ depends explicitly on time, it is no longer possible to solve the time-dependent Schrödinger equation

$$\imath\hbar \frac{\partial \psi}{\partial t} = \mathcal{H}\psi \tag{4.1.2}$$

by means of a separation ansatz

$$\psi_n(x,\ t) = \exp\left(-\frac{\imath}{\hbar} E_n t\right)\psi_n(x). \tag{4.1.3}$$

Consequently the energy E is no longer a constant of motion. But $\mathcal{H}$ is still invariant against a *discrete* shift τ. This in turn means that it is possible to find solutions $\psi_n(x,\ t)$ of the time-dependent Schrödinger equation which are simultaneously eigenfunctions of the corresponding shift operator T_τ,

$$T_\tau \psi_n(x,\ t) = \psi_n(x,\ t + \tau) = \lambda_n \psi_n(x,\ t). \tag{4.1.4}$$

For the solution to be stationary λ_n must be a pure phase factor,

$$\psi_n(x,\ t + \tau) = e^{-\imath\phi_n}\psi_n(x,\ t). \tag{4.1.5}$$

It follows that $\psi_n(x,\ t)$ can be written as

$$\psi_n(x,\ t) = e^{-\imath\omega_n t} u_n(x,\ t), \tag{4.1.6}$$

where $u_n(x,\ t)$ is periodic in time, $u_n(x,\ t+\tau) = u_n(x,t)$, and where $\omega_n = \phi_n/\tau$. This is the *Floquet theorem*. It is completely analogous to the Bloch theorem in solid state physics, where the periodicity of the crystal lattice leads to a corresponding relation for the spatial coordinates. Comparing Eqs. (4.1.3) and (4.1.6) we see that in Floquet systems the role of the energy is taken by the so-called quasi-energy

$$\overline{E}_n = \hbar\omega_n = \frac{\hbar}{\tau}\phi_n. \qquad (4.1.7)$$

As the Floquet phase ϕ_n is defined only up to integer multiples of 2π, quasi-energies are defined only up to integer multiples of h/τ. An analogous situation is again found in crystals, where the wave numbers are defined only up to multiples of reciprocal lattice vectors with the consequence that we can restrict the wavenumbers to the first Brillouin zone. A relevant example has already been given in Section 3.2.2. In Floquet systems the range $\langle -h/2\tau,\, h/2\tau\rangle$ is conventionally taken as the first Brillouin zone.

The time-evolution operator $U(t)$ is defined via the equation

$$\psi(x,\, t) = U(t)\psi(x,\, 0). \qquad (4.1.8)$$

Inserting this expression into the time-dependent Schrödinger equation (4.1.2), we get the operator equation

$$\imath\hbar\dot{U} = \mathscr{H}U, \qquad (4.1.9)$$

which has to be solved with the initial condition $U(0) = 1$. From the condition that the Hamiltonian is Hermitian, it follows immediately that U is unitary. For the proof we take the adjoint of Eq. (4.1.9),

$$-\imath\hbar\dot{U}^\dagger = U^\dagger\mathscr{H}, \qquad (4.1.10)$$

multiply both sides of Eq. (4.1.9) from the left with $U^\dagger$, both sides of Eq. (4.1.10) from the right with U, and take the difference. We thus find $d(U^\dagger U)/dt = 0$, i.e. $U^\dagger U$ is constant. From the initial condition $U(0) = 1$ it follows that the constant must be one,

$$U^\dagger U = 1. \qquad (4.1.11)$$

With $t = \tau$ Eq. (4.1.8) reads

$$\psi(x,\, \tau) = U(\tau)\psi(x,\, 0), \qquad (4.1.12)$$

which can be generalized to

$$\psi(x,\, n\tau) = [U(\tau)]^n\psi(x,\, 0). \qquad (4.1.13)$$

It follows for the time-evolution operator

$$U(n\tau) = [U(\tau)]^n. \qquad (4.1.14)$$

For a stroboscopic observation of the system at times $n\tau$ ($n = 0,\, 1,\, \ldots$) the knowledge of $U(\tau)$ is thus sufficient. As $U(\tau)$ is unitary, it can be diagonalized with the help of a unitary transformation

$$U = V^\dagger U_D V, \qquad (4.1.15)$$

where U_D is a diagonal matrix with matrix elements

$$(U_D)_{nn} = e^{-\imath\phi_n}. \qquad (4.1.16)$$

The eigenphases of U are exactly the Floquet phases introduced above. The study of Floquet systems is thus reduced to the study of the eigenphases of $U(\tau)$. This operator is therefore called the Floquet operator, and will be denoted by F in the following.

The determination of the eigenphases of F is usually a difficult task. To illustrate this, we take as an example a Hamilton operator with a time dependence which is given by one single cosine term:

$$\mathscr{H}(t) = \mathscr{H}_0 + 2V_0 \cos(\omega t). \tag{4.1.17}$$

The factor 2 is introduced for later convenience. Using the Floquet theorem (4.1.6) every stationary solution of the time-dependent Schrödinger equation can be written as

$$\psi(x, t) = e^{-\imath\Omega t} \sum_k u_k(x) e^{\imath k\omega t}, \tag{4.1.18}$$

where we have denoted the as yet unknown Floquet phase by Ω, and have expanded the periodic function $u_n(x, t)$ into a Fourier series. The index n has been omitted for the sake of convenience. Entering this ansatz into the Schrödinger equation and equating the coefficients of $e^{\imath k\omega t}$ for all k, we get a coupled equation system for the u_k:

$$\hbar(\Omega - k\omega)u_k = \mathscr{H}_0 u_k + V_0(u_{k-1} + u_{k+1}). \tag{4.1.19}$$

This can be written in matrix notation as

$$Hu = \hbar\Omega u, \tag{4.1.20}$$

where

$$H = \begin{pmatrix} \ddots & \ddots & & & \\ \ddots & \mathscr{H}_0 + (k-1)\hbar\omega 1 & V_0 & & \\ & V_0 & \mathscr{H}_0 + k\hbar\omega 1 & V_0 & \\ & & V_0 & \mathscr{H}_0 + (k+1)\hbar\omega 1 & \ddots \\ & & & \ddots & \ddots \end{pmatrix}, \tag{4.1.21}$$

and

$$u = \begin{pmatrix} \vdots \\ u_{k-1} \\ u_k \\ u_{k+1} \\ \vdots \end{pmatrix}. \tag{4.1.22}$$

The eigenphases of the Floquet operator are thus obtained as the eigenvalues of the matrix H. The procedure described implies an obvious problem. The matrix is unbounded in both directions and must be truncated for a numerical calculation. The problem is even more severe because the matrix elements themselves are already matrices of infinite rank.

For these reasons periodically *driven* systems are not very popular among theoreticians. They prefer by far the periodically *kicked* systems described by the Hamiltonian

$$\mathscr{H}(t) = \mathscr{H}_0 + V_0 \sum_n \delta(t - n\tau). \tag{4.1.23}$$

The experimentalist, on the other hand, prefers driven systems, which are much easier to realize! For kicked systems the associated Floquet operator is easily obtained. To this end we replace the delta function for a moment by a pulse of finite width $\Delta\tau$ and height $(\Delta\tau)^{-1}$, and write

$$\mathscr{H}(t) = \begin{cases} \mathscr{H}_0 & n\tau < t < (n+1)\tau - \Delta\tau \\ \mathscr{H}_0 + \dfrac{1}{\Delta\tau} V_0 & (n+1)\tau - \Delta\tau < t < (n+1)\tau \end{cases}. \tag{4.1.24}$$

As the Hamiltonian is piecewise constant, the time-evolution operator $U(t)$ can be obtained by direct integration. For $0 < t < \tau - \Delta\tau$ we have

$$U(t) = \exp\left(-\frac{\imath}{\hbar}\mathscr{H}_0 t\right), \tag{4.1.25}$$

and for $\tau - \Delta\tau < t < \tau$

$$U(t) = \exp\left[-\frac{\imath}{\hbar}\left(\mathscr{H}_0 + \frac{1}{\Delta\tau}V_0\right)(t - \tau + \Delta\tau)\right] U(\tau - \Delta\tau), \tag{4.1.26}$$

whence follows for the Floquet operator $F = U(\tau)$,

$$F = \exp\left[-\frac{\imath}{\hbar}\left(\mathscr{H}_0 + \frac{1}{\Delta\tau}V_0\right)\Delta\tau\right] \exp\left(-\frac{\imath}{\hbar}\mathscr{H}_0(\tau - \Delta\tau)\right). \tag{4.1.27}$$

Now the limit $\Delta\tau \to 0$ can be performed, yielding

$$F = \exp\left(-\frac{\imath}{\hbar}V_0\right) \exp\left(-\frac{\imath}{\hbar}\mathscr{H}_0 \tau\right). \tag{4.1.28}$$

The kicked tops represent a very popular class of these systems. The second important system is the kicked rotator, which has been studied extensively especially in the context of dynamical localization. Both systems will be discussed in the following sections.

4.1.2 Circular ensembles

Symmetry considerations are very important for Floquet systems, too. We can be rather brief here, since the argumentation follows essentially the same lines as for the conservative systems. First we note that the time-evolution operator U is uniquely defined by the operator Schrödinger equation

$$\imath\hbar\dot{U} = \mathscr{H}U, \tag{4.1.29}$$

and the initial condition $U(0) = 1$. This is a general property of linear differential equation systems. Now assume that there is a symmetry operator R commuting with $\mathscr{H}$. Multiplying both sides of Eq. (4.1.29) from the left with R and from the right with R^{-1}, we obtain

$$\imath\hbar\frac{d(RUR^{-1})}{dt} = \mathscr{H}(RUR^{-1}). \tag{4.1.30}$$

$\hat{U} = RUR^{-1}$ thus obeys the same equation as U. It also obeys the initial condition $\hat{U}(0) = 1$, whence follows $\hat{U} = U$, or

$$UR = RU. \tag{4.1.31}$$

If the eigenfunctions of R are taken as basis functions, the matrix representations of both U and $\mathscr{H}$ take block form, i.e. the differential equation system (4.1.29) decomposes into disjunct subsystems. We therefore restrict the following discussion to sub-Hamiltonians lacking any symmetry, exactly as we proceeded in Section 3.1.1 in the case of the conservative systems.

The time-reversal symmetry again needs special treatment. If there is no time-reversal symmetry, no further information can be obtained in addition to the fact that the time-evolution operator must be unitary. If there is time-reversal symmetry, we must once more discriminate between systems without and with a spin-$\frac{1}{2}$ interaction. Let us begin with the first case. Taking on both sides of Eq. (4.1.29) the time reversal and the complex conjugate, we get

$$\imath\hbar\frac{dU(-t)^*}{dt} = \mathscr{H}U(-t)^*. \tag{4.1.32}$$

$U(-t)^*$ obeys the same differential equation system as U, including the initial condition, whence follows

$$U(t) = U(-t)^*. \tag{4.1.33}$$

This holds for arbitrary times. For $t = \tau$ we have in particular $U(-\tau) = U(\tau)^{-1} = U(\tau)^\dagger = F^\dagger$, since U is unitary. Inserting this into Eq. (4.1.33), we obtain

$$F = (F^\dagger)^* = F^T. \tag{4.1.34}$$

For Floquet systems with time-reversal symmetry and no spin-$\frac{1}{2}$ interaction the Floquet operator can thus be represented by a symmetric matrix.

In the presence of spin-$\frac{1}{2}$ interactions $\mathscr{H}$ no longer commutes with the complex conjugation operator C due to the presence of the spin matrices. We have, however, learnt that in this case $\mathscr{H}$ commutes with ZC where Z is the diagonal matrix with the quaternion τ_y in the diagonal (see Eq. (3.1.56)). Repeating the above calculation we now obtain the relation

$$F = ZF^T Z^{-1} = -ZF^T Z. \tag{4.1.35}$$

It should be remembered that the matrix $Q^R = ZQ^T Z^{-1}$ is the dual matrix of Q. Matrices with the property $Q^R = Q$ are called self-dual. In Floquet systems with time-reversal symmetry and a spin-$\frac{1}{2}$ interaction the Floquet operator can therefore be represented by a self-dual matrix.

In Section 3.1.3 we defined the three Gaussian matrix ensembles by the conditions that (i) all matrix elements are uncorrelated, and that (ii) the ensembles remain invariant under orthogonal, unitary, and symplectic transformations, respectively. We now proceed in an analogous way for the Floquet systems.

First we look for the transformations preserving the unitary, symmetric, and self-dual properties of the matrix representations respectively, of the Floquet operators for the three symmetry cases discussed above. For systems without time-reversal symmetry the most general transformation, not changing the unitarity property of a matrix, is given by

$$S' = USV, \tag{4.1.36}$$

where U, V are arbitrary unitary matrices. It is evident that S' is unitary if S is unitary. For systems with time-reversal symmetry but no spin-$\frac{1}{2}$ interaction a matrix S remains symmetric under all transformations

$$S' = USU^T, \tag{4.1.37}$$

where U is an arbitrary unitary matrix. For systems with time-reversal symmetry and an additional spin-$\frac{1}{2}$ interaction the transformation

$$S' = USU^R \tag{4.1.38}$$

preserves the self-duality of S.

We are now ready to define the so-called *circular ensembles*. The circular unitary ensemble (CUE) is the ensemble of all unitary matrices remaining invariant under transformations described by Eq. (4.1.36). Analogously the circular orthogonal ensemble (COE) is the ensemble of all symmetric matrices remaining invariant under transformations obeying Eq. (4.1.37), and the circular symplectic ensemble (CSE) is the ensemble of self-dual matrices being invariant under transformations described by Eq. (4.1.38).

The ensembles are called 'circular', as the eigenvalues of the Floquet operator are unimodular, i.e. all of them are on the unit circle in the complex plane. The invariance properties formulated above are sufficient to define the circular ensembles uniquely. It is not necessary to impose the statistical independence of the matrix elements as an additional condition.

The joint eigenphase distribution function for the circular ensemble takes the extraordinarily simple form

$$P(\phi_1, \ldots, \phi_N) \sim \prod_{n>m} |e^{\imath\phi_n} - e^{\imath\phi_m}|^\nu, \tag{4.1.39}$$

where ν is the universality index taking the values 1, 2, and 4 for the COE, CUE, and CSE, respectively. The proof can be found in Chapter 9 of Ref. [Meh91].

The calculation of the n-point correlation functions is essentially performed in the same way as for the Gaussian ensembles (see Chapter 10 of Ref. [Meh91]). For the one-point correlation function, i.e. the ensemble averaged density of states, we obtain

$$R_1(\phi) = \frac{N}{2\pi}. \tag{4.1.40}$$

There is a constant probability of finding an eigenphase on the unit circle. The circular ensembles have the great advantage that finite size effects are absent, since the energy axis is so to speak closed to a circle. In the Gaussian ensembles finite size effects have been responsible in particular for the unphysical semicircle law for the density of states.

In the limit of large N the n-point correlation functions of the circular and the Gaussian ensembles become identical. Therefore all quantities calculated in Chapter 3, especially the nearest neighbour spacing distribution and the spectral rigidity, hold unchanged for the circular ensembles, too.

As an example we consider a periodically kicked top described by the Hamiltonian [Haa87]

$$\mathscr{H}(t) = \frac{\hbar p}{\tau} I_y + \frac{\hbar k}{2I} I_z^2 \sum_{n=-\infty}^{\infty} \delta(t - n\tau). \tag{4.1.41}$$

The first term may be interpreted as the interaction of a nuclear spin with a magnetic field, leading to a precession with the Larmor angular frequency p/τ about the y axis. The second term describes periodic kicks proportional to I_z^2 with strength k/I, where I is the spin quantum number. It is a bit difficult to imagine realizations of kicks proportional to I_z^2, and not to I_z. One possibility is a periodically switched quadrupole interaction between a nuclear quadrupole moment and an electric field gradient.

Using Eq. (4.1.28) we get for the associated Floquet operator

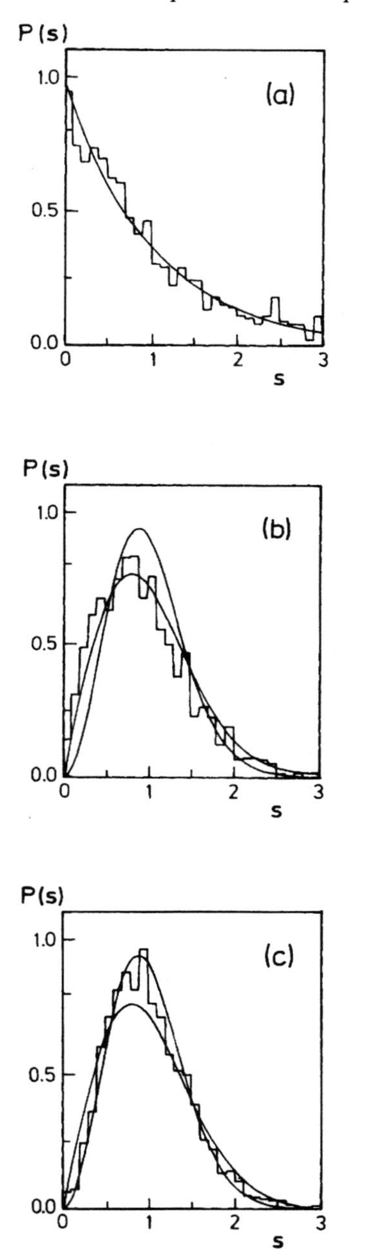

Figure 4.1. Level spacing distribution for a kicked top with time-reversal symmetry (see Eq. (4.1.42)) with $p = 2$, $0.1 \leqslant k \leqslant 0.3$, where the classical motion is essentially regular (a), with $p = 1.7$, $10.0 \leqslant k \leqslant 10.5$, where the motion is chaotic (b), and for a kicked top with broken time-reversal symmetry (see Eq. (4.1.43)) with $p = 1.7$, $10.0 \leqslant k \leqslant 10.5$ in the chaotic range (c) [Haa87] (Copyright 1987 by Springer Verlag).

$$F = \exp\left(-\imath \frac{k}{2I} I_z^2\right) \exp(-\imath p I_y). \tag{4.1.42}$$

The Hamiltonian (4.1.41) is invariant with respect to a rotation by an angle of π about the y axis. Therefore the eigenphases fall into two disjunct subspectra. The corresponding classical system behaves more or less regularly for kick strengths $k < 2$, whereas for $k > 6$ it is completely chaotic [Haa87]. This is also seen in the level spacing distribution. Figure 4.1(a) shows the result with k values taken in the interval $0.1 \leqslant k \leqslant 0.3$. To improve the statistics the results of both symmetry classes and five different kick strengths have been superimposed. The resulting spacing distribution is in excellent agreement with the expected Poissonian behaviour. For k in the chaotic region a GOE Wigner distribution is expected and observed. Figure 4.1(b) shows the result with k taken from the interval $10.0 \leqslant k \leqslant 10.5$ and $p = 1.7$. Altogether 20 spectra have been superimposed.

The kicked top discussed belongs to the COE class as it is possible to find real matrix representations both for I_z and I_y (see the discussion in Section 3.1.2). This is no longer the case for the Floquet operator

$$F = \exp\left(-\imath \frac{k'}{2I} I_x^2\right) \exp\left(-\imath \frac{k}{2I} I_z^2\right) \exp\left(-\imath p I_y\right), \tag{4.1.43}$$

differing from the operator in Eq. (4.1.42) by an additional kick proportional to I_x^2. As it is impossible to find real representations for all three components of the angular momentum, this system belongs to the CUE class. This is verified by the level spacing distribution shown in Fig. 4.1(c), which is in good accordance with the Wigner distribution for the unitary case. k values and p have been chosen as in Fig. 4.1(b).

An example for a kicked top belonging to the symplectic case has been given by the same authors [Sch88] (see also the introduction of the book by Haake [Haa91]). Again a perfect agreement with the corresponding Wigner distribution for the nearest neighbour spacing has been found.

4.2 Dynamical localization

4.2.1 *Kicked rotator*

Dynamical localization was found for the first time in quantum mechanical studies of the kicked rotator [Cas79, Chi81, Chi88]. The phenomenon is presented in the introduction to the *Quantum Chaos* monograph by Casati and Chirikov [Cas95] and the review by Izrailev [Izr90]. The following presentation is based on these two articles.

The Hamiltonian of the kicked rotator is given by

$$\mathscr{H}(t) = \frac{L^2}{2} + k \cos\theta \sum_n \delta(t-n). \tag{4.2.1}$$

It describes the free rotation of a pendulum with angular momentum L, which is periodically kicked by a gravitational potential of strength k. The moment of inertia I and the kick period T have been normalized to one.

It has already been discussed in the introduction that the equations of motion classically reduce to the famous standard map

$$\theta_{n+1} = \theta_n + L_{n+1}, \tag{4.2.2}$$

$$L_{n+1} = L_n + k \sin\theta_n, \tag{4.2.3}$$

where θ_n, L_n are the values of the dynamical variables immediately after the nth kick. The system is conceptually simple, but it already shows the full complexity of real systems such as Kolmogorov–Arnol'd–Moser tori, or breaking of resonant tori under the influence of a perturbation [Chi79]. Details can be found in monographs of classical nonlinear dynamics such as Refs. [Ott93, Sch84]. For small k values the rotator shows regular behaviour for most initial values of θ and L, but with increasing k more and more tori break up, until above a critical value of $k = k_c = 0.9716\ldots$ the last invariant torus will have been destroyed (see Fig. 1.3).

In the introduction the results by Geisel *et al.* [Gei86] on the spread of the momentum exactly at k_c have been discussed. Tori and cantori, acting as diffusion barriers in the classical system, have been found to influence the quantum mechanical behaviour as well. With increasing k the classical phase space becomes more and more chaotic, until for $k > 5$ most regular parts have disappeared (see Fig. 1.3(c)).

In this range the transport of angular momentum becomes diffusive. Qualitatively this can be seen by an inspection of Eqs. (4.2.2) and (4.2.3): as soon as $|L_n|$ is of the order of 2π or larger, which will be the case after a few kicks if k is large, successive θ_n are uncorrelated. Then the sign of $\sin\theta_n$ is random, and the sequence $\{L_n\}$ describes a random walk. For a more quantitative argumentation we define $f_N(\Delta L)$ as the probability that after N kicks $(L_N - L_0)$ takes the value ΔL or

$$f_N(\Delta L) = \langle \delta(\Delta L - L_N + L_0) \rangle, \tag{4.2.4}$$

where the brackets denote an averaging over different initial conditions. For the calculation of the average the delta function is expressed in terms of its Fourier integral,

$$f_N(\Delta L) = \left\langle \frac{1}{2\pi} \int_{-\infty}^{\infty} dt\, e^{\imath(\Delta L - L_N + L_0)t} \right\rangle$$
$$= \left\langle \frac{1}{2\pi} \int_{-\infty}^{\infty} dt\, e^{\imath \Delta L t} \prod_{n=0}^{N-1} e^{-\imath(L_{n+1} - L_n)t} \right\rangle$$
$$= \left\langle \frac{1}{2\pi} \int_{-\infty}^{\infty} dt\, e^{\imath \Delta L t} \prod_{n=0}^{N-1} e^{-\imath kt \sin\theta_n} \right\rangle. \qquad (4.2.5)$$

This result is still exact. Now we apply the approximation that for large k values the θ_n are uncorrelated. Then the averages on the right hand side can be calculated independently. With

$$\langle e^{-\imath kt \sin\theta} \rangle = \frac{1}{2\pi} \int_0^{2\pi} e^{-\imath kt \sin\theta}\, d\theta = J_0(kt), \qquad (4.2.6)$$

where $J_0(kt)$ is a Bessel function, we obtain from Eq. (4.2.5)

$$f_N(\Delta L) = \frac{1}{2\pi} \int_{-\infty}^{\infty} dt\, e^{\imath \Delta L t} [J_0(kt)]^N. \qquad (4.2.7)$$

For large N values we have

$$[J_0(kt)]^N = e^{N \ln J_0(kt)} = \exp\left(-\frac{N}{4}(kt)^2 + \cdots \right). \qquad (4.2.8)$$

In the limit $N \to \infty$ the unwritten terms can be neglected. Inserting expression (4.2.8) into Eq. (4.2.7) we end with

$$f_N(L) = \frac{1}{k\sqrt{\pi N}} \exp\left(-\frac{(\Delta L)^2}{k^2 N} \right). \qquad (4.2.9)$$

For the quadratic average of ΔL it follows

$$\langle (\Delta L)^2 \rangle = \frac{k^2}{2} N. \qquad (4.2.10)$$

$\langle (\Delta L)^2 \rangle$ therefore increases linearly with N. This, together with the Gaussian distribution for $f_N(L)$, is typical for a diffusion process with a diffusion constant

$$D = \frac{k^2}{2}. \qquad (4.2.11)$$

This is a manifestation of the *central limit theorem*: in the limit of large N the sum of N random numbers is Gaussian distributed with a second moment proportional to N. The details of the distribution of the random numbers are irrelevant.

We now come to the quantum mechanical description of the kicked rotator. Using Eq. (4.1.28) we obtain for the Floquet operator

$$F = \exp\left(-\frac{\imath}{\hbar}k\cos\theta\right)\exp\left(-\frac{\imath}{\hbar}\frac{\tau}{2}L^2\right), \tag{4.2.12}$$

where L^2 now has to be interpreted as the quantum mechanical operator $-\hbar^2\partial^2/\partial\theta^2$. In the eigenbasis of L,

$$|n\rangle = \frac{1}{\sqrt{2\pi}}\, e^{\imath n\theta}, \tag{4.2.13}$$

the matrix elements of F are given by

$$\begin{aligned} F_{nm} &= \langle n|F|m\rangle \\ &= \frac{1}{2\pi}\int_0^{2\pi} e^{-\imath n\theta}\exp\left(-\frac{\imath}{\hbar}k\cos\theta\right)\exp\left(-\frac{\imath}{\hbar}\frac{\tau}{2}L^2\right)e^{\imath m\theta}\, d\theta \\ &= \exp\left(-\frac{\imath}{\hbar}\frac{\tau}{2}m^2\right)\frac{1}{2\pi}\int_0^{2\pi}\exp\left(-\frac{\imath}{\hbar}k\cos\theta\right)\mathrm{e}^{\imath(m-n)\theta}\, d\theta \\ &= \exp\left(-\frac{\imath}{\hbar}\frac{\tau}{2}m^2\right)\imath^{m-n}J_{m-n}\left(\frac{k}{\hbar}\right). \end{aligned} \tag{4.2.14}$$

In this basis only the matrix elements close to the diagonal are large, as the Bessel functions decrease rapidly with increasing order. This motivated Izrailev and others to study banded matrices [Izr90, Cas91] (see Section 3.2.2). Equation (4.2.14) allows a straightforward calculation of the quantum mechanical evolution of the system from a given initial condition: let $a(\theta)$ be the probability amplitude to find the rotator at time $t = 0$ at the angle θ. From this the components of the state vector

$$A_0 = \begin{pmatrix} a_1 \\ a_2 \\ \vdots \end{pmatrix} \tag{4.2.15}$$

in the basis of eigenfunctions of L are obtained by a simple Fourier transformation

$$a_n = \frac{1}{2\pi}\int_0^{2\pi} a(\theta)e^{-\imath n\theta}\, d\theta, \tag{4.2.16}$$

whence follows for the state vector after the Nth kick

$$A_N = F^N A_0. \tag{4.2.17}$$

For a comparison of the quantum mechanical with the classical behaviour we calculate the quantum mechanical expectation value of L^2 after the Nth kick:

$$\langle L^2\rangle = A_N^\dagger L^2 A_N. \tag{4.2.18}$$

Inserting expression (4.2.17) for A_N we get

$$\langle L^2 \rangle = \hbar^2 \sum_{l,m,n} l^2 (F^N)_{lm} (F^N)^*_{ln} a_m a^*_n. \tag{4.2.19}$$

From this equation $\langle L^2 \rangle$ can be calculated from a given initial distribution $a(\theta)$. In Fig. 4.2(a) $\langle L^2 \rangle$ is plotted as a function of the kick number for $k = 5$, in a region where the classical phase space is nearly completely chaotic. The classical rotator exactly follows the expected diffusion law described by Eq. (4.2.10), but quantum mechanically a different behaviour is found [Cas79]. For kick numbers below about 50 the classical and quantum mechanical probabilities are still close together, but then the quantum mechanical diffusion begins to saturate until, after the 1000th kick, the quantum diffusion has completely stopped. The same findings are presented in Fig. 4.2(b) from another point of view. Here the logarithm of the distribution function $f_N(L)$ after the 1000th kick is shown. Classically the expected Gaussian distribution (4.2.9) is found (remember that in the logarithmic plot a Gaussian function becomes an

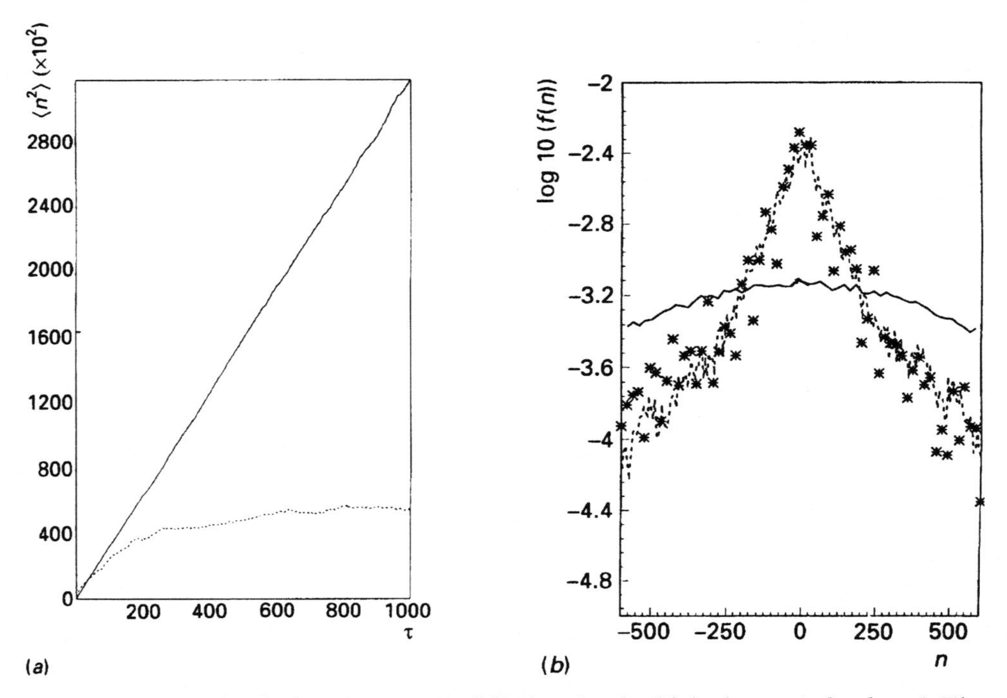

Figure 4.2. (a) Classical and quantum diffusion for the kicked rotator for $k = 5$. The quadratically averaged angular momentum $\langle n^2 \rangle$ is plotted versus the kick number for the classical (solid line) and the quantum mechanical (dotted line) cases. (b) Classical and quantum mechanical probability distributions after 1000 kicks. For the stars see the original work [Cas95].

inverted parabola!), but quantum mechanically a distribution is obtained which suggests an exponential localization,

$$f_N(L) = \frac{1}{l_s} \exp\left(-\frac{2|L|}{l_s}\right), \tag{4.2.20}$$

with a localization length l_s.

Qualitatively it is rather easily seen that quantum diffusion must saturate [Chi88]. To this end we diagonalize the Floquet operator by means of a unitary transformation,

$$F_{ln} = \sum_k e^{-\imath\phi_k} U^*_{kl} U_{kn}, \tag{4.2.21}$$

where the ϕ_k are the Floquet phases. Let us assume that prior to the first kick the rotator is in the $l = 0$ state. Then a_0 is one, and all other a_m are zero. Equation (4.2.19) now reads

$$\langle L^2 \rangle = \hbar^2 \sum_{l,k,k'} l^2\, e^{\imath N(\phi_k - \phi_{k'})} U^*_{kl} U_{k0} U_{k'l} U^*_{k'0}. \tag{4.2.22}$$

We have seen that in the eigenbasis of L the Floquet operators have a band structure. Consequently the operator U, diagonalizing F, is banded too. This means that the sums in Eq. (4.2.22) are essentially truncated as soon as the summation indices k, k' exceed a maximum value l_s. The Floquet phases associated with these states will be more or less equally distributed on the unit circle with a mean level density of $l_s/2\pi$. The smallest phase difference $\phi_k - \phi_{k'}$ occuring in the sum is thus of the order of the reciprocal level density $2\pi/l_s$. For kick numbers $N \leqslant l_s$ the system does not 'feel' that the involved subspectrum is discrete. This is the range of classical diffusion. For $N \gg l_s$, on the other hand, the phase factors oscillate rapidly, and on the average only the terms with $k = k'$ survive,

$$\langle L^2 \rangle = \hbar^2 \sum_{l,k} l^2 |U_{kl}|^2 |U_{k0}|^2. \tag{4.2.23}$$

The expectation value of L^2 is now independent of N, the diffusion has stopped.

The cross-over from one region to the other takes place at $N \sim l_s$. Equation (4.2.23) shows that l_s is also of the order of the maximum l quantum number obtained after many kicks, whence follows $\langle (\Delta L)^2 \rangle \sim l_s^2$. Note that in the present example we have $\langle L^2 \rangle = \langle (\Delta L)^2 \rangle$, since we started with an angular momentum of $l = 0$. Inserting the results into the diffusion relation (4.2.10), we approximately get

$$l_s^2 \sim \frac{k^2}{2} l_s = D l_s, \tag{4.2.24}$$

whence follows the remarkable result [Chi81, Chi88]

$$l_s = \alpha D. \tag{4.2.25}$$

The quantum mechanical localization length is proportional to the classical diffusion constant! The proportionality factor α cannot be determined from this qualitative argument. For a somewhat different system, the Lloyd model, where the potential in the kick term is given by

$$V(\theta) = V_0 \arctan(E - 2k\cos\theta), \tag{4.2.26}$$

the factor could be analytically determined as $\alpha = \frac{1}{2}$ [She86, Fis89]. For the kicked rotator only numerical results are available. Figure 4.3 shows the localization length as a function of the classical diffusion constant in a double logarithmic plot, demonstrating that the proportionality between both quantities is valid over five orders of magnitude. The solid straight line corresponds to a factor $\alpha = \frac{1}{2}$ suggesting that this value also holds for the kicked rotator [She86, She87].

We have not yet discussed the finding that the localization shows an exponential behaviour. This point is postponed to Section 4.3.1, where we shall come back to the question of localization in kicked systems from a completely different point of view.

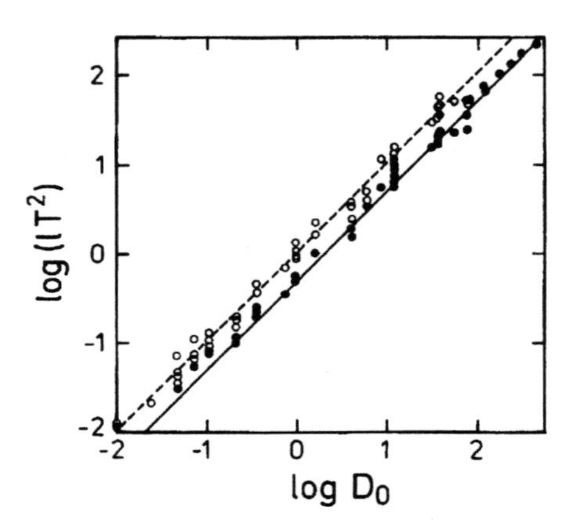

Figure 4.3. Localization length l versus classical diffusion constant D_0 for the standard map. The open and closed circles are obtained from numerical simulations. The solid line corresponds to the theoretical expectation $l = D/2$ [She86] (Copyright 1986 by the American Physical Society).

4.2.2 Hydrogen atoms in strong radiofrequency fields

The studies of microwave ionization of excited hydrogen atoms belong to the key experiments of quantum chaos. In a pioneer work Bayfield and Koch [Bay74] studied the microwave ionization of hydrogen atoms in excited states with main quantum numbers in the range of $n = 60$ to 70. States with main quantum numbers up to about 100 can be excited in a controlled way [Bel96]. The present state of the art is reviewed in two articles by Koch [Koc95a] and by Koch and van Leeuwen [Koc95b]. The theoretical background is provided in two reports by Casati *et al.* [Cas87a] and Jensen *et al.* [Jen91]. The most recent results are compiled in the introduction of the *Quantum Chaos* monograph by Casati and Chirikov [Cas95]. The monograph also contains reprints of a number of important papers in the field. Both the classical and the quantum mechanical aspects of the hydrogen atom in a strong microwave field are treated in the monograph by Blümel and Reinhardt [Blü97].

The hydrogen atom in a strong microwave field is the first experimental example showing dynamical localization. Both aspects discussed in the last section, namely classical diffusion in the semiclassical region, and quantum suppression of diffusion in the dynamical localization range can also be found in this system. The Schrödinger equation describing the dynamics of an electron under the combined influence of the Coulomb interaction with the nucleus and the electromagnetic interaction with the microwaves is given by

$$\imath\hbar\frac{\partial\psi}{\partial t} = \left(\frac{p^2}{2m_e} - \frac{e}{r} + zF\cos(\omega t)\right)\psi. \tag{4.2.27}$$

m_e is the reduced mass of the electron, and e is the elementary charge. The last term describes the interaction of a microwave field oscillating in the z direction with the electric dipole moment of the electron. The Schrödinger equation (4.2.27) is invariant with respect to the following scaling transformation:

$$\begin{aligned} &\hat{r} = rn^{-2}, \quad \hat{p} = pn, \quad \hat{t} = tn^{-3}, \quad \hat{\omega} = \omega n^3, \\ &\hat{E} = En^2, \quad \hat{F} = Fn^4, \quad \hat{\hbar} = \hbar n^{-1}. \end{aligned} \tag{4.2.28}$$

If the units are fixed by the convention $\hbar = e = m_e = 1$, radius, momentum, and angular frequency for the nth shell equal 1 in the rescaled units, and the energy is given by $-\frac{1}{2}$. The scaling relations (4.2.28) show that in terms of the rescaled variables the classical dynamics becomes independent of n. Quantum mechanically the situation is different. The scaling of $\hbar$ shows that an increase of n is equivalent to a decrease of $\hat{\hbar}$. An increase of n therefore corresponds to an approach towards the semiclassical limit. Observation of a scaling behaviour with n is hence a clear indication that we are in the range of classical dynamics.

Figure 4.4 shows a sketch of the experimental set-up used by Koch and coworkers [Koc95a]. A beam of hydrogen atoms passes from the left through a number of electric fields. In this region the atoms are excited by the pulse of a CO_2 laser to the $n = 10$ state. With a second laser pulse the wanted final state with principal quantum numbers in the range $n = 24$ to 98 is obtained. The fine tuning is achieved by a variation of the electric fields via the Stark effect. In this way every wanted n state in the mentioned range can be populated in a controlled way. The excited atoms pass through a cavity generating a longitudinally oscillating electric field with frequencies in the range of 7.6 to 36 GHz, and electric field strengths of up to 100 V cm^{-1}. Part of the atoms is further excited by absorption of the microwaves and eventually ionized. The fraction of ionized atoms can be determined either by detecting the surviving atoms, the ions produced, or the electrons released. Figure 4.5(a) shows the fraction of hydrogen atoms surviving the microwave ionization as a function of the microwave field amplitude F [Koc95a]. The atoms were prepared in three different n states. Principal quantum numbers and the corresponding microwave frequencies were chosen in such a way that the rescaled angular frequency was identical in all three cases. The survival probability declines sharply at a critical field amplitude which depends on the n quantum number. If, however, the data are plotted as a function of the *scaled* field amplitude $\hat{F} = Fn^4$, a perfect scaling behaviour is found. Thus the ionization can be understood in a purely classical manner. We should note that the number of microwave photons necessary to ionize the hydrogen atom is between 50 and 75 for the three examples shown in the figure. If the ionization probability were calculated quantum mechanically using high-order perturbation theory, a completely different picture would be obtained, which is not in accordance with the observation.

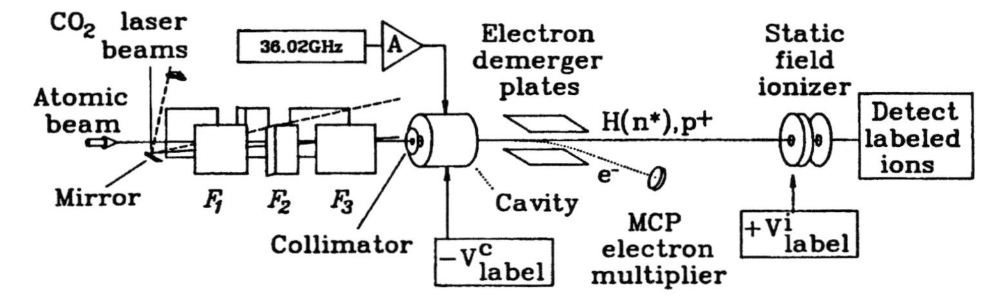

Figure 4.4. Schematic view of an apparatus to study the microwave ionization of hydrogen atoms. The atoms enter from the left, are laser excited into a Rydberg state close to the ionization threshold and ionized by microwaves in a resonance cavity. For details see text [Koc95a] (with kind permission from Elsevier Science).

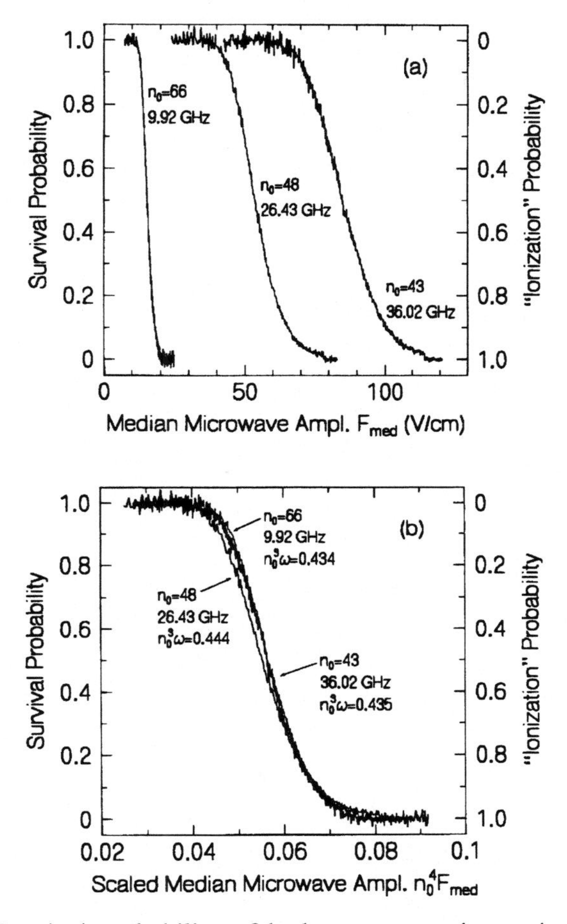

Figure 4.5. (a) Survival probability of hydrogen atoms in a microwave cavity as a function of the microwave field strength F. The atoms have been prepared in states with three different principal quantum numbers n_0. The microwave frequencies $\omega/2\pi$ have been chosen in such a way that in all three cases the scaled frequencies $n_0^3\omega$ have the identical value of 0.439(5). (b) Survival probability as a function of the scaled field strength $n_0^4 F$ [Koc95a] (with kind permission from Elsevier Science).

The exact quantum mechanical treatment of the Schrödinger equation (4.2.27) is difficult for reasons discussed in Section 4.1.1. The complexity of the system is considerably reduced, if the motion of the electron is restricted to one dimension by replacing the r^{-1} term in Eq. (4.2.27) by z^{-1} [Cas87a]. This procedure is less crude than we might expect. In Section 2 of Ref. [Cas87a] the limits of this approximation are discussed with the result that under the conditions of the experiments the errors are only marginal. Alternatively r^{-1} may be substituted by the potential [Jen91]

$$V(z) = \begin{cases} -\dfrac{1}{z} & z > 0 \\ \infty & z < 0 \end{cases}. \qquad (4.2.29)$$

This choice is somewhat closer to the original Kepler problem. Classically the electron falls freely in this potential towards the nucleus, is totally reflected, and so on. This may be interpreted as the motion on a Kepler ellipse with an eccentricity $\epsilon = 1$.

A 'black box' quantum mechanical calculation is not very instructive. We therefore apply a further simplification and come to the perhaps surprising findings that the hydrogen atom in a strong microwave field is closely related to the kicked rotator discussed in the last section. The oscillatory behaviour of the electron suggests a stroboscopic view of the motion, i.e. a replacement of the continuous motion by a discrete map. The classical equations of motion are obtained from the Hamilton–Jacobi equations:

$$\dot{z} = \frac{\partial \mathcal{H}}{\partial p} = p, \quad \dot{p} = -\frac{\partial \mathcal{H}}{\partial z} = -V'(z) + F\cos(\omega t). \qquad (4.2.30)$$

For the construction of the map we introduce the phase of the microwave field and the actual energy of the system as new variables

$$\Phi(t) = \omega t,$$

$$E(t) = \frac{p^2}{2} + V(z) + zF\cos(\omega t). \qquad (4.2.31)$$

From the equations of motion we obtain for the time derivative of the energy

$$\dot{E} = -\omega z F \sin(\omega t). \qquad (4.2.32)$$

Let Φ_n, E_n now be the values of phase and energy, respectively, at the beginning of the nth cycle. We choose the moment of largest distance from the nucleus as the starting point. We obtain the new values for the dynamical variables at the end of the next cycle from Eqs. (4.2.31) and (4.2.32) as

$$\Phi_{n+1} = \Phi_n + \omega T, \qquad (4.2.33)$$

$$E_{n+1} = E_n - \omega F \oint z(t)\sin(\omega t + \Phi_n)\, dt, \qquad (4.2.34)$$

where T is the cycle time and the loop integral is over one cycle:

$$\oint z(t)\sin(\omega t + \Phi_n)\, dt = \int_{-T/2}^{T/2} z(t)\sin(\Phi_n + \omega t)\, dt$$
$$= \int_{0}^{T/2} z(t)[\sin(\Phi_n + \omega t) + \sin(\Phi_n - \omega t)]\, dt$$
$$= 2\sin\Phi_n \int_0^{T/2} z(t)\cos(\omega t)\, dt. \tag{4.2.35}$$

Equations (4.2.33) and (4.2.35) are still exact, but for further calculations some approximations have to be applied. A relation between T and E is obtained from the integration of the equations of motion. For the case $\omega T \gg 1$, which will be considered now, the influence of the microwaves on the motion of the electron is small, and the equation of motion is simply given by Newton's law

$$\ddot{z} = -V'(z). \tag{4.2.36}$$

The integration can be performed by means of standard techniques and yields t as a function of z:

$$t = \frac{1}{\sqrt{2}} \int_0^z \left(\frac{1}{z} - |E|\right)^{-1/2} dz, \tag{4.2.37}$$

whence follows for the relation between the cycle time T and the energy E:

$$T = \sqrt{2} \int_0^{1/|E|} \left(\frac{1}{z} - |E|\right)^{-1/2} dz = \frac{2\pi}{(2|E|)^{3/2}}, \tag{4.2.38}$$

which could have also been obtained directly from Kepler's third law. The integration in Eq. (4.2.37) can be performed analytically but yields a somewhat inconvenient expression. In the limit of large principal quantum numbers, however, we may put $E = 0$ in the integral and obtain

$$z(t) = \left(\frac{3t}{\sqrt{2}}\right)^{2/3}. \tag{4.2.39}$$

This is inserted into the integral on the right hand side of Eq. (4.2.35). Moreover we replace the upper limit of the integration by ∞. This is admissible in the limit $\omega T \gg 1$ and is therefore no additional approximation. We then have

$$2\int_0^{T/2} z(t)\cos(\omega t)\,dt \approx 2\int_0^{\infty}\left(\frac{3t}{\sqrt{2}}\right)^{2/3}\cos(\omega t)\,dt$$
$$= -(48)^{1/6}\Gamma\left(\frac{2}{3}\right)\omega^{-5/3}$$
$$\approx -2.58\omega^{-5/3}. \tag{4.2.40}$$

Collecting the results we obtain from Eqs. (4.2.33) and (4.2.34) the so-called *Kepler map* [Cas87b, Cas90]

$$\Phi_{n+1} = \Phi_n + 2\pi\omega(2|E_{n+1}|)^{-3/2}, \tag{4.2.41}$$

$$E_{n+1} = E_n + \omega k \sin\Phi_n, \tag{4.2.42}$$

with $k = 2.58\omega^{-5/3}F$. Somewhat arbitrarily we have inserted E_{n+1} for E on the right hand side of Eq. (4.2.41). We could also have inserted E_n or any value between E_n and E_{n+1}. The choice applied here is in accordance with the cited references.

The Kepler map bears a strong resemblance to the standard map (4.2.2). The agreement becomes perfect if E_n is linearized about its initial value E_0. To this end we turn to the rescaled variables (4.2.28). Since in rescaled variables the initial energy is given by $\hat{E}_0 = -\frac{1}{2}$, we may put $\hat{E}_n = -\frac{1}{2} + N_n\hat{\hbar}\hat{\omega}$, where N_n is the number of absorbed photons. Expanding the root in Eq. (4.2.41) and taking only the terms which are linear in the photon number we obtain

$$\Phi_{n+1} = \Phi_n + L_{n+1}, \tag{4.2.43}$$

$$L_{n+1} = L_n + K\sin\Phi_n, \tag{4.2.44}$$

where we have introduced the new variable

$$L_n = 2\pi\hat{\omega}(1 + 3N_n\hat{\hbar}\hat{\omega}). \tag{4.2.45}$$

Equations (4.2.43) and (4.2.44) are identical with the standard map, where the kick strength is given by

$$K = 6\pi\hat{\omega}^2\hat{k} = 48.6\hat{\omega}^{1/3}\hat{F}. \tag{4.2.46}$$

This correspondence between the Kepler and the standard maps suggests that the phenomenology found for the kicked rotator is also observed for hydrogen atoms in strong microwave fields. We begin with a discussion of the classical regime. For the standard map unbounded diffusion in phase space becomes possible for kick strengths larger than about one. In this range the electron diffuses through the phase space until it eventually reaches the ionization threshold. In the measurements shown in Fig. 4.5 the scaled microwave frequency was given by $\hat{\omega} = 0.439$. Inserting this value into Eq. (4.2.46) and taking $K = 1$, ionization would be expected for scaled field amplitudes larger than $\hat{F} = 0.027$. The actual ionization threshold of $\hat{F} = 0.04$ is somewhat higher (see Fig. 4.5(b)). A quantitative agreement cannot be expected in view of the approximations applied in the derivation of the Kepler map, which definitely becomes invalid for $\hat{\omega} < 1$.

According to Eq. (4.2.46) the threshold amplitude should drop off proportional to $\hat{\omega}^{-1/3}$ with the frequency. Figure 4.6 shows the experimental result. The scaled field amplitude necessary to ionize 10% of the atoms is plotted as a function of $\hat{\omega}$. For $\hat{\omega} < 1$ the expected behaviour is actually found. The triangles are the result of a Monte-Carlo simulation of classical diffusion, showing perfect agreement with the measurements including some fine structures superimposed onto the general drop-off.

For $\hat{\omega} > 1$ the threshold amplitudes found experimentally are larger than expected for classical diffusion up to a factor of two. This is the first experimental realization of the dynamical localization phenomenon [Gal88, Bay89]. The threshold amplitude in the localization range can be estimated by means of the expressions derived in Section 4.2.1. The essential relation is the proportionality between localization length and classical diffusion constant, $l_s = \alpha D$, where a typical value for α is $\frac{1}{2}$ (see Eq. (4.2.25)). In Section 4.2.1 l_s has been introduced as the number of angular momentum states occupied by the kicked

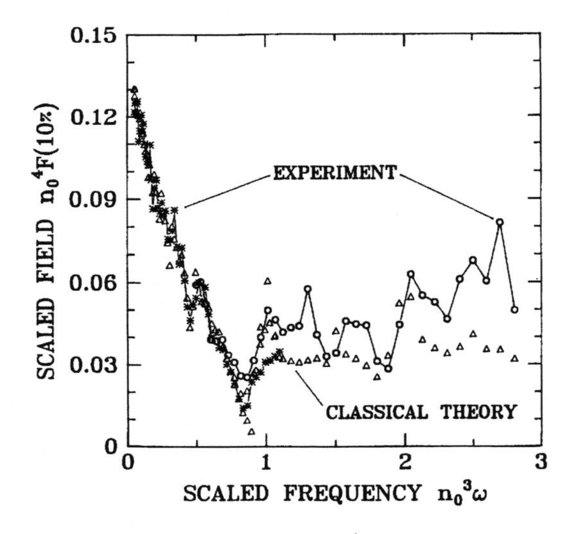

Figure 4.6. Scaled microwave field strength $\hat{F} = n_0^4 F$ needed to ionize 10% of the hydrogen atoms versus the scaled frequency $\hat{\omega} = n_0^3\omega$. For $\hat{\omega} < 1$ the excitation of the hydrogen atoms by the microwaves resembles a diffusive process which eventually leads to ionization. This is the cause of the sharp decline of the ionization field strength in this range. The increase of the ionization field strength above the values expected from classical theory in the range $\hat{\omega} > 1$ is an experimental verification of the dynamical localization [Koc95a] (with kind permission from Elsevier Science).

rotator until dynamical localization stops the diffusion. In the hydrogen atom the role of the angular momentum is taken by the number of photons absorbed or emitted, and consequently we have to take for D the photon number diffusion constant. Because of $L_n = \text{const.} - 6\pi\hat{\hbar}\hat{\omega}^2 N_n$ (see Eq. (4.2.45)) the photon number diffusion constant is smaller than the L diffusion constant by a factor of $(6\pi\hat{\hbar}\hat{\omega}^2)^{-2}$. Thus relation (4.2.25) between the localization length and the classical diffusion constant in the present case yields

$$\begin{aligned} l_s &= \frac{\alpha}{2}\left(\frac{K}{6\pi\hat{\hbar}\hat{\omega}^2}\right)^2 \\ &= 3.3\alpha\hat{F}^2\hat{\hbar}^{-2}\hat{\omega}^{-10/3} \end{aligned} \tag{4.2.47}$$

for the average number of photons, which can be absorbed until dynamical localization stops the classical diffusion. Ionization is observed if this number is sufficient to reach the ionization threshold. In rescaled units the energy prior to the microwave excitation is $\frac{1}{2}$. Therefore the number of photons necessary for ionization is given by $(2\hat{\hbar}\hat{\omega})^{-1}$. Taking $\alpha = \frac{1}{2}$ the threshold field amplitude is obtained as

$$\begin{aligned} \hat{F} &= 0.55\hat{\hbar}^{1/2}\hat{\omega}^{7/6} \\ &= 0.55n^{-1/2}\hat{\omega}^{7/6}, \end{aligned} \tag{4.2.48}$$

since in rescaled units $\hat{\hbar}$ is given by n^{-1} (see Eq. (4.2.28)). In the experiments shown in Fig. 4.6 the principal quantum numbers were in the range of 45 to 80. Inserting this into Eq. (4.2.48) ionization field strengths between 0.06 and 0.08 are found for $\hat{\omega} = 1$, somewhat larger than the experimental value. With increasing $\hat{\omega}$ the experimentally found ionization thresholds increase more slowly than predicted. These deviations can be explained by applying more refined techniques [Cas90].

In another experiment, performed with Rb atoms, the predicted dependence (2.48) between the scaled ionization field amplitude and the scaled angular frequency was really found [Ben95]. The scaling relations, which proved to be extremely useful for the hydrogen atoms, only approximately hold for Rb atoms due to the presence of large quantum defects. Therefore the authors had to apply a modified scaling. The details as well as the references to the original works can be found in a review article by Benson *et al.* [Ben95].

We may imagine situations where the atoms are never ionized even if the microwave field strength is large enough in view of the above argumentation. This is the case if the atom is prepared in a Floquet eigenstate (see Section 4.1.1). Then the original state is totally restored after a complete period of up to a phase factor, and ionization will never be observed. In calculations of the wave packet dynamics in hydrogen atoms in microwave fields such eternally living states have indeed been found [Buc95a, Zak95]. The authors speculated that the considerable enhancement of the ionization threshold at a scaled angular frequency of $\hat{\omega} = 1.3$ (see Fig. 4.6) might be caused by the fact that here the atoms have been prepared incidentally in a state close to a Floquet eigenstate [Buc95b]. There are, however, alternative attempts to explain this enhancement by scarring of the eigenfunctions of the involved states [Jen91].

4.2.3 Ultra-cold atoms in magneto-optical traps

A realization of the dynamical localization of atoms has been achieved by Raizen and coworkers [Moo94] following a suggestion by Graham *et al.* [Gra92]. The experiments were performed with sodium atoms at temperatures in the μK region. The atoms were trapped and simultaneously cooled in a magneto-optical trap by the combined interaction of an inhomogeneous magnetic field and the electromagnetic field of a strong laser. A more detailed description of the trap and the cooling technique cannot be given here. A very nice introduction can be found in an article by Chu [Chu91].

In the experiments about 10^5 atoms were trapped within a volume of diameter 0.3 mm at a temperature of 17 μK. At the end of the preparation step the trap was turned off, and a modulated standing light field was switched on for typically 10 μs. The Hamiltonian for the interaction of the sodium atom with the light field is given by

$$\mathscr{H}_0 = \mathscr{H}_{el} + \frac{p^2}{2M} + eF\cos\{k_L[x - \Delta L\sin(\omega t)]\}\cos(\omega_L t). \qquad (4.2.49)$$

Here $\mathscr{H}_{el}$ describes the interaction of the valence electron with the atom, $p^2/2M$ is the kinetic energy of the atom, and the last term describes the electric dipole interaction of the light field with the electron, where ω_L and ω are laser and modulation angular frequency, respectively. The standing waves were generated by directing two laser beams from opposite directions into the trap, the modulation was achieved by passing one beam through an electro-optical phase modulator [Moo94]. The laser frequency was chosen close to the $(3S_{1/2}, F = 2) \to (3P_{3/2}, F = 3)$ transition of the D_2 line ($\lambda = 589$ nm) of the sodium atom. The electronic Hamiltonian can be reduced to a two-level system, containing the ground state $|g\rangle$ and the excited state $|e\rangle$. Taking the energy average of the two states as the energy zero, the matrix elements of $\mathscr{H}_{el}$ and eF are given by

$$\langle e|\mathcal{H}_{el}|e\rangle = \frac{\hbar\omega_0}{2}, \qquad \langle e|\mathcal{H}_{el}|g\rangle = 0,$$
$$\langle g|\mathcal{H}_{el}|e\rangle = 0, \qquad \langle g|\mathcal{H}_{el}|g\rangle = -\frac{\hbar\omega_0}{2}, \tag{4.2.50}$$

and

$$\langle e|eF|e\rangle = 0, \qquad \langle e|eF|g\rangle = \hbar\Omega,$$
$$\langle g|eF|e\rangle = \hbar\Omega, \qquad \langle g|eF|g\rangle = 0, \tag{4.2.51}$$

where the nondiagonal matrix elements of eF have been expressed in terms of the *Rabi frequency* Ω. The diagonal matrix elements of eF vanish since the electric dipole operator has nonvanishing matrix elements only between states differing in their angular quantum numbers by one. The Hamilton operator can be written very compactly using the Pauli spin matrices as

$$\mathcal{H}_0 = \frac{p^2}{2M} + \frac{\hbar\omega_0}{2}\sigma_z + \hbar\Omega\cos\{k_L[x - \Delta L\sin(\omega t)]\}\cos(\omega_L t)\sigma_x. \tag{4.2.52}$$

This is a demonstration of the fact that any two-level system is equivalent to a spin-$\frac{1}{2}$ system [Fey57].

The following steps are familiar to everybody acquainted with magnetic resonance. First the Hamiltonian is transferred to a coordinate system rotating with angular frequency ω_L about the z axis of the spin space. From the wave function ψ and the Hamiltonian $\mathcal{H}_0$ in the laboratory frame the corresponding quantities in the rotating frame are obtained as

$$\psi_{\text{rot}} = \exp\left(\frac{\imath\omega_L\sigma_z t}{2}\right)\psi, \tag{4.2.53}$$

and

$$\mathcal{H}_{\text{rot}} = -\hbar\omega_L\sigma_z + \exp\left(\frac{\imath\omega_L\sigma_z t}{2}\right)\mathcal{H}_0\exp\left(-\frac{\imath\omega_L\sigma_z t}{2}\right). \tag{4.2.54}$$

Using the rotation property

$$e^{\imath\alpha\sigma_z}\sigma_x e^{-\imath\alpha\sigma_z} = \sigma_x\cos 2\alpha - \sigma_y\sin 2\alpha, \tag{4.2.55}$$

of the Pauli matrices, $\mathcal{H}_{\text{rot}}$ can be written as

$$\mathcal{H}_{\text{rot}} = \frac{p^2}{2M} + \frac{\hbar(\omega_0 - \omega_L)}{2}\sigma_z + \frac{\hbar\Omega}{2}\cos\{k_L[x - \Delta L\sin(\omega t)]\}$$
$$\times [\sigma_x(1 + \cos(2\omega_L t)) - \sigma_y\sin(2\omega_L t)]. \tag{4.2.56}$$

The factor 2 in the arguments of the sine and the cosine functions on the right hand side of Eq. (4.2.55) results from the fact that the spinor rotations have a period of 4π. The relation is proved with help of the Euler relation for spinors,

$$e^{\imath\alpha\sigma_z} = \cos\alpha + \imath\sigma_z\sin\alpha. \tag{4.2.57}$$

Corresponding relations hold for σ_x and σ_y.

In the *rotating wave approximation* the terms on the right hand side of Eq. (4.2.56) oscillating with the angular frequency $2\omega_L$, are neglected. The remaining Hamiltonian has only a slow time-dependence due to the modulation of the field. It can be written as

$$\mathscr{H}_{\text{rot}} = \frac{p^2}{2M} + \hbar\Omega_{\text{eff}}(\sigma_z \cos\alpha + \sigma_x \sin\alpha), \tag{4.2.58}$$

where Ω_{eff} and the angle α are given by

$$\Omega_{\text{eff}} = \frac{1}{2}\sqrt{(\omega_0 - \omega_L)^2 + \Omega^2 \cos^2\{k_L[x - \Delta L \sin(\omega t)]\}}, \tag{4.2.59}$$

$$\tan\alpha = \frac{\Omega \cos\left[k_L(x - \Delta L \sin(\omega L))\right]}{(\omega_0 - \omega_L)}. \tag{4.2.60}$$

By a further rotation in the spin space by the angle $-\alpha$ about the y axis we end with the effective Hamiltonian

$$\begin{aligned}\mathscr{H}_{\text{eff}} &= \hbar\dot{\alpha}\sigma_y + e^{-\imath\alpha\sigma_y}\mathscr{H}\, e^{\imath\alpha\sigma_y} \\ &= \frac{p^2}{2M} + \hbar(\dot{\alpha}\sigma_y + \Omega_{\text{eff}}\sigma_z).\end{aligned} \tag{4.2.61}$$

If α changes slowly compared to all other time variations in the system, the term $\hbar\dot{\alpha}\sigma_y$ can be neglected. This is the *adiabatic approximation*. The effective Hamiltonian is now diagonal in the spin space, and we may restrict ourselves to one spin component, say the $-\frac{1}{2}$-state. If, moreover, the detuning frequency $(\omega_0 - \omega_L)$ is large compared to the Rabi frequency Ω, the square root in Eq. (4.2.59) can be expanded, and we get

$$\mathscr{H}_{\text{eff}} = \frac{p^2}{2M} - \frac{\hbar(\omega_0 - \omega_L)}{2}\left\{1 + \frac{\Omega^2}{2(\omega_0 - \omega_L)^2}\cos^2[k_L(x - \Delta L \sin(\omega t))]\right\}. \tag{4.2.62}$$

Rescaling the variables and the Hamiltonian according to

$$t \rightarrow \frac{t}{\omega}, \quad x \rightarrow \frac{x}{2k_L}, \quad p \rightarrow \frac{M\omega}{2k_L}p, \quad \mathscr{H}_{\text{eff}} \rightarrow \frac{M\omega^2}{4k_L^2}\mathscr{H}, \tag{4.2.63}$$

and removing the constant term from $\mathscr{H}_{\text{eff}}$ by a shift of the energy zero, we end with the standard form for the Hamiltonian [Gra92]

$$\mathscr{H} = \frac{p^2}{2} - K\cos(x - \lambda \sin t), \tag{4.2.64}$$

where

$$K = \frac{\hbar k_L^2}{2M}\frac{\Omega^2}{(\omega_0 - \omega_L)\omega^2}, \quad \lambda = 2k_L\Delta L. \tag{4.2.65}$$

We have again obtained a Hamiltonian with a periodic time dependence. We are now going to approximate classical dynamics by a discrete map, exactly as we proceeded with the hydrogen atom in a strong microwave field. From Eq. (4.2.64) we obtain the following classical equations of motion for the dynamical variables x and p:

$$\dot{x} = \frac{\partial \mathcal{H}}{\partial p} = p \tag{4.2.66}$$

$$\dot{p} = -\frac{\partial \mathcal{H}}{\partial x} = -K \sin(x - \lambda \sin t). \tag{4.2.67}$$

For large values of λ the phase of the sine function,

$$\Phi(t) = x(t) - \lambda \sin t, \tag{4.2.68}$$

varies rapidly. Therefore the force acting on the atom is zero on average. The only exceptions are the stationary phase points t_s, where $\Phi'(t_s) = 0$, whence follows

$$p(t_s) = \lambda \cos t_s. \tag{4.2.69}$$

This can be realized only for $|p| \leqslant \lambda$. For $|p| > \lambda$ the atoms are not influenced by the light field. We can thus conclude that the motion of the atoms may be chaotic at most for $|p| \leqslant \lambda$. Only at the stationary phase point the force acts for some time in the same direction leading to a net change of the momentum. Because of this more or less discontinuous behaviour it is again possible to approximate the equations of motion by a discontinuous map. From Eq. (4.2.66) we obtain for the change of the momentum over one period

$$\Delta p = -K \int_0^{2\pi} \sin \Phi(t)\, dt. \tag{4.2.70}$$

We approximate the integral by assuming that the change of $x(t)$ over one period is small. Now the integration can be carried out and yields

$$\begin{aligned}
\Delta p &= -K \int_0^{2\pi} \sin(x - \lambda \sin t)\, dt \\
&= -K \sin x \int_0^{2\pi} \cos(\lambda \sin t)\, dt \\
&= -2\pi K \sin x J_0(\lambda) \\
&\approx -2K \sqrt{\frac{2\pi}{\lambda}} \cos\left(\lambda - \frac{\pi}{4}\right) \sin x,
\end{aligned} \tag{4.2.71}$$

where the asymptotic representation of the Bessel function has been used in the last step. Graham *et al.* [Gra92] have instead obtained by a somewhat different approximation $\Delta p = -2K\sqrt{\pi/\lambda} \sin x$.

From the first equation of motion (4.2.66) we obtain for the change of x over one period

$$\Delta x = 2\pi p. \tag{4.2.72}$$

With Eqs. (4.2.71) and (4.2.72) we have obtained a discrete map for the dynamical variables x and p:

$$x_{n+1} = x_n + 2\pi p_{n+1},$$
$$p_{n+1} = p_n - 2K\sqrt{\frac{2\pi}{\lambda}}\cos\left(\lambda - \frac{\pi}{4}\right)\sin x_n. \tag{4.2.73}$$

The derivation, however, contains one flaw. There are *two* stationary phase points in the interval 0 to 2π. For large λ values they are close to $\pi/2$ and $3\pi/2$. The atom is therefore kicked twice within one period. It has not been considered that x and p change their values after the first kick. In a more careful treatment we would have divided the map into two successive steps, one for each kick. But since we are only interested in the qualitative behaviour, we have simplified the discussion, following Ref. [Gra92] in this respect.

Equations (4.2.73) are identical with the standard map (4.2.2) if we substitute L_n by $2\pi p_n$. The control parameter k is given by

$$k = 4\pi K\sqrt{\frac{2\pi}{\lambda}}\cos\left(\lambda - \frac{\pi}{4}\right). \tag{4.2.74}$$

We can now apply our previous knowledge of the standard map. Classical diffusion is observed for $k > 1$. According to Eq. (4.2.74) this is the case if $K > 0.03\sqrt{\lambda}$. The diffusion constant for the momentum p is obtained from Eq. (4.2.11) as

$$D = \frac{1}{2}\left(\frac{k}{2\pi}\right)^2 = \frac{2\pi K^2}{\lambda}, \tag{4.2.75}$$

where we have replaced the term $\cos^2(\lambda - \pi/4)$ by its average value $\frac{1}{2}$. The factor 2π in the denominator on the right hand side results from the substitution $L_n = 2\pi p_n$ (compare the analogous procedure in Eq. (4.2.47)).

Classical diffusion will be stopped only for $|p| \geqslant \lambda$ where the motion ceases to be chaotic (see Eq. (4.2.69)). In the range of classical diffusion a spreading of momenta with a variance

$$\sqrt{\langle p^2\rangle} \sim \lambda \tag{4.2.76}$$

is therefore expected. In the range of dynamical localization, on the other hand, an exponential momentum distribution is expected with a localization length proportional to the classical diffusion constant D, whence follows

$$\sqrt{\langle p^2 \rangle} \sim \frac{1}{\lambda}. \tag{4.2.77}$$

Equations (4.2.76) and (4.2.77) allow easy discrimination between the regimes of classical diffusion and dynamical localization.

To determine the distribution of momenta, a refined technique has been applied [Moo94]. If the standing laser field is turned off, the atoms begin to expand freely in the dark for some ms. Thus the spread of momenta is turned into a spread of positions. Now the final positions of the atoms are frozen by turning on the cooling laser once more. The technique works since the cooling force is proportional to the velocity of the atoms. The cooling laser thus acts exactly as a viscous liquid. It has, so to speak, the effect of 'optical' molasses [Chu91].

Figure 4.7 shows the experimental results for the momentum spread as a function of λ [Rob95]. The light solid lines denote the behaviour expected from the standard map approximation. The experimental results follow more or less the predictions from Eqs. (4.2.76) and (4.2.77). We observe both the linear increase for small λ values, and the decrease with λ^{-1} for large λ values. We

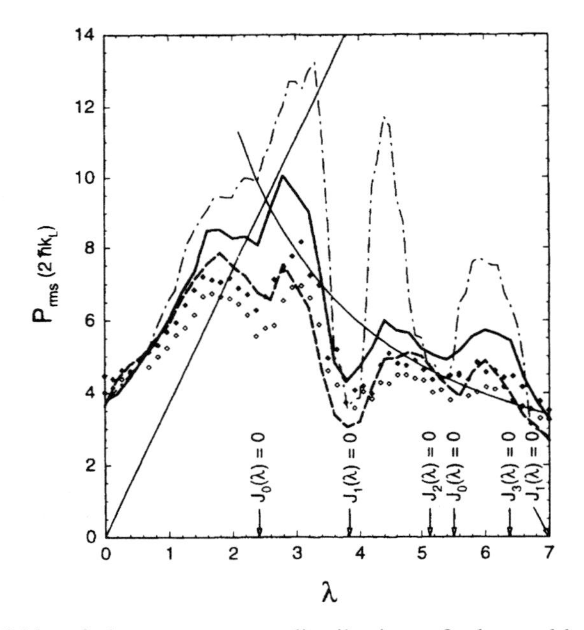

Figure 4.7. Width of the momentum distribution of ultra-cold sodium atoms in a spatially modulated standing light field as a function of the modulation strength λ. The light solid lines denote the behaviour expected from the standard map approximation (see Eqs. (4.2.76) and (4.2.77)). The heavy solid and dashed lines result from two quantum mechanical calculations, the dashed dotted line is from a classical simulation. For $\lambda \geqslant 2$ dynamical localization becomes manifest [Rob95] (Copyright 1995 by the American Physical Society).

even find the oscillatory behaviour predicted by Eq. (4.2.74). Nevertheless, a more detailed analysis shows that some care has to be taken [Bar95]. The light dashed-dotted line in Fig. 4.7 is obtained from a classical simulation, showing that the overall trend of the experimentally found behaviour can be explained classically. The heavy solid and broken lines are the results of two different quantum mechanical calculations. The measurements closely follow the latter. This shows that in the region of linear increase of the momentum spread, found for small λ values, as well as in the minima of the oscillations, the dynamics is exclusively classic. Only in the maxima of the oscillations is dynamical localization expected.

These findings are supported by the results shown in Fig. 4.8 [Rob95]. The upper panel shows phase portraits for different λ values. In the middle panel the classical momentum distribution is shown as obtained from the phase portraits. The bottom panel finally shows the experimentally measured momentum distribution together with the prediction from a Floquet calculation. A comparison of the two lower rows shows that all results, except for $\lambda = 3$, agree with the classical prediction. For $\lambda = 3$ dynamical localization is found. Whereas classically the momentum is spread more or less uniformly over the

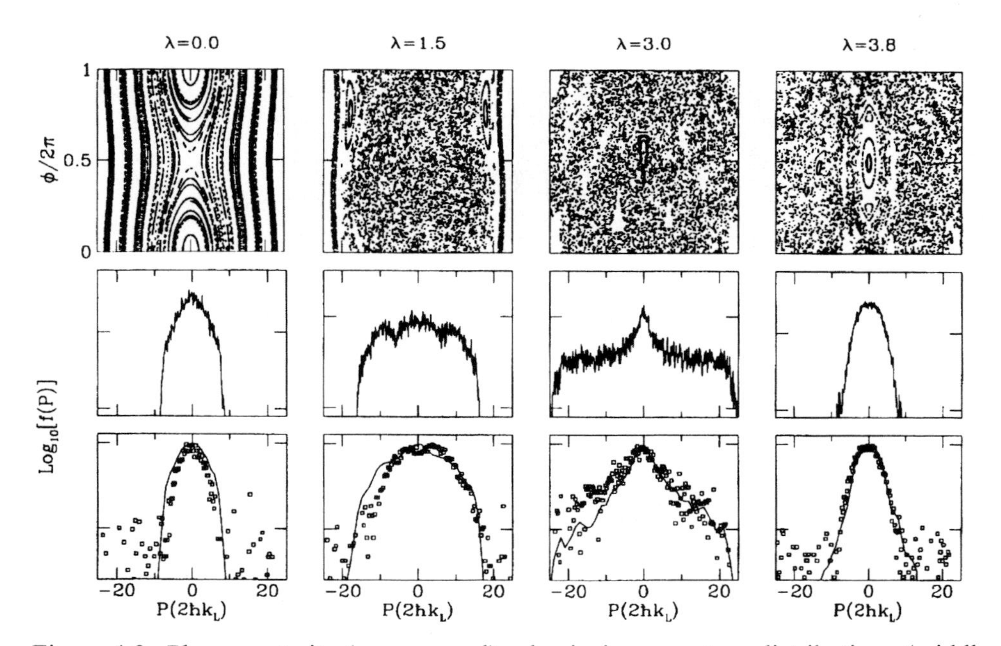

Figure 4.8. Phase portraits (upper panel), classical momentum distributions (middle panel) and experimental momentum distributions (lower panel) for ultra-cold sodium atoms. The solid lines in the bottom panel are predictions from the Floquet theory. For $\lambda = 3$ exponential localization is clearly visible [Rob95] (Copyright 1995 by the American Physical Society).

accessible phase space (the small central peak is caused by a small surviving regular island), experimentally, as well as in the Floquet calculation, an exponential localization of the momentum is found. Recently, even the transition from classical diffusion to quantum mechanical localization in dependence of the kick number (see Fig. 4.2(a)) could be verified experimentally [Moo95].

4.3 Tight-binding systems

4.3.1 Anderson model

We leave the Floquet systems for a moment, and approach an apparently unrelated topic, namely the study of electronic wave functions on a disordered crystalline lattice. We start with a one-dimensional lattice of equally spaced atoms with a distance d. The Hamiltonian is given by

$$\mathscr{H} = -\frac{\hbar^2}{2m_e}\frac{d^2}{dx^2} + V(x), \tag{4.3.1}$$

where $V(x)$ is the lattice potential. To solve the Schrödinger equation we apply the *tight-binding approximation* and assume that the electron interacts strongly with one single atom at a given time, while the interaction with all other atoms is weak. Then the electronic wave function can be expressed in terms of a linear combination of single atom eigenfunctions $\psi_0(x)$ as

$$\psi(x) = \sum_n a_n \psi_0(x - nd), \tag{4.3.2}$$

where the sum is over all lattice positions. Usually only one atomic level is considered. Entering with ansatz (4.3.2) the Schrödinger equation, and assuming that the overlap of the wave functions between different sites can be neglected,

$$\int \psi_0^*(x - nd)\psi_0(x - md)\, dx = \delta_{nm}, \tag{4.3.3}$$

we obtain

$$\sum_{k \neq n} W_{nk} a_k + E_n^0 a_n = E a_n, \tag{4.3.4}$$

where

$$E_n^0 = \int \psi_0^*(x - nd)\mathscr{H}\psi_0(x - nd)\, dx,$$

$$W_{nk} = \int \psi_0^*(x - nd)\mathscr{H}\psi_0(x - kd)\, dx. \tag{4.3.5}$$

For periodic potentials E_n^0 is independent of n, and the W_{nk} values exclusively depend on the difference $(n-k)$. In this case the wave function $\psi(x)$ can be written as a Bloch function

$$\psi(x) = e^{\imath kx} u_k(x), \tag{4.3.6}$$

leading to the well-known electronic band model of a crystalline solid. If the periodic arrangement is disturbed by disorder, we arrive at the *Anderson model* [And58]. Disorder may be introduced in two different ways: For bond disorder the bond strengths W_{nk} are statistically distributed, and the E_n^0 are constant. For site disorder the bond strengths W_{nk} are constant and are usually restricted to nearest neighbour interactions, whereas the E_n^0 vary from site to site.

There are numerous review articles on the Anderson model [Ish73, Tho74, And78, Lee85]. The most recent results have been reviewed by Kramer and McKinnon [Kra93]. Only the main result will be recapitulated: in one-dimensional systems *all* electronic wave functions are localized, even with vanishingly small disorder. The localization is caused by destructive quantum mechanical interference due to coherent backscattering from the impurities.

The Anderson localization and the dynamical localization discussed in the last sections are closely related to each other. The following presentation is based on the papers by Fishman, Grempel, and Prange [Fis82, Gre84] who were the first to notice this relation. The very similar structure of the Anderson equation system (4.3.4) and of the equation system (4.1.19) we obtained from the Fourier expansion of the Floquet system, suggests that there *is* a relation between the two systems.

We start with the Floquet Hamiltonian

$$\mathscr{H}(t) = K(L) + V(\theta) \sum_n \delta(t-n), \tag{4.3.7}$$

where $K(L)$ and $V(\theta)$ are arbitrary functions of angular momentum and angle, respectively. The kicked rotator (4.2.1) is obtained as a special case with

$$K(L) = \frac{L^2}{2}, \qquad \mathrm{V}(\theta) = k\cos\theta. \tag{4.3.8}$$

The Floquet operator corresponding to the Hamiltonian (4.3.7) is given by

$$F = e^{-\imath V(\theta)}\, e^{-\imath K(L)}. \tag{4.3.9}$$

In the eigenbasis of L the matrix elements of F are given by

$$F_{nm} = e^{-\imath K(m)} J_{n-m}, \tag{4.3.10}$$

with

$$J_n = \frac{1}{2\pi} \int_0^{2\pi} e^{-\imath V(\theta)}\, e^{\imath n\theta}\, d\theta. \tag{4.3.11}$$

The notation indicates that J_n may be considered as a generalized Bessel function. For the kicked rotator J_n reduces, except for some phase factor, to an ordinary Bessel function. Let us now assume that

$$A = \begin{pmatrix} a_1 \\ a_2 \\ \vdots \end{pmatrix} \tag{4.3.12}$$

is an eigenvector of F to the eigenphase ϕ,

$$FA = e^{-\imath V(\theta)}\, e^{-\imath K(L)} A = e^{-\imath\phi} A. \tag{4.3.13}$$

Substituting

$$e^{-\imath V(\theta)} = \frac{1 + \imath W(\theta)}{1 - \imath W(\theta)}, \tag{4.3.14}$$

where $W(\theta)$ is given by

$$W(\theta) = -\tan\frac{V(\theta)}{2}, \tag{4.3.15}$$

Eq. (4.3.13) may be expressed as

$$(1 + \imath W) e^{\imath(\phi - K)} A = (1 - \imath W) A. \tag{4.3.16}$$

Introducing the new state vector

$$\overline{A} = (e^{\imath(\phi - K)} + 1) A, \tag{4.3.17}$$

Eq. (4.3.16) may be written as

$$\frac{e^{\imath(\phi - K)} - 1}{e^{\imath(\phi - K)} + 1} \overline{A} + \imath W \overline{A} = 0, \tag{4.3.18}$$

or

$$\tan\left(\frac{\phi - K}{2}\right) \overline{A} + W \overline{A} = 0. \tag{4.3.19}$$

This may be equivalently expressed as

$$\sum_{k \neq n} W_{n-k} \overline{a}_k + E_n^0 \overline{a}_n = E \overline{a}_n, \tag{4.3.20}$$

where W_n is the Fourier transform of $W(\theta)$,

$$W_n = \frac{1}{2\pi} \int_0^{2\pi} W(\theta) e^{-\imath n\theta}\, d\theta \tag{4.3.21}$$

and where E_n^0 and E are given by

$$E_n^0 = \tan\left(\frac{\phi - K(n)}{2}\right), \tag{4.3.22}$$

and

$$E = -W_0. \tag{4.3.23}$$

Equation (4.3.20) corresponds to a tight-binding Hamiltonian with bond strengths W_{n-m} depending only on the distance between neighbours at sites n and m, and site-dependent energies E_n^0. The correspondence between kicked systems described by the Hamiltonian (4.3.7), in particular the kicked rotator, and the Anderson model with site disorder has thus been proved. There remains, however, one open question. The sequence $\{E_n^0\}$ defined by Eq. (4.3.22) is not random, but at most pseudo-random. There is one attempt to close this gap [Alt96]. It uses the same approach as applied in the proof of the Bohigas–Giannoni–Schmit conjecture, and consequently faces the sane problems as discussed in Section 3.2.1 [Cas98]. Nobody really doubts, however, that for questions of localization it is irrelevant whether the site energies are described by random or pseudo-random sequences. Any numerical evidence shows that pseudo-random sequences also lead to localization [Gre84, Gri88].

4.3.2 Transfer matrix method

Using the equivalence between the kicked rotator and Anderson localization, it is rather easy to prove that wave functions in disordered systems are exponentially localized. We restrict ourselves to a model with site disorder and constant nearest neighbour interactions. Normalizing the interaction strength W to one, the equation system (4.3.4) reads

$$a_{n+1} + E_n^0 a_n + a_{n-1} = E a_n. \tag{4.3.24}$$

It can be rewritten in matrix form as

$$\begin{pmatrix} a_{n+1} \\ a_n \end{pmatrix} = T_n \begin{pmatrix} a_n \\ a_{n-1} \end{pmatrix}, \tag{4.3.25}$$

where T_n is the transfer matrix given by

$$T_n = \begin{pmatrix} E - E_n^0 & -1 \\ 1 & \cdot \end{pmatrix}. \tag{4.3.26}$$

The transfer matrix technique is very useful especially for one-dimensional models, as the components of the state vector at the end of a chain are obtained from the starting values by simple matrix multiplication,

$$\begin{pmatrix} a_{N+1} \\ a_N \end{pmatrix} = T \begin{pmatrix} a_1 \\ a_0 \end{pmatrix}, \tag{4.3.27}$$

with

$$T = T_N T_{N-1} \ldots T_1. \qquad (4.3.28)$$

Let us first consider the situation that all T_n are equal and given by

$$\overline{T} = \begin{pmatrix} E - E_0 & -1 \\ 1 & \cdot \end{pmatrix}. \qquad (4.3.29)$$

Two different situations have to be discriminated. For $|E - E_0| < 2$ we may write

$$\cos\alpha = \frac{E - E_0}{2}, \qquad (4.3.30)$$

and obtain

$$T = \cos\alpha \cdot 1 + \imath \sin\alpha \cdot \sigma, \qquad (4.3.31)$$

with

$$\sigma = \frac{1}{\imath \sin\alpha} \begin{pmatrix} \cos\alpha & -1 \\ 1 & -\cos\alpha \end{pmatrix}. \qquad (4.3.32)$$

Since σ obeys the relation $\sigma^2 = 1$, expression (4.3.31) may be written as

$$\overline{T} = e^{\imath\alpha\sigma}. \qquad (4.3.33)$$

The corresponding expression for $|E - E_0| > 2$ is obtained by replacing α by $\imath\lambda$ everywhere, whence follows

$$\overline{T} = e^{-\lambda\sigma}. \qquad (4.3.34)$$

From Eqs. (4.3.33) and (4.3.34) we get for the transfer matrix $T = \overline{T}^N$ of the complete chain

$$T = \begin{cases} e^{\imath N\alpha\sigma}, & \text{if } |E - E_0| < 2 \\ e^{-N\lambda\sigma}, & \text{if } |E - E_0| > 2 \end{cases}. \qquad (4.3.35)$$

For $|E - E_0| < 2$ the components of the state vector are periodically modulated. This corresponds to a delocalized state. For $|E - E_0| > 2$, on the other hand, the components of T increase or decrease exponentially. This situation corresponds to exponentially localized states, since only state vectors with exponentially decreasing components are physically allowed.

We have thus obtained, by means of a possibly unfamiliar approach, the electronic band model of the solid. The energies with $|E - E_0| < 2$ correspond to the allowed Bloch bands, and the energies with $|E - E_0| > 2$ to the forbidden regions. We must not be surprised at having obtained only one allowed band. This follows from the fact that only one single atom wave function has been considered in the ansatz (4.3.2). The approximation is justified as long as the band width is small compared to the gap between neighbouring bands.

The trace of the transfer matrix contains all the information needed to decide whether a state is localized or not. For localized states Eq. (4.3.35) yields $\mathrm{Tr}(T) = \cosh N\lambda$. From this we obtain for the average localization length per site in the limit of large N

$$\bar{\lambda} = \frac{1}{N} \ln[\mathrm{Tr}(T)]. \tag{4.3.36}$$

Subsequently the situation of site disorder will be considered. We choose an energy E such that for all site energies $|E - E_n^0|$ is smaller than 2. Then the transfer matrix T_n can again be written in terms of a spinor rotation operator

$$T_n = e^{\imath \alpha_n \sigma_n} \tag{4.3.37}$$

where now

$$\cos \alpha_n = \frac{E - E_n^0}{2}, \tag{4.3.38}$$

and

$$\sigma_n = \frac{1}{\imath \sin \alpha_n} \begin{pmatrix} \cos \alpha_n & -1 \\ 1 & -\cos \alpha_n \end{pmatrix}. \tag{4.3.39}$$

An alternative parametrization of σ_n is obtained by introducing the new parameter β_n related to α_n via the relations

$$\sinh \beta_n = \frac{\cos \alpha_n}{\sin \alpha_n}, \qquad \cosh \beta_n = \frac{1}{\sin \alpha_n}, \tag{4.3.40}$$

whence follows

$$\sigma_n = \imath \begin{pmatrix} -\sinh \beta_n & \cosh \beta_n \\ -\cosh \beta_n & \sinh \beta_n \end{pmatrix}. \tag{4.3.41}$$

Using well-known spinor rotation properties (see Eq. (4.2.55)) this can be written as

$$\sigma_n = -\exp\left(\frac{\beta_n}{2}\sigma_x\right) \sigma_y \exp\left(-\frac{\beta_n}{2}\sigma_x\right) \tag{4.3.42}$$

where σ_x and σ_y are ordinary Pauli spin matrices. Applying this result to Eq. (4.3.37) we obtain for the transfer matrix

$$T_n = \exp\left(\frac{\beta_n}{2}\sigma_x\right) e^{-\imath \alpha_n \sigma_y} \exp\left(-\frac{\beta_n}{2}\sigma_x\right). \tag{4.3.43}$$

This shows that T_n may be interpreted as the time-evolution operator of a spin-$\frac{1}{2}$ system being kicked by an imaginary magnetic field parallel to the x axis at the beginning and the end, and precessing about the y axis between the kicks. The trace of the transfer matrix for the complete chain reads

$$\begin{aligned}\mathrm{Tr}(T) &= \mathrm{Tr}\left(\prod_{n=1}^{N} T_n\right) \\ &= \mathrm{Tr}\left[\prod_{n=1}^{N} e^{\delta_n \sigma_x} e^{-i\alpha_n \sigma_y}\right]\end{aligned} \tag{4.3.44}$$

where δ_n is given by

$$\delta_n = \frac{\beta_n - \beta_{n+1}}{2}, \tag{4.3.45}$$

with $\beta_{N+1} = \beta_1$. The product is ordered such that n increases from the right to the left. Introducing γ_n as the sum over all rotation angles α_k with $k \leqslant n$,

$$\gamma_n = \sum_{k=1}^{n} \alpha_k, \quad \gamma_0 = 0, \tag{4.3.46}$$

Equation (4.3.44) reads

$$\begin{aligned}\mathrm{Tr}(T) &= \mathrm{Tr}\left[\prod_{n=1}^{N} e^{\delta_n \sigma_x} e^{-i(\gamma_n - \gamma_{n-1})\sigma_y}\right] \\ &= \mathrm{Tr}\left[e^{-i\gamma_N \sigma_y} \prod_{n=1}^{N} (e^{i\gamma_n \sigma_y} e^{\delta_n \sigma_x} e^{-i\gamma_n \sigma_y})\right] \\ &= \mathrm{Tr}\left[e^{-i\gamma_N \sigma_y} \prod_{n=1}^{N} e^{\delta_n(\sigma_x \cos 2\gamma_n + \sigma_z \sin 2\gamma_n)}\right]\end{aligned} \tag{4.3.47}$$

where again the spinor rotation relation (4.2.55) has been used.

Equation (4.3.47) is still exact. But to proceed further we have to assume that the δ_n are small compared to one. This does not necessarily mean that the disorder is small. Equation (4.3.45) shows that it is sufficient to assume that the *changes* from site to site are small. Then we can apply the exponential operator relation

$$\exp(\delta A)\exp(\delta B) = \exp\left(-\frac{\delta}{2}[A, B]\right)\exp[\delta(A + B)], \tag{4.3.48}$$

which is correct up to $\mathcal{O}(\delta^2)$. The relation is easily proved by expanding the exponentials on both sides into their Taylor series. By repeated application of the relation we get

$$\mathrm{Tr}(T) = \mathrm{Tr}\Bigg\{ \exp(-\imath\gamma_N\sigma_y)$$

$$\times \exp\left(-\frac{1}{2}\sum_{n>m} \delta_n\delta_m[\sigma_x \cos 2\gamma_n + \sigma_z \sin 2\gamma_n,\ \sigma_x \cos 2\gamma_m + \sigma_z \sin 2\gamma_m] \right)$$

$$\times \exp\left(\sum_n \delta_n(\sigma_x \cos 2\gamma_n + \sigma_z \sin 2\gamma_n) \right) \Bigg\}. \qquad (4.3.49)$$

This may be written as

$$\mathrm{Tr}(T) = \mathrm{Tr}[e^{-\imath\alpha\sigma_y}\, e^{\delta(\sigma_x \cos 2\gamma + \sigma_z \sin 2\gamma)}], \qquad (4.3.50)$$

where we have introduced the abbreviations

$$\delta^2 = \left(\sum_n \delta_n \cos 2\gamma_n \right)^2 + \left(\sum_n \delta_n \sin 2\gamma_n \right)^2$$

$$= \sum_{n,m} \delta_n\delta_m \cos 2(\gamma_n - \gamma_m), \qquad (4.3.51)$$

$$\alpha = \gamma_N + \sum_{n>m} \delta_n\delta_m \sin 2(\gamma_n - \gamma_m), \qquad (4.3.52)$$

and

$$\sin 2\gamma = \frac{1}{\delta}\sum_n \delta_n \sin 2\gamma_n, \qquad \cos 2\gamma = \frac{1}{\delta}\sum_n \delta_n \cos 2\gamma_n. \qquad (4.3.53)$$

Applying a last spinor rotation about the y axis we finally get

$$\mathrm{Tr}(T) = \mathrm{Tr}\big(e^{-\imath\alpha\sigma_y}\, e^{\delta\sigma_x}\big)$$

$$= 2\cos\alpha \cosh\delta. \qquad (4.3.54)$$

This is the central result of this section. The disorder has led to a localization of all states. The localization length per site is obtained by means of Eq. (4.3.36) as

$$\bar{\lambda} = \frac{1}{N}[\,\ln(2\cos\alpha) + \ln(\cosh\delta)]. \qquad (4.3.55)$$

The cosine term vanishes in the limit $N \to \infty$. Expanding the second term into a Taylor series and taking only terms up to δ^2 we obtain for the localization length per site

$$\bar{\lambda} = \frac{\delta^2}{2N}. \qquad (4.3.56)$$

Now we specialize the calculation further to the case that the site energies E_n^0

are small and uncorrelated. A relation between the site energies and the δ_n is obtained from Eqs. (4.3.38) and (4.3.40) as follows:

$$\begin{aligned}\delta_n &= \frac{1}{2}(\beta_n - \beta_{n+1}) \approx \frac{1}{2}\sinh(\beta_n - \beta_{n+1}) \\ &= \frac{1}{2}(\sinh\beta_n \cosh\beta_{n+1} - \sinh\beta_{n+1}\cosh\beta_n) \\ &= \frac{1}{2}\frac{\cos\alpha_n - \cos\alpha_{n+1}}{\sin\alpha_n \sin\alpha_{n+1}} \\ &\approx \frac{E^0_{n+1} - E^0_n}{4\sin^2\alpha}.\end{aligned} \tag{4.3.57}$$

In the last step we have replaced α_n and α_{n+1} by α in the denominator. As the difference between the α_n and α is of $\mathcal{O}(E^0_n)$, expression (4.3.57) is correct up to the first order in the site energies. In the same approximation we may replace the argument $2(\gamma_n - \gamma_m)$ in the cosine function in Eq. (4.3.51) by $2(n-m)\alpha$ (see definition (4.3.46) of the γ_n). We then obtain from Eqs. (4.3.51) and (4.3.56) for the ensemble average of the localization length per site

$$\langle\bar{\lambda}\rangle = \frac{1}{32N\sin^4\alpha}\left\langle\sum_{n,m}(E^0_{n+1} - E^0_n)(E^0_{m+1} - E^0_m)\cos 2(n-m)\alpha\right\rangle. \tag{4.3.58}$$

For uncorrelated site energies all averages $\langle E^0_n E^0_m\rangle$ with $n \neq m$ vanish. So we end with

$$\langle\bar{\lambda}\rangle = \frac{1}{8\sin^2\alpha}\left\langle\left(E^0_n\right)^2\right\rangle. \tag{4.3.59}$$

The same result has been derived in the literature in a completely different way [Tho72]. It also holds for the case that the site energies are described by the pseudo-random sequence [Gri88, Tho88]

$$E^0_n = \lambda\cos(\alpha|n|^\nu) \quad \text{with } \nu > 1. \tag{4.3.60}$$

It should be noted that for the derivation of expression (4.3.56) it has been sufficient to assume that the site energies vary slowly. The result (4.3.59), however, only holds under the more stringent assumption that the site energies are small.

4.3.3 Harper equation

The calculation of the localization length performed in the last section for the Anderson model with site disorder becomes obsolete if the site energies are correlated, e.g. for site energies varying periodically with the site. This is the case for an electron propagating on a two-dimensional square lattice with a perpendicularly applied magnetic field. Its Schrödinger equation is given by

$$\left[\frac{1}{2m_e}\left(\boldsymbol{p}-\frac{e}{c}\boldsymbol{A}\right)^2+V(x, y)\right]\psi = E\psi, \tag{4.3.61}$$

where the potential $V(x, y)$ is periodic in x and y with a period length d. The level spacing distributions of this system have already been discussed in Section 3.2.2. For the vector potential $\boldsymbol{A}$ we now choose, in contrast to the former practice, a gauge asymmetric in x and y,

$$\boldsymbol{A} = B(0, -x, 0). \tag{4.3.62}$$

By means of the gauge transformation

$$\hat{\psi} = \exp\left(-\frac{\imath e}{\hbar c}\int \boldsymbol{A}\, d\boldsymbol{x}\right)\psi \tag{4.3.63}$$

the vector potential term is removed from the Schrödinger equation. After propagation in the y direction by one lattice constant, the wave function acquires an additional phase $-2\pi\imath\alpha x/d$ from the vector potential, where α is given by

$$\alpha = \frac{eBd^2}{hc}. \tag{4.3.64}$$

In Section 3.2.2 we have seen that α corresponds to the number of magnetic flux quanta per unit cell. After propagation in the x direction the wave function experiences no phase shift. This is a consequence of the asymmetry in the applied gauge (4.3.62).

As in Section 4.3.1 we express $\hat{\psi}$ as a superposition of single atom wave functions,

$$\hat{\psi}(x, y) = \sum_{n,m} a_{nm}\, e^{-2\pi\imath m\alpha x/d}\psi_0(x - nd, y - md). \tag{4.3.65}$$

Entering with this ansatz the Schrödinger equation, we again end with a tight-binding Hamiltonian:

$$\sum_{k\neq n, l\neq m} W_{nm,kl}\, a_{kl} + E^0_{nm} a_{nm} = E a_{nm}, \tag{4.3.66}$$

where now

$$E^0_{nm} = \int \psi_0(x - nd, y - md)\mathscr{H}_0\psi_0(x - nd, y - md)\, dx\, dy,$$

$$W^0_{nm,kl} = \int \psi_0(x - nd, y - md)e^{2\pi\imath(m-l)\alpha x/d}\mathscr{H}_0 \times \psi_0(x - kd, y - ld)\, dx\, dy. \tag{4.3.67}$$

$\mathscr{H}_0$ is the Hamiltonian entering the Schrödinger equation (4.3.61), but with the vector potential term absent. E^0_{nm} is independent of the lattice site because of the translational invariance of $\mathscr{H}_0$, and may therefore be taken as zero without loss of generality. $W_{nm,kl}$ can be written as

$$W_{nm,kl} = e^{2\pi\imath(m-l)k\alpha} W_{n-k,m-l}, \tag{4.3.68}$$

where

$$W_{nm} = \int \psi_0(x - nd, y - md)e^{2\pi\imath m\alpha x/d}\mathscr{H}_0\psi_0(x, y)\, dx\, dy. \tag{4.3.69}$$

This is again a consequence of the translational invariance. Taking only the nearest neighbour interactions, Eq. (4.3.66) yields

$$W_{10}(a_{n+1,m} + a_{n-1,m}) + W_{01}\left(a_{n,m+1}\, e^{-2\pi\imath n\alpha} + a_{n,m-1}\, e^{2\pi\imath n\alpha}\right) = Ea_{nm}. \tag{4.3.70}$$

In the derivation we have used $W_{10} = W_{-10}$ and $W_{01} = W_{0\,-1}$. This follows from the fact that the single atom wave functions have well-defined parities, $\psi_0(-x, -y) = (-1)^{\pi_a}\psi_0(x, y)$.

As the coefficients of equation system (4.3.70) do not depend on m, the m dependence can be separated by putting

$$a_{nm} = e^{\imath m\phi} a_n, \tag{4.3.71}$$

where ϕ is the phase shift per lattice constant in the y direction. This corresponds to a free propagation of the particle in the y direction with a wave number $k_y = \phi/d$. Equation (4.3.70) now reduces to

$$a_{n+1} + a_{n-1} + \lambda\cos(2\pi n\alpha - \phi)a_n = \epsilon a_n, \tag{4.3.72}$$

where the abbreviations $\epsilon = E/W_{10}$ and $\lambda = 2W_{10}/W_{01}$ have been introduced. For the special case $W_{01} = W_{10}$, i.e. $\lambda = 2$, we obtain the *Harper equation*. This is a tight-binding equation, too, but now with a periodic site dependence of the energy,

$$E^0_n = 2\cos(2\pi n\alpha - \phi). \tag{4.3.73}$$

The system behaviour is qualitatively different for rational and irrational values of α. For rational values of $\alpha = p/q$ the sequence $\{E^0_n\}$ has a period length q. Assuming for reasons of simplicity that the total number of sites is an integer multiple Nq of q, the transfer matrix for the complete chain is given by

$$T = [T^{(q)}]^N, \tag{4.3.74}$$

where

$$T^{(q)} = \prod_{n=1}^{q} T_n. \tag{4.3.75}$$

In the last section we learnt that the states are delocalized for $|\mathrm{Tr}(T^{(q)})| < 2$. As an example we consider the simplest of all situations, the case $\alpha = \frac{1}{2}$. Here the period length is two, and the corresponding transfer matrix is given by

$$T^{(2)} = \begin{pmatrix} \epsilon - 2\cos\phi & -1 \\ 1 & \cdot \end{pmatrix} \begin{pmatrix} \epsilon + 2\cos\phi & -1 \\ 1 & \cdot \end{pmatrix}, \tag{4.3.76}$$

with a trace given by

$$\mathrm{Tr}\big(T^{(2)}\big) = \epsilon^2 - 4\cos^2\phi - 2. \tag{4.3.77}$$

The condition $\mathrm{Tr}\big(T^{(2)}\big) < 2$ holds for two energy ranges limited by

$$-2\sqrt{1+\cos^2\phi} < \epsilon < -2|\cos\phi|,$$

$$2|\cos\phi| < \epsilon < 2\sqrt{1+\cos^2\phi}. \tag{4.3.78}$$

As ϕ can vary between 0 and 2π, the conditions (4.3.78) define two bands with edges given by

$$-2\sqrt{2} < \epsilon < 0, \quad 0 < \epsilon < 2\sqrt{2}. \tag{4.3.79}$$

In the general case of a rational $\alpha = p/q$ there are q different bands. Calculating the band structure of the Harper equation for different α values, we obtain the pattern shown in Fig. 4.9, which is known as the Hofstadter butterfly [Hof76].

Typical crystalline lattice spacings are in the range of about 0.2 nm. Hence magnetic inductions of some 10^5 T would be needed for an experimental realization of the butterfly (see Eq. (4.3.64)). This is far beyond the technical limits. Alternatively the lattice constant can be increased by using artificial superlattices. In such systems the first results showing commensurability effects in the magnetoconductance have already been reported [Sch96]. Further indications for fractional numbers of flux quanta per unit cell have been found in the magnetic field dependence of the critical temperature of superconducting aluminum [Pan84].

Microwave analogue experiments present another possibility to study the Hofstadter butterfly. The propagation of microwaves through a one-dimensional array of scatterers is governed by a transfer matrix equation very similar to that discussed for the one-dimensional tight-binding system in the last section. By varying the lengths of the scatterers periodically, Kuhl and Stöck-

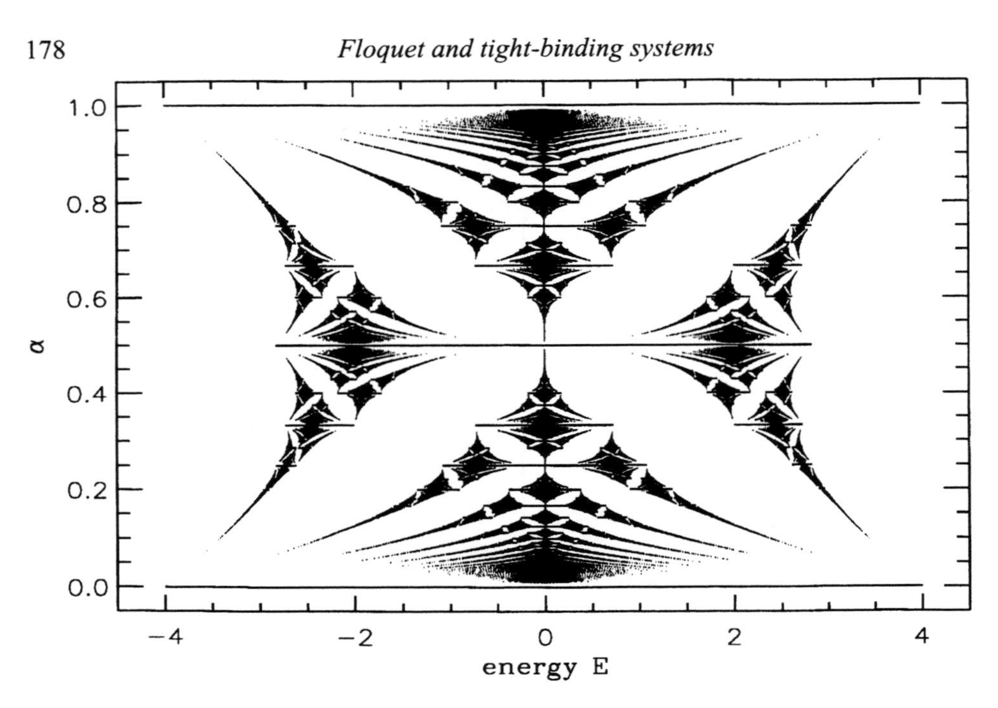

Figure 4.9. The Hofstadter butterfly, showing the eigenvalue spectrum of an electron on a square lattice with a perpendicularly applied magnetic field. The energies are plotted on the abscissa and range from −4 to +4. On the ordinate the number of flux quanta per unit cell α is shown between 0 and 1. For rational numbers $\alpha = p/q$ the spectrum is made up of q bands, for irrational α the eigenvalues form a Cantor set.

mann [Kuh98] succeeded in reproducing the Hofstadter butterfly in the microwave transmission spectrum of such a waveguide (see Fig. 4.10).

Theoretical studies of spectral statistics of the butterfly have been performed in particular by Geisel and coworkers. The following presentation is essentially based on two reports by the group [Gei95, Fle95]. We start with the nearest neighbour spacing distribution. It is not evident how to define this quantity for the Harper equation. For irrational values of α the spectrum forms a Cantor set. A textbook example of a Cantor set is obtained from the following prescription (see Chapter 3.3 of Ref. [Sch84]). Take a line of length one and remove one third from its centre. In the next step remove one third from the centres of the remaining two segments, and so on. The remnants form a Cantor set with a fractal dimension somewhere between 0 and 1. The characteristic property of this set is its self-similarity on different scales, exactly as found for the Hofstadter butterfly.

This construction suggests a way to define nearest neighbour spacing distributions in the butterfly for an irrational α. We approximate α by a set of rational approximants $\alpha_n = p_n/q_n$, and perform the limit $n \to \infty$. This is best

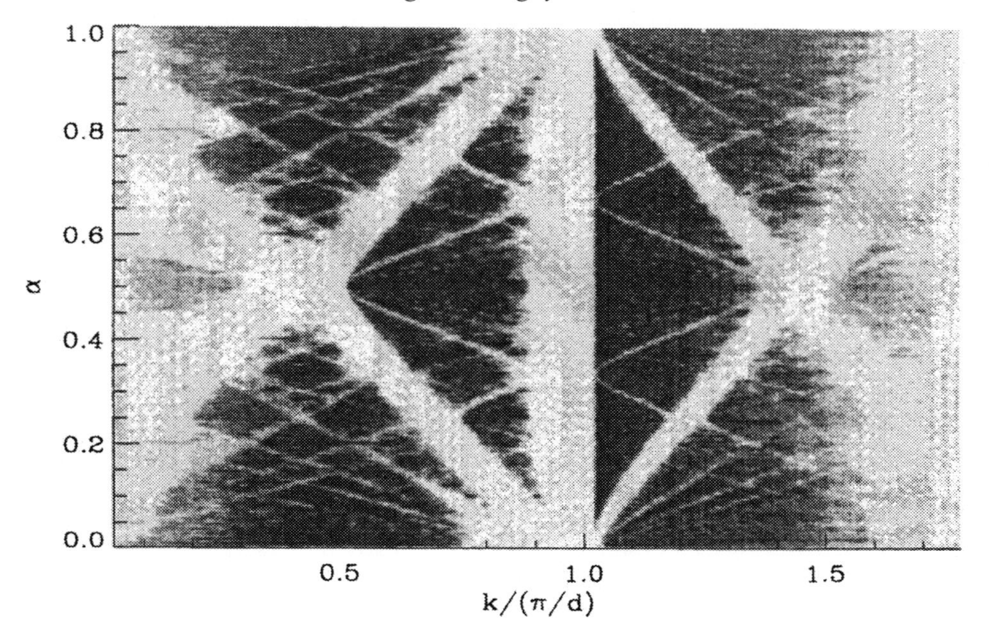

Figure 4.10. Microwave transmission spectrum in a one-dimensional waveguide containing 100 scatterers with periodically varying lengths $l_n = l_0 \Theta[\cos(2\pi n\alpha)]$, where $\Theta(x)$ is the Heaviside step function. The k axis corresponds to a frequency range from 7.5 to 15 GHz, on the ordinate α is plotted between 0 and 1 [Kuh98] (Copyright 1998 by the American Physical Society).

achieved by means of continued fractions. The golden mean $\alpha_G = (\sqrt{5} - 1)/2 = 0.6180 \ldots$ is of special importance in this context. It has the continued fraction

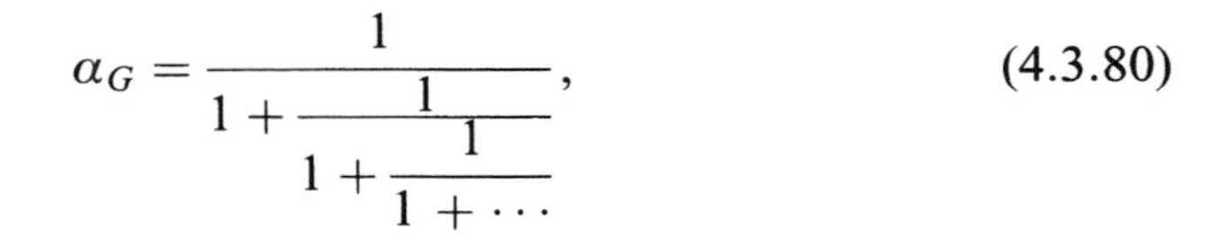

$$\alpha_G = \cfrac{1}{1 + \cfrac{1}{1 + \cfrac{1}{1 + \cdots}}}, \tag{4.3.80}$$

containing only ones. It can be considered as the most irrational of all numbers, as its continued fraction has the slowest convergence to the correct value of all numbers between 0 and 1.

The set of approximants $\{(\alpha_G)_n\}$ of α_G, obtained by truncating the continued fraction at different depths, is given by

$$(\alpha_G)_n = \frac{a_{n-1}}{a_n}, \tag{4.3.81}$$

where the a_n are the Fibonacci numbers defined by the recursion relation

$$a_0 = 1, \quad a_1 = 1,$$

$$a_n = a_{n-1} + a_{n-2}, \qquad n = 2, 3, \ldots . \tag{4.3.82}$$

Figure 4.11 shows the integrated level spacing distribution $I(s) = \int_0^s p(s)\, ds$ obtained for two different approximants of the golden mean [Gei91]. The data obey an inverse power,

$$I(s) \sim s^{1+\nu}, \tag{4.3.83}$$

whence follows for the level spacing distribution

$$p(s) \sim s^{\nu} \tag{4.3.84}$$

with a numerically found exponent of $\nu = -1.5009(10)$. The situation is completely different compared to the Poisson and Gaussian ensembles formerly discussed, where only positive exponents between 0 and 4 have been found. In the Hofstadter butterfly we have an extreme level clustering, which is even stronger than that observed for the Poissonian ensemble with $\nu = 0$. This is a direct consequence of the self-similar structure of the spectrum, which also affects the spectral rigidity. The latter is shown in Fig. 4.12 for one approximant of the golden mean. Again a power law

$$\Delta_3(L) \sim L^{\gamma} \tag{4.3.85}$$

is observed with $\gamma = 1.48(6)$. This has to be compared with the linear behaviour found for the Poissonian, and the logarithmic behaviour found for the Gaussian ensembles (see Section 3.2.4). The exponents ν and γ can be associated with the Hausdorff dimensions D_0 and D_2, respectively [Gei91, Gei95]. Readers not familiar with fractal dimensions will find an elementary introduction in Chapter 5 of Ref. [Sch84].

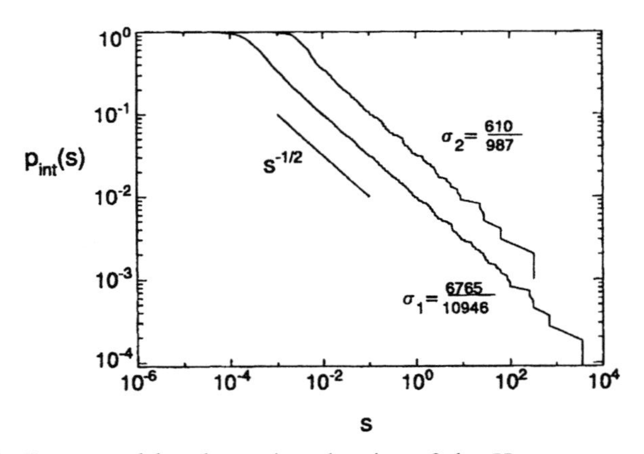

Figure 4.11. Integrated level spacing density of the Harper equation for two approximants of the golden mean, taken from the Fibonacci sequence (4.3.82) [Gei91] (Copyright 1991 by the American Physical Society).

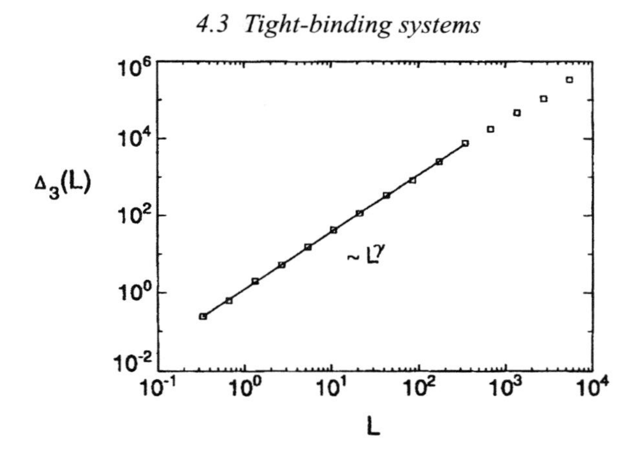

Figure 4.12. Spectral rigidity for the Harper equation for the approximant $(\alpha_G)_{20} = 6765/10946$ of the golden mean [Gei91] (Copyright 1991 by the American Physical Society).

Geisel *et al.* give a heuristic explanation for the exponent $\nu = -3/2$ [Gei91, Gei95]. Take a rational approximant $\alpha_n = p_n/q_n$ of an irrational number α, not necessarily the golden mean. Now consider a segment of the chain of length q_n. Along this distance a particle is not yet aware of the periodicity of the lattice, and we may assume that a wave packet performs a diffusive motion with an averaged squared displacement $\langle x^2(t)\rangle \sim t$. From this a maximum period $\tau_n \sim q_n^2$ is obtained for which the periodicity is not yet evident. This period, on the other hand, can be related via the uncertainty relation to a smallest energy distance $s_n \sim \hbar/\tau_n$ between neighbouring states. The number of states associated with the q_n sites, on the other hand, is exactly q_n, since each site brings in one level. But if the smallest energy difference between neighbouring states is s_n, then q_n must equal the number of states with an energy spacing larger than or equal to s_n. Collecting the results we have

$$\int_{s_n}^{\infty} p(s)\, ds_n \sim q_n \sim \sqrt{\tau_n} \sim s_n^{-1/2}, \tag{4.3.86}$$

whence follows $p(s) \sim s^{-3/2}$. For a refined energy spectrum the next approximant $\alpha_{n+1} = p_{n+1}/q_{n+1}$ is considered. Repeating the steps we conclude that relation (4.3.86) should hold on all scales. A completely analogous argument has been applied in Section 4.2.1, where we derived the relation between diffusion constant and localization length for the kicked rotator.

The above argument is based on the assumption that the spread of a wave packet is diffusive. To check this Geisel *et al.* [Gei91] have studied the spreading of a wave packet on a finite lattice with 987 sites for different values of λ. For $\lambda = 2$, corresponding to the Harper equation, the authors have indeed found a diffusion behaviour according to $\langle x^2(t)\rangle \sim t^\beta$ with $\beta = 1$. In classical mechanics this is the normal behaviour, in quantum mechanics, however, this is not self-evident, as we have learnt in this chapter. In a more recent work Ketzmerick *et al.* [Ket97] could relate the exponent β to the Hausdorff dimensions of the energy spectrum and the eigenfunctions. They obtained a value of β close (but not equal) to 1. Consequently the classical diffusion behaviour observed has to be considered as a mere coincidence.

5

Eigenvalue dynamics

In many spectroscopic experiments the energy levels of a system are determined as a function of an external parameter. Take as an example a hydrogen atom in a magnetic field. As long as the magnetic interaction is small compared to the spacing between neighbouring levels, a level with angular momentum quantum number L splits equidistantly into its $2L + 1$ magnetic sublevels. A qualitatively different behaviour is found if the magnetic Zeeman interaction is of the same order of magnitude as the mean level spacing. In this range any regularity in the magnetic splitting is lost, and the energy levels move erratically as a function of the magnetic field. In billiard systems a shape parameter such as one length may take the role of the magnetic field. In chaotic systems degeneracies no longer occur, apart from accidental ones having the measure zero. Consequently crossings of eigenvalues are never observed when the parameter is changed, and they are converted into anti-crossings. For two-level systems this is an effect well-known in elementary quantum mechanics. The resulting motion of the eigenvalues dependent on the external parameter strongly resembles the dynamics of the particles of a one-dimensional gas with repulsive interaction.

The dynamical concept was introduced by Pechukas [Pec83] and further developed by Yukawa [Yuk85], and is therefore called the *Pechukas–Yukawa model* in the following. The major part of this chapter will treat the consequences of this model. We shall see that ordinary statistical mechanics can describe the eigenvalue dynamics of the spectra of chaotic systems. Random matrix theory will then result as a direct consequence of the Boltzmann ansatz of statistical mechanics. The Pechukas–Yukawa model has been discussed in several textbooks [Nak93, Haa91], but these do not include a number of important recent results.

If the Hamiltonian depends on two parameters, there is a new aspect of quality. In the two-dimensional parameter space closed loops are possible with the surprising consequence that the phases of the eigenfunctions may have changed. These geometrical phases are known as *Berry's* phases, after M. Berry who was the first to study the influence of adiabatic changes on the phase of wave functions systematically [Ber84a].

5.1 Pechukas–Yukawa model

5.1.1 Equations of motion

The determination of the eigenvalues of quantum mechanical systems dependent on an external parameter is an everyday problem of practical importance. Typical Hamiltonians are of the form

$$\mathscr{H} = \mathscr{H}_0 + tV, \tag{5.1.1}$$

where $\mathscr{H}_0$ is the Hamiltonian of the unperturbed system, and tV is a perturbation. Often $\mathscr{H}_0$ corresponds to an integrable Hamiltonian which gradually becomes chaotic, if the strength t of the perturbation is increased. A typical example is the hydrogen atom in a strong magnetic field.

Pechukas [Pec83] recognized that quantum mechanical perturbation theory allows a dynamical interpretation, if the perturbation strength t is interpreted as 'time'. The following derivation is essentially based on his paper. We denote the eigenvalues by x_n and the corresponding eigenfunctions by $|n\rangle$,

$$\mathscr{H}|n\rangle = x_n|n\rangle. \tag{5.1.2}$$

The eigenfunctions obey the orthogonality relation

$$\langle n|m\rangle = \delta_{nm}, \tag{5.1.3}$$

whence follows after differentiation with respect to t

$$\left(\frac{\partial}{\partial t}\langle n|\right)|m\rangle + \langle n|\left(\frac{\partial}{\partial t}|m\rangle\right) = 0. \tag{5.1.4}$$

By means of the operator

$$H = \imath\frac{\partial}{\partial t}, \tag{5.1.5}$$

with the matrix elements

$$H_{nm} = \langle n|H|m\rangle, \tag{5.1.6}$$

Equation (5.1.4) can be written as $H_{nm} = H^*_{mn}$, i.e. H is Hermitian. Let us now derive an equation system for the 'time' evolution of the matrix elements of an arbitrary operator A. Using the completeness relation

$$\sum_k |k\rangle\langle k| = 1 \tag{5.1.7}$$

we have

$$\frac{d}{dt}\langle n|A|m\rangle = \left(\frac{\partial}{\partial t}\langle n|\right)A|m\rangle + \langle n|\frac{\partial A}{\partial t}|m\rangle + \langle n|A\left(\frac{\partial}{\partial t}|m\rangle\right)$$
$$= \sum_k \left[\left(\frac{\partial}{\partial t}\langle n|\right)|k\rangle\langle k|A|m\rangle + \langle n|A|k\rangle\langle k|\left(\frac{\partial}{\partial t}|m\rangle\right)\right]$$
$$+ \langle n|\frac{\partial A}{\partial t}|m\rangle. \tag{5.1.8}$$

Using definition (5.1.5), we get

$$\frac{dA_{nm}}{dt} = \imath\sum_k (H_{nk}A_{km} - A_{nk}H_{km}) + \left(\frac{\partial A}{\partial t}\right)_{nm}. \tag{5.1.9}$$

In compact matrix notation this can be written as

$$\frac{dA}{dt} = \imath[H, A] + \frac{\partial A}{\partial t}. \tag{5.1.10}$$

This is exactly the equation known from quantum mechanics for the time evolution of operators, if H is interpreted as a Hamilton operator and t as a time. This allows us to derive a nearly one-to-one correspondence between the eigenvalue dynamics and the dynamics of a one-dimensional interacting gas.

In the next step we insert in Eq. (5.1.9) the Hamiltonian $\mathscr{H}$ for A. We then obtain

$$\dot{x}_n\delta_{nm} = \imath H_{nm}(x_m - x_n) + V_{nm}, \tag{5.1.11}$$

where the dot means differentiation with respect to t, whence follows

$$\dot{x}_n = V_{nn} \tag{5.1.12}$$

for $n = m$, and

$$H_{nm} = \frac{V_{nm}}{\imath(x_n - x_m)} \tag{5.1.13}$$

for $n \neq m$. Next we insert V for A and obtain

$$\dot{V}_{nn} = 2\sum_{k\neq n} \frac{V_{nk}V_{kn}}{x_n - x_k}, \tag{5.1.14}$$

for $n = m$, and

$$\dot{V}_{nm} = \sum_{k\neq n,m} V_{nk}V_{km}\left(\frac{1}{x_n - x_k} + \frac{1}{x_m - x_k}\right)$$
$$+ \imath(H_{nn} - H_{mm})V_{nm} - \frac{V_{nm}(V_{nn} - V_{mm})}{x_n - x_m}, \tag{5.1.15}$$

for $n \neq m$. The first term in the second row can be removed by means of the substitution

$$V_{nm} = \hat{V}_{nm} \exp\left[\imath \int (H_{nn} - H_{mm})\, dt\right]. \tag{5.1.16}$$

The $\hat{V}_{nm}$ obey the same equation system as the V_{nm} but without the first term in the second row of Eq. (5.1.15). We may therefore assume $H_{nn} = 0$ without loss of generality, and again write V_{nm} for $\hat{V}_{nm}$.

The second term in the second row can be removed by introducing the new variables

$$p_n = V_{nn}, \tag{5.1.17}$$

$$f_{nm} = (x_n - x_m) V_{nm}, \qquad n \neq m. \tag{5.1.18}$$

We then obtain the closed dynamical equation system

$$\dot{x}_n = p_n, \tag{5.1.19}$$

$$\dot{p}_n = 2 \sum_{k \neq n} \frac{|f_{nk}|^2}{(x_n - x_k)^3}, \tag{5.1.20}$$

$$\dot{f}_{nm} = \sum_{k \neq n,m} f_{nk} f_{km} \left(\frac{1}{(x_n - x_k)^2} - \frac{1}{(x_m - x_k)^2} \right), \tag{5.1.21}$$

for the dynamical variables x_n, p_n, f_{nm}. This is the Pechukas–Yukawa equation system. In Pechukas' original version [Pec83] x_n and V_{nm} were used as dynamical variables. The replacement of the dynamical variables V_{nm} with $n \neq m$ by f_{nm} is due to Yukawa [Yuk85]. The underlying motivation will become clear in the next section.

If we assume for a moment that the f_{nm} were constant, Eqs. (5.1.19) and (5.1.20) would describe the motion of particles in a one-dimensional gas with a mutual repulsive force according to an r^{-3} distance law. In this analogy the x_n take the role of the positions, and the p_n that of the momenta. For the case of constant f_{nm} the Pechukas–Yukawa equation system reduces to the Calogero–Moser system [Cal71, Mos75], one of the rare examples of many-body systems allowing a closed solution. For details see Ref. [Hop92]. In the Pechukas–Yukawa system the f_{nm}, describing the interaction strengths between 'particles' n and m, are 'time-dependent' themselves.

The Hamiltonian (5.1.1) has one serious disadvantage, which becomes evident if we square $\mathcal{H}$ and take the trace:

$$\mathrm{Tr}(\mathcal{H}^2) = \mathrm{Tr}\left(\mathcal{H}_0^2\right) + 2t\, \mathrm{Tr}(\mathcal{H}_0 V) + t^2\, \mathrm{Tr}(V^2). \tag{5.1.22}$$

We assume that the matrix representation of $\mathcal{H}$ is truncated to a finite rank N to allow the calculation of the trace. If $\mathcal{H}_0$ and V are uncorrelated, $\mathrm{Tr}(\mathcal{H}_0 V)$ is small compared to $\mathrm{Tr}(\mathcal{H}_0^2)$ and $\mathrm{Tr}(V^2)$ in the limit of large N, and the mixed term in Eq. (5.1.22) can be neglected. If further $\mathscr{H}_0$ and V are normalized according to $\mathrm{Tr}(\mathscr{H}_0^2) = \mathrm{Tr}(V^2)$, which is always possible by a proper definition of the perturbation parameter t, we have

$$\mathrm{Tr}(\mathscr{H}^2) = (1 + t^2)\mathrm{Tr}\left(\mathscr{H}_0^2\right). \tag{5.1.23}$$

From the invariance properties of the trace we get $\mathrm{Tr}(\mathscr{H}^2) = \sum x_n^2$. The increase of the perturbation strength t therefore leads to inflation of the eigenvalue gas. To eliminate this problem we have to rescale the Hamiltonian. The most general ansatz to achieve this is given by

$$\mathscr{H} = g(\lambda)(\mathscr{H}_0 + \lambda V), \tag{5.1.24}$$

where λ is a function of t which has still to be determined. Haake and Lenz [Haa90] have shown that $g(\lambda)$ and $\lambda(t)$ cannot be chosen arbitrarily to obtain a well-behaved dynamical equation system (the term 'well-behaved' will be explained in Section 5.1.3), but must obey the relation

$$g^2\dot{\lambda} = \mathrm{const.} \tag{5.1.25}$$

Because of this condition, direct removal of the factor $(1 + t^2)$ in Eq. (5.1.23) by rescaling $\mathscr{H}$ by a factor $(1 + t^2)^{-1/2}$ is not possible. But we may put alternatively

$$g(\lambda) = \frac{1}{\sqrt{1 + \lambda^2}}, \tag{5.1.26}$$

and determine $\lambda(t)$ from Eq. (5.1.25). In this way we obtain, after suitable fixing of the constants, $\lambda(t) = \tan t$. This results in the modified Hamiltonian

$$\mathscr{H} = \mathscr{H}_0 \cos t + V \sin t, \tag{5.1.27}$$

which has been used in a number of level dynamics studies such as those performed by Zakrzewski and Delande [Zak93a] which will be discussed later on. With the Hamiltonian (5.1.27) we again obtain a Pechukas–Yukawa equation system where only the f_{nm} have to be redefined according to

$$f_{nm} = \frac{(x_n - x_m)V_{nm}}{\cos t}. \tag{5.1.28}$$

Furthermore Eq. (5.1.20) for $\dot{p}_n$ is modified to

$$\dot{p}_n = -x_n + 2\sum_{k \neq n} \frac{|f_{nk}|^2}{(x_n - x_k)^3}. \tag{5.1.29}$$

The redefinition of the Hamiltonian has generated an additional harmonic force term $-x_n$. This term is a direct consequence of the periodic parameter dependence of the Hamiltonian (5.1.27).

The inflation problem does not exist in Floquet systems, as here all

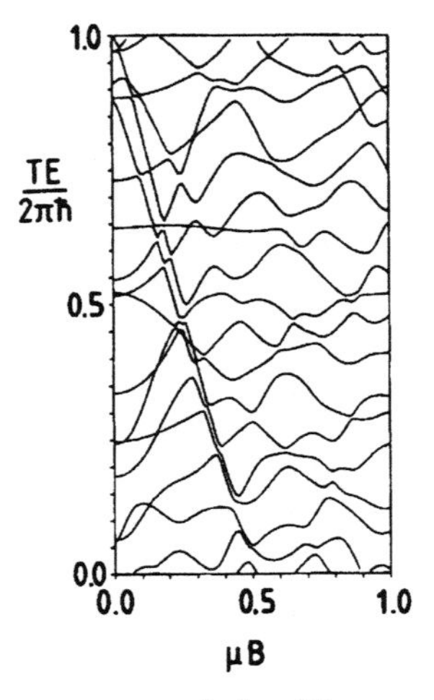

Figure 5.1. Quasi-energy spectrum of the Floquet operator (5.1.30) dependent on $k = \mu B$ for a spin quantum number $S = 16$. The quasi-energies are plotted in units of h/T, where T is the kick period. Only states with even parity are shown [Gas90] (Copyright 1990 by the American Physical Society).

eigenvalues are confined to the unit circle. The Hamiltonian defined in Eq. (5.1.1) has then to be replaced by the Floquet operator

$$F = e^{\imath \mathscr{H}_0 T} e^{\imath k V}, \qquad (5.1.30)$$

where now the kick strength k takes the role of the level dynamics parameter. A numerical example for Floquet level dynamics with $\mathscr{H}_0 = -AS_z^2$ and $V = S_x$ is shown in Fig. 5.1 [Gas90]. In the figure the level dynamics parameter is denoted by μB suggesting a kick by a magnetic field. The dynamical equation system for the Floquet system closely resembles the original Pechukas–Yukawa equations [Kuś87]. Details can be found in Ref. [Haa91].

5.1.2 Constants of motion

The Pechukas–Yukawa equation system has a considerable number of constants of motion. We shall even see that the equation system is completely integrable. To obtain the constants of motion we start with the differential equation (5.1.10) for the time evolution of operators. Taking the trace on both sides we have

$$\frac{d}{dt}[\mathrm{Tr}(A)] = \mathrm{Tr}\left(\frac{\partial A}{\partial t}\right). \qquad (5.1.31)$$

The trace of the commutator $[\mathscr{H}, \mathrm{A}]$ vanishes because of the trace property

$\mathrm{Tr}(AB) = \mathrm{Tr}(BA)$. It is again assumed that the Hamiltonian has been truncated to a matrix of finite rank N. Equation (5.1.31) shows that the trace of any operator which does not explicitly depend on the 'time' is a constant of motion. Two operators of this type are directly obtained from the Hamiltonian (5.1.1), namely V and $F = [\mathscr{H}, V]$ with the matrix elements V_{nm} and $f_{nm} = (x_n - x_m)V_{nm}$, respectively. Again we come across the dynamical variable f_{nm} introduced in the last section.

It is evident from Eq. (5.1.31) that any expression of the type

$$\mathrm{Tr}(F^{n_1} V^{n_2} F^{n_3} \ldots) \tag{5.1.32}$$

is a constant of motion. Not all of these expressions are independent of each other, however. To determine the number of independent constants of motion, we change over to an arbitrary basis $|\alpha\rangle$ independent of t [Kuś88]. In this basis the matrix elements of $\mathscr{H}$ read

$$\mathscr{H}_{\alpha\beta} = (\mathscr{H}_0)_{\alpha\beta} + tV_{\alpha\beta}. \tag{5.1.33}$$

These equations can be considered as the solution of the differential equation system

$$\dot{\mathscr{H}}_{\alpha\beta} = V_{\alpha\beta}. \tag{5.1.34}$$

In this interpretation the $(\mathscr{H}_0)_{\alpha\beta}$ take the roles of integration constants. The number of independent constants of motion is thus given by the number of independent components of $\mathscr{H}_0$.

As the x_n are obtained from $\mathscr{H}$ by diagonalization, the simple differential equation system (5.1.34) must be equivalent to the Pechukas–Yukawa equations. We might get the impression that a triviality has been artificially complicated in the Pechukas–Yukawa approach. But this is the wrong view. Nobody would calculate the level dynamics of a quantum mechanical spectrum by integrating the Pechukas–Yukawa equations. The linking of two seemingly unrelated fields, namely the eigenvalue dynamics of quantum mechanical spectra and classic statistical mechanics, makes up the very success of the Pechukas–Yukawa approach.

The integrability of the Pechukas–Yukawa equations has been proved in a completely different way in two independent, but essentially equivalent papers by Yukawa [Yuk86] and Nakamura and Lashmaman [Nak86]. The dimension of the phase space spanned by the dynamical variables x_n, p_n, f_{nm} is $2N + M$ where $M = \nu N(N-1)/2$ is the number of variables f_{nm}. ν is again the universality index, i.e. $\nu = 1, 2, 4$ for the GOE, the GUE, and the GSE, respectively. According to the above argumentation the number of constants of motion is $N + M$. For all three ensembles the trajectories in phase space thus span an N-dimensional subspace, just as for a classically integrable system with N degrees of freedom.

The constants of motion of the type (5.1.32), which contain V and F only up to the second power, are of particular importance. Since the traces of F and VF vanish, there are only three of them left. We begin with

$$P = \text{Tr}(V) = \sum_n V_{nn} = \sum_n p_n, \tag{5.1.35}$$

which can be interpreted as the 'total momentum' of the eigenvalues. In the 'centre of mass' system

$$\bar{x}_n = x_n - \frac{t}{N}\,\text{Tr}(V) \tag{5.1.36}$$

the 'total momentum' vanishes. Without loss of generality we may therefore assume $P = 0$. The next constant of motion is

$$W = \frac{1}{2}\,\text{Tr}(V^2) = \frac{1}{2}\sum_n (V_{nn})^2 + \frac{1}{2}\sum_{n,m}{}' V_{nm}V_{mn}. \tag{5.1.37}$$

Using the definitions (5.1.17) and (5.1.18) this may be written as

$$W = \frac{1}{2}\sum_n p_n^2 + \frac{1}{2}\sum_{n,m}{}' \frac{|f_{nm}|^2}{(x_n - x_m)^2}. \tag{5.1.38}$$

This expression allows an obvious interpretation: W corresponds to the 'total energy' of the eigenvalue gas with a 'kinetic energy'

$$W_{\text{kin}} = \frac{1}{2}\sum_n p_n^2, \tag{5.1.39}$$

and a 'potential energy'

$$W_{\text{pot}} = \frac{1}{2}\sum_{n,m}{}' \frac{|f_{nm}|^2}{(x_n - x_m)^2}. \tag{5.1.40}$$

With the modified Hamiltonian (5.1.27) the 'potential energy' contains an additional spring term resulting from the additional force $-x_n$ which enters into the equation of motion (5.1.29). In this case the 'total energy' is given by

$$W = \frac{1}{2}\sum_n (p_n^2 + x_n^2) + \frac{1}{2}\sum_{n,m}{}' \frac{|f_{nm}|^2}{(x_n - x_m)^2}. \tag{5.1.41}$$

The other constant of motion quadratically depending on the dynamical variables is

$$Q = \frac{1}{2}\,\text{Tr}(F^2) = \frac{1}{2}\sum_{n,m}{}' |f_{nm}|^2, \tag{5.1.42}$$

which is sometimes called 'angular momentum', a somewhat questionable notation, since the eigenvalue dynamics is one-dimensional.

5.1.3 Phase space density

The phase space density $\rho(x_n, p_n, f_{nm})$ describes the equilibrium distribution of the dynamical variables x_n, p_n, f_{nm} in the $(2N+M)$-dimensional phase space. $\rho(x_n, p_n, f_{nm})$ obeys the equation

$$\frac{\partial}{\partial t}\int_{dV} \rho \, dV + \int_{dS} \rho \boldsymbol{v} \, \boldsymbol{dS} = 0, \tag{5.1.43}$$

where dV is an arbitrary small phase space volume element, and $\boldsymbol{dS}$ its surface. The first term gives the probability per time that the phase space trajectory enters or leaves the volume element dV, and the second term gives the probability current passing through the boundary $\boldsymbol{dS}$ into the volume element.

$$\boldsymbol{v} = (\dot{x}_n, \dot{p}_n, \dot{f}_{nm}) \tag{5.1.44}$$

is the vector of the time derivatives of the dynamical variables. Applying Stokes' theorem to the second term, we obtain a continuity equation

$$\frac{\partial \rho}{\partial t} + \boldsymbol{\nabla}(\rho \boldsymbol{v}) = 0, \tag{5.1.45}$$

for ρ, just as in ordinary statistical mechanics. The total time derivative of the phase space density is obtained as

$$\begin{aligned} \dot{\rho} &= \frac{\partial \rho}{\partial t} + \sum_n \dot{x}_n \frac{\partial \rho}{\partial x_n} + \sum_n \dot{p}_n \frac{\partial \rho}{\partial p_n} + \sum_{n,m} \dot{f}_{nm} \frac{\partial \rho}{\partial f_{nm}} \\ &= \frac{\partial \rho}{\partial t} + (\boldsymbol{v}\boldsymbol{\nabla})\rho. \end{aligned} \tag{5.1.46}$$

Combining Eqs. (5.1.45) and (5.1.46) we get

$$\dot{\rho} + (\boldsymbol{\nabla}\boldsymbol{v})\rho = 0. \tag{5.1.47}$$

From the Pechukas–Yukawa equations we have

$$(\boldsymbol{\nabla}\boldsymbol{v}) = \sum_n \frac{\partial \dot{x}_n}{\partial x_n} + \sum_n \frac{\partial \dot{p}_n}{\partial p_n} + \sum_{n,m} \frac{\partial \dot{f}_{nm}}{\partial f_{nm}} = 0, \tag{5.1.48}$$

as can be directly verified by means of Eqs. (5.1.19) to (5.1.21). Yukawa [Yuk85] introduced the f_{nm} for the very purpose of obtaining a divergence-free flow of the dynamical variables. With the V_{nm} as variables there would be contributions from the term $\partial \dot{V}_{nm}/\partial V_{nm}$ to the divergence of the velocity, as is evident from Eq. (5.1.15). The Haake–Lenz condition (5.1.25), too, is enforced by the demand to obtain a divergence-free flow of the dynamical variables.

Again we have found a parallel with classical mechanics where the vanishing divergence of the velocities of the dynamical variables is a consequence of the Hamilton–Jacobi equations

$$\frac{\partial \mathcal{H}}{\partial p} = \dot{q}, \quad \frac{\partial \mathcal{H}}{\partial q} = -\dot{p}. \tag{5.1.49}$$

Equation (5.1.47) now reduces to $\dot{\rho} = 0$, i.e. ρ is a constant of motion. ρ can therefore be expressed in terms of the constants of motion of the Pechukas–Yukawa equation system. As the equations are completely integrable, the phase space variables must obey the constraints

$$C_n = c_n^0, \quad n = 1, \ldots, N + M, \tag{5.1.50}$$

where the C_n are the constants of motion, and the c_n^0 are the respective initial values. From this we arrive at the micro-canonical distribution

$$\rho \sim \prod_n \delta(C_n - c_n^0). \tag{5.1.51}$$

The proportionality factor has to be chosen in such a way that the integral of ρ over all phase space variables is normalized to one. Calculations with micro-canonical quantities pose difficult technical problems. Therefore we resort to the corresponding macro-canonical quantity

$$\rho = \frac{1}{Z} \exp\left(-\sum_n \lambda_n C_n \right). \tag{5.1.52}$$

The normalization factor

$$Z = \int \exp\left(-\sum_n \lambda_n C_n \right) dx_n \, dp_n \, df_{nm} \tag{5.1.53}$$

is called the partition function and plays a central role in statistical mechanics. For the macro-canonical ensemble the constraints (5.1.50) are replaced by the weaker conditions

$$\int_n C_n(x_n, p_n, f_{nm}) \rho(x_n, p_n, f_{nm}) \, dx_n \, dp_n \, df_{nm} = c_n^0, \quad n = 1, \ldots, N + M, \tag{5.1.54}$$

which can be written more compactly by means of Eqs. (5.1.52) and (5.1.53) as

$$c_n^0 = -\frac{\partial \ln Z}{\partial \lambda_n}. \tag{5.1.55}$$

The partition function alone is thus sufficient to determine the parameters λ_n from the initial values c_n^0.

For the micro-canonical phase space density (5.1.51) the constants of motion C_n are constant by definition, but for the corresponding macro-canonical expression (5.1.52) only the average values of the C_n are fixed. Therefore the macro-canonical phase space density can be at most qualitatively correct. But in the limit of a large number of eigenvalues the fluctuations about the average values rapidly decrease, and the micro-canonical and the macro-canonical approach should come to the same results asymptotically.

We might also argue that the calculation of the constants of motion such as W or Q (see Eqs. (5.1.38) and (5.1.42)) is necessarily restricted to a finite number of levels, for which the macro-canonical description would be more appropriate anyway.

The macro-canonical distribution (5.1.52) can be obtained alternatively using an argument from information theory. We find that the distribution (5.1.52) contains the least information

$$I = \int \rho(x_n, p_n, f_{nm}) \ln \rho(x_n, p_n, f_{nm}) \, dx_n \, dp_n \, df_{nm} \tag{5.1.56}$$

under the constraints imposed by Eqs. (5.1.54). In this approach the λ_n take the roles of the Lagrange multipliers. A similar procedure was applied in Section 3.1.3 to motivate the distribution of matrix elements in the Gaussian ensemble.

We have thus obtained an expression which is completely analogous to the Boltzmann ansatz of classical mechanics. The complete integrability of the Pechukas–Yukawa equations casts some doubts on the argumentation. Obviously the motion in phase space cannot be really ergodic. But experiments as well as calculations yield only a very small portion of all the dynamical variables, i.e. only low-dimensional subspaces of the complete phase space are really explored. It is at least plausible to assume that by this projection from the high-dimensional phase space to a low dimensional subspace quasi-ergodicity is obtained.

It should be remembered that the same problem already exists in ordinary statistical mechanics. Here too, the number of systems where ergodicity has been proved is extremely small, and the Boltzmann ansatz is primarily justified by its success.

5.1.4 Pechukas–Yukawa model and random matrix theory

The macro-canonical expression (5.1.52) for the stationary phase space distribution, though already an approximation, is still too complicated to be tractable. Therefore we go one step further and restrict the calculation to only two constants of motion, namely the 'total energy' W and the 'angular momentum' Q introduced in Section 5.1.2. We then arrive at the ansatz originally introduced by Yukawa [Yuk85]:

$$\rho = \frac{1}{Z} e^{-\beta W - \gamma Q}. \tag{5.1.57}$$

For W we take the expression (5.1.41), including the spring potential $\frac{1}{2}\sum x_n^2$. The stationary phase space distribution then reads

$$\rho = \frac{1}{Z} \exp\left\{ -\frac{\beta}{2} \left[\sum_n (p_n^2 + x_n^2) + \sum_{n,m}{}' \frac{|f_{nm}|^2}{(x_n - x_m)^2} \right] - \frac{\gamma}{2} \sum_{n,m}{}' |f_{nm}|^2 \right\}. \tag{5.1.58}$$

The restriction to the two constants of motion W and Q can be made plausible by the following arguments [Has93]:

– they are the only constants of motion quadratic in the p_n and the f_{nm},
– of all possible distributions the Yukawa ansatz for given x_n contains the least information with respect to the p_n and the f_{nm}.

But the main justification of ansatz (5.1.57) is its simplicity. The influence of the other constants of motion is discussed in Chapter 6 of Ref. [Haa91].

The correlated energy distribution function is obtained from Eq. (5.1.58) by integrating ρ over the variables p_n and f_{nm},

$$P(x_1, \ldots, x_n) = \int \rho(x_n, p_n, f_{nm})\, dp_n\, df_{nm}. \tag{5.1.59}$$

Performing the integrations we obtain

$$P(x_1, \ldots, x_n) \sim \prod_{n>m} \left| \frac{(x_n - x_m)^2}{1 + \frac{\gamma}{\beta}(x_n - x_m)^2} \right|^{\nu/2} \exp\left(-\frac{\beta}{2} \sum_n x_n^2 \right), \tag{5.1.60}$$

where ν is the universality index.

In the limiting case $\gamma \gg \beta$ the distribution reduces to

$$P(x_1, \ldots, x_n) \sim \exp\left(-\frac{\beta}{2} \sum_n x_n^2 \right). \tag{5.1.61}$$

This is identical with the energy distribution function for the Poissonian ensemble. In the other limiting case $\beta \gg \gamma$ we have

$$P(x_1, \ldots, x_n) \sim \prod_{n>m} |x_n - x_m|^\nu \exp\left(-\frac{\beta}{2} \sum_n x_n^2 \right). \tag{5.1.62}$$

This is exactly the correlated eigenenergy distribution for the Gaussian ensembles (see Eq. (3.1.95)).

As a function of β/γ the correlated eigenenergy distribution derived from the Yukawa ansatz thus yields a continuous transition from the Poisson to one of the Gaussian ensembles. By this we have linked the spectral statistics of chaotic systems to ordinary statistical mechanics. This may explain the great success of random matrix theory in the description of the spectra of chaotic systems.

Comparatively little has been done to study the implications of the correlated eigenenergy distribution function (Eq. (5.1.60)) in the transition region. This is somewhat astonishing as the Lagrange parameters β, γ are not free fit parameters of the model, but can be calculated from the partition function by means of Eq. (5.1.55).

Some results are available in Floquet systems. Here the correlated eigenphase distribution function is given by [Ma94]

$$P(\phi_1, \dots, \phi_N) \sim \prod_{n>m} \left| \frac{1 - \cos(\phi_n - \phi_m)}{\beta + 2\gamma[1 - \cos(\phi_n - \phi_m)]} \right|^{\nu/2} . \qquad (5.1.63)$$

Again we recover the eigenphase distribution functions for the Poissonian and the Gaussian circular ensembles in the limits $\beta \ll \gamma$ and $\beta \gg \gamma$, respectively (see Eq. (4.1.39)). With the substitution

$$\lambda = \left(1 + \frac{\beta}{2\gamma}\right) - \sqrt{\left(1 + \frac{\beta}{2\gamma}\right)^2 - 1} \qquad (5.1.64)$$

Eq. (5.1.63) can be alternatively written as

$$P(\phi_1, \dots, \phi_N) \sim \prod_{n>m} \left| \frac{e^{\imath\phi_n} - e^{\imath\phi_m}}{e^{\imath\phi_n} - \lambda e^{\imath\phi_m}} \right|^{\nu} . \qquad (5.1.65)$$

For the unitary case $\nu = 2$ Eq. (5.1.65) describes the eigenphase distribution for the Gaudin ensemble, for which a number of analytical results were obtained in the pioneer times of random matrix theory [Gau66]. In a recent paper Ma calculated the two-point correlation function and the number variance for the Gaudin ensemble [Ma95]. A comparison with numerical results, however, which would be highly desirable to check the validity of the Yukawa conjecture in the transition region, is still lacking.

5.2 Billiard level dynamics

5.2.1 Billiard and Pechukas–Yukawa level dynamics

In one respect the eigenvalue dynamics observed in billiard systems differs from the Pechukas–Yukawa level dynamics previously discussed. Whereas in the latter case the Hamilton operator changes as a function of a given parameter, in billiards usually the shape is varied [Tak91, Tak92]. There are, however, also cases where the spectra of billiards are studied as a function of an external magnetic field [Nak88]. Sieber *et al.* [Sie95] have introduced mixed

Dirichlet and Neumann boundary conditions, and have studied the spectra as a function of the mixing angle.

An experimental example where the eigenvalues of a quarter Sinai microwave billiard have been determined as a function of the length is shown in Fig. 5.2 [Kol94]. The observed overall decrease of the eigenvalues with increasing length results from the fact that the mean level spacing decreases with the area (see Section 2.2.1). From the dynamical point of view this has to be interpreted as a nonvanishing centre of mass velocity. The peculiar vertical structures seen in the figure can be associated with bouncing-ball wave functions, i.e. standing waves between the long sides of the billiard. Examples of such wave functions have already been presented in Chapter 2 (see Figs. 2.5, 2.9 and 2.13). The bouncing ball is responsible for a number of deviations from the universal behaviour in various spectral correlations. For a better understanding of these features periodic orbit theory is indispensable. A detailed discussion will be presented in Section 7.3.3.

In billiard level dynamics it is the shape which is parameter-dependent and not the Hamiltonian. It is by no means clear whether the Pechukas–Yukawa model can be applied at all to billiard systems. Therefore we first have to establish a correspondence between the two situations.

We may of course describe the billiard boundary by a potential being zero within and ∞ outside the billiard. Then again the Hamiltonian is parameter-dependent, but at the cost of introducing a discontinuous potential. This can be avoided by means of the conformal mapping technique. In two dimensions this is best achieved within the framework of complex function theory. It is well known that any simply connected region in the complex plane can be mapped

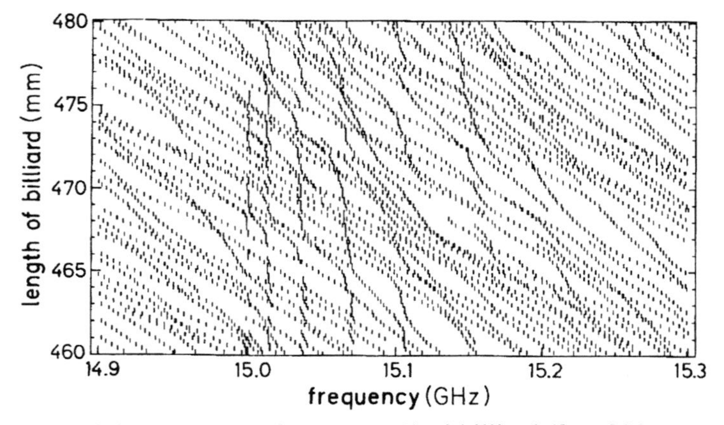

Figure 5.2. Part of the spectrum of a quarter Sinai billiard ($b = 200$ mm, $r = 70$ mm) as a function of length a [Kol94] (Copyright 1994 by the American Physical Society).

to other arbitrary simply connected regions, e.g. to the unit circle. As an example we take the stadium billiard. Let us assume that $w = f(z)$ is the function mapping the stadium to the unit circle, where $z = x + \imath y$, and $w = u + \imath v$. It is a simple exercise to show that the time-independent Schrödinger equation

$$\left(\frac{\partial^2}{\partial x^2} + \frac{\partial^2}{\partial y^2} + E \right) \psi = 0 \tag{5.2.1}$$

is transformed to

$$\left[|f'(z)|^2 \left(\frac{\partial^2}{\partial u^2} + \frac{\partial^2}{\partial v^2} \right) + E \right] \psi = 0 \tag{5.2.2}$$

by a substitution of variables from x, y to u, v. Instead of solving the original Schrödinger equation Eq. (5.2.1) for the stadium we may alternatively solve the transformed equation Eq. (5.2.2) for the unit circle. As the function $f(z)$ depends on the length-to-radius ratio, the parameter dependence is transferred by the mapping from the boundary condition to the Hamiltonian without the complication of introducing a discontinuous potential (though the conformal mapping technique may produce singularities as well, especially close to corners). It is usually difficult to find the correct mapping function. Only in rare cases, among them the polygonal billiards, can analytical expressions be obtained (see Chapter 16 of Ref. [Hen86]). For billiards constructed by a conformal mapping, such as the Robnik billiard (see Section 3.2.2), the correct mapping function is of course automatically known.

If only one billiard wall is shifted, whereas the residual part of the boundary remains unchanged, there is a more direct method to derive the corresponding dynamical equation system [Stö97]. We choose a coordinate system where the movable wall is parallel to the y axis, and is shifted towards the positive x direction (see Fig. 5.3). A and S are area and circumference of the billiard at the beginning, and x_m and ψ_m the corresponding eigenvalues and eigenfunctions. They obey the Schrödinger equation

$$-\Delta \psi_m = x_m \psi_m, \tag{5.2.3}$$

with the Dirichlet boundary condition $\psi_m|_S = 0$. After a shift of the wall by Δl, area and boundary take the new values A_1 and S_1. The new eigenvalues x_n^1 and the eigenfunctions ψ_n^1 obey the Schrödinger equation

$$-\Delta \psi_n^1 = x_n^1 \psi_n^1, \tag{5.2.4}$$

with the Dirichlet boundary condition $\psi_n^1|_{S_1} = 0$ on the new surface S_1. Multiplying both sides of Eq. (5.2.3) by ψ_n^1, both sides of Eq. (5.2.4) by ψ_m and taking the difference, we have

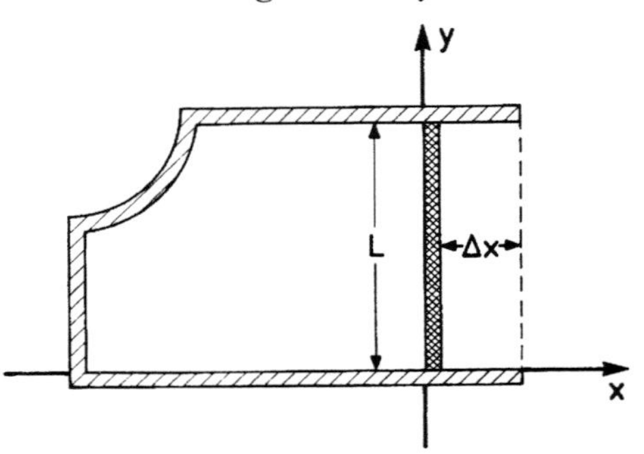

Figure 5.3. The coordinate system applied for the level dynamics calculation described in the text.

$$\begin{aligned}(x_n^1 - x_m)\psi_n^1\psi_m &= \psi_n^1\Delta\psi_m - \psi_m\Delta\psi_n^1 \\ &= \nabla(\psi_n^1\nabla\psi_m - \psi_m\nabla\psi_n^1).\end{aligned} \tag{5.2.5}$$

Integrating both sides over the area A of the original billiard and applying Green's theorem, we get

$$\begin{aligned}(x_n^1 - x_m)\int_A \psi_n^1\psi_m\, dA &= \int_S (\psi_n^1\nabla_\perp\psi_m - \psi_m\nabla_\perp\psi_n^1)\, dS \\ &= \int_0^L \psi_n^1(0, y)\frac{\partial\psi_m(0, y)}{\partial x}\, dy,\end{aligned} \tag{5.2.6}$$

where $\nabla_\perp$ denotes the normal derivative directed outwards. In the second step we have used the fact that ψ_m vanishes everywhere on S, whereas ψ_n^1 vanishes on S with the exception of the movable wall (see Fig 5.3).

In the next step we perform the limit $\Delta l \to 0$. As ψ_n^1 vanishes on S_1, we have $\psi_n^1(\Delta l, y) = 0$. Expanding $\psi_n^1(\Delta l, y)$ into a Taylor series at $x = 0$ we get

$$\begin{aligned}0 &= \psi_n^1(\Delta l, y) \\ &= \psi_n^1(0, y) + \Delta l\frac{\partial\psi_n^1(0, y)}{\partial x} + \mathcal{O}[(\Delta l)^2],\end{aligned} \tag{5.2.7}$$

whence follows

$$\psi_n^1(0, y) = -\Delta l\frac{\partial\psi_n(0, y)}{\partial x} + \mathcal{O}[(\Delta l)^2]. \tag{5.2.8}$$

In the second step we have used the fact that the difference between $\psi_n^1(0, y)$ and $\psi_n(0, y)$ is of $\mathcal{O}(\Delta l)$. This results in an additional error for $\psi_n^1(0, y)$, which again is only of $\mathcal{O}[(\Delta l)^2]$. Inserting expression (5.2.8) for $\psi_n^1(0, y)$ into the right hand side of Eq. (5.2.6) and performing the limit $\Delta l \to 0$ we end with

$$\dot{x}_n = V_{nn} \tag{5.2.9}$$

for $n = m$, and

$$(x_n - x_m)\int \dot{\psi}_n \psi_m = V_{nm} \tag{5.2.10}$$

for $n \neq m$. Here the dot means differentiation with respect to the billiard length, and the V_{nm} are given by

$$V_{nm} = -\int_0^L \frac{\partial \psi_n(0, y)}{\partial x}\frac{\partial \psi_m(0, y)}{\partial x}\, dy. \tag{5.2.11}$$

It follows from Eqs. (5.2.9) and (5.2.11) that all eigenvalue velocities are negative. For the averaged velocities this is an immediate consequence of the fact that the billiard area is increased by the shift of the wall, but for the individual eigenvalues this is by no means self-evident. For an experimental demonstration see Fig 5.2.

To obtain a dynamical equation for the V_{nm} as well, we differentiate Eq. (5.2.11) with respect to the length:

$$\dot{V}_{nm} = -\int_0^L \left(\frac{\partial \dot{\psi}_n(0, y)}{\partial x}\frac{\partial \psi_m(0, y)}{\partial x} + \frac{\partial \psi_n(0, y)}{\partial x}\frac{\partial \dot{\psi}_m(0, y)}{\partial x}\right) dy. \tag{5.2.12}$$

The $\dot{\psi}_n$ can be expressed in terms of the ψ_n using the completeness relation and Eq. (5.2.10):

$$\begin{aligned}\dot{\psi}_n &= \sum_l \left(\int \dot{\psi}_n \psi_l \, dA\right)\psi_l \\ &= \sum_{l \neq n} \frac{V_{nl}}{x_n - x_l}\psi_l.\end{aligned} \tag{5.2.13}$$

Inserting this expression into Eq. (5.2.12) we end with

$$\dot{V}_{nn} = 2\sum_{l \neq n}(V_{nl})^2 \frac{1}{x_n - x_l} \tag{5.2.14}$$

for $n = m$, and

$$\dot{V}_{nm} = \sum_{l \neq n,m} V_{nl}V_{lm}\left(\frac{1}{x_n - x_l} + \frac{1}{x_m - x_l}\right) - V_{nm}\frac{V_{nn} - V_{mm}}{x_n - x_m} \tag{5.2.15}$$

for $n \neq m$. Equations (5.2.9), (5.2.14), and (5.2.15) are identical with the Pechukas–Yukawa equations derived in Section 5.1.1. Billiard and Pechukas–Yukawa level dynamics are thus equivalent when the billiard length is taken as the parameter.

5.2.2 Tests of the Yukawa conjecture

We have seen that in chaotic systems the Yukawa conjecture (5.1.57) reduces to

$$\rho(x_n, p_n, f_{nm}) = \frac{1}{Z}\exp(-\beta W), \tag{5.2.16}$$

where

$$W = \frac{1}{2}\sum_n \left(p_n^2 + x_n^2\right) + \frac{1}{2}\sum_{n,m}{}' \frac{|f_{nm}|^2}{(x_n - x_m)^2} \tag{5.2.17}$$

is the 'total energy'. This is exactly the Boltzmann ansatz of classical mechanics, where β^{-1} corresponds to the 'temperature' of the eigenvalue gas. In view of the somewhat questionable assumptions applied in the derivation of the conjecture it is indispensable to test the consequences of the conjecture by means of experimental or numerical spectra.

We have already seen in Section 5.1.4 that the correlated energy distribution function resulting from Eq. (5.2.16) is identical to the energy distribution function of the Gaussian ensembles. For tests going beyond random matrix theory, we therefore have to study distributions of the dynamical variables p_n and f_{nm} as well.

We start with the velocity distribution function. If Eq. (5.2.16) is integrated over the variables x_n and f_{nm}, we obtain a Gaussian distribution for the centre of mass velocities $v_n = p_n - \langle p_n \rangle$,

$$P_{\text{vel}}(v) = \sqrt{\frac{\beta}{2\pi}}\exp\left(-\frac{\beta}{2}v^2\right), \tag{5.2.18}$$

exactly as in classical statistical mechanics. Figure 5.4(a) shows a velocity distribution for a Sinai billiard with mixed boundary conditions, where the mixing angle between Dirichlet and Neumann conditions is the level dynamics parameter [Sie95]. The observed distribution deviates significantly from the expected Gaussian behaviour. Similar discrepancies are found in Sinai microwave billiards [Kol94]. They are again caused by the bouncing-ball structures as shown in Fig 5.2. If the eigenvalues influenced by the bouncing ball are excluded from the distribution, the expected Gaussian behaviour is actually found (see Fig 5.4(b)).

The same velocity distribution has been analytically derived using supersymmetry techniques for the spectra of disordered systems [Sim93]. Here an external parameter such as the strength of magnetic field applied has been varied.

A frequently studied quantity is the curvature distribution, defined by

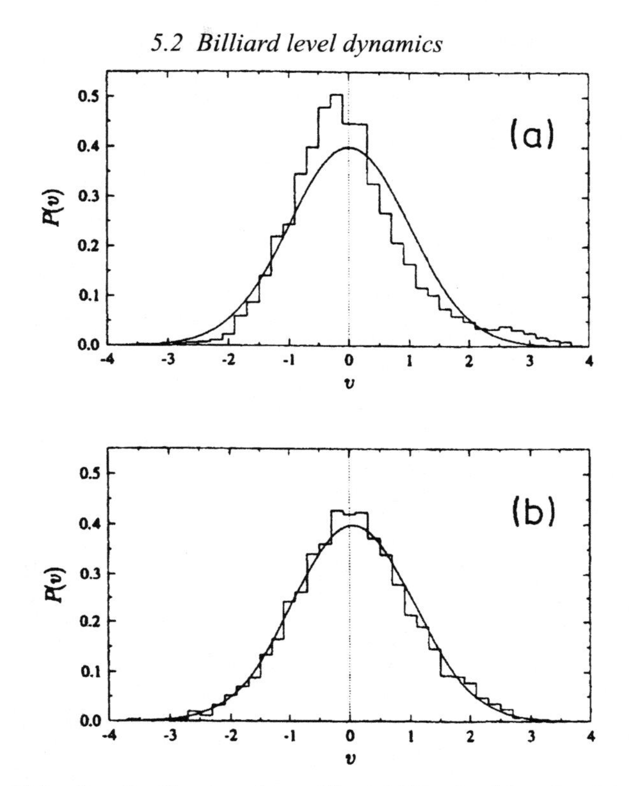

Figure 5.4. (a) Velocity distribution for a Sinai billiard with mixed boundary conditions, where the mixing angle between Dirichlet and Neumann condition is the level dynamics parameter. The solid line is a Gaussian. (b) Velocity distribution for the same Sinai billiard, but with the regions disturbed by the bouncing ball omitted from the histogram [Sie95].

$$K_n = \ddot{x}_n \frac{\langle x_n - x_{n-1} \rangle}{\langle \dot{x}_n^2 \rangle - \langle \dot{x}_n \rangle^2} . \tag{5.2.19}$$

The brackets denote a local average over the energy. Up to the normalization factor the curvature corresponds to the eigenvalue acceleration. The largest curvatures are observed close to avoided crossings. For the calculation of the asymptotic curvature distribution we may therefore restrict the dynamics to a 'two-particle' interaction at the moment of a close encounter. For an encounter of the eigenvalues x_{n-1} and x_n the Pechukas–Yukawa equations (5.1.19) to (5.1.21) reduce to

$$\ddot{x}_n = -\ddot{x}_{n-1} = 2 \frac{|f_{n-1,n}|^2}{(x_n - x_{n-1})^3} . \tag{5.2.20}$$

From this we obtain for the asymptotic curvature distribution

$$P_{\text{curv}}(K) \sim$$
$$\int \delta\left(K - \frac{2c|f_{n-1,n}|^2}{(x_n - x_{n-1})^3}\right) \exp\left(-\frac{\beta}{2}\frac{|f_{n-1,n}|^2}{(x_n - x_{n-1})^2}\right) dx_{n-1}\, dx_n\, df_{n,n-1}$$
$$\sim \int \delta\left(K - \frac{2cf^2}{x^3}\right) \exp\left(-\frac{\beta}{2}\frac{f^2}{x^2}\right) dx\, df, \quad (5.2.21)$$

where c is an abbreviation of the normalization constant entering Eq. (5.2.19). With the substitutions $x = \hat{x}/K$, $f = \hat{f}/K$ we get

$$P_{\text{curv}}(K) = \frac{1}{K^3} \int \delta\left(1 - \frac{2c\hat{f}^2}{\hat{x}^3}\right) \exp\left(-\frac{\beta}{2}\frac{\hat{f}^2}{\hat{x}^2}\right) d\hat{x}\, d\hat{f}$$
$$\sim \frac{1}{K^3}. \quad (5.2.22)$$

Equation (5.2.21) holds for real symmetric matrices. For Hermitian matrices we have to replace f^2 by ff^*, and to integrate independently over Re(f) and Im(f). For the general case we obtain

$$P_{\text{curv}}(K) \sim \frac{1}{K^{2+\nu}}, \quad (5.2.23)$$

where ν is the universality index. Figure 5.5 shows curvature distributions for the Hamiltonian $\mathscr{H} = \mathscr{H}_0 + Vt$ calculated by Gaspard *et al.* [Gas90]. Both $\mathscr{H}_0$ and V have been taken from one of the Gaussian ensembles. The asymptotic behaviour predicted from Eq. (5.2.23) is found for all three Gaussian ensembles. Zakrzewski and Delande [Zak93a] have studied curvature distributions of kicked tops of different universality classes and have found that they are described for all K values by the expression

$$P_{\text{curv}}(K) = \frac{N_\nu}{\left[1 + \left(\frac{K}{\gamma_\nu}\right)^2\right]^{(\nu+2)/2}}, \quad (5.2.24)$$

where N_ν is a normalization constant, and where

$$\gamma_\nu = \frac{\nu\pi}{\beta}. \quad (5.2.25)$$

The conjecture has been subsequently proved by v. Oppen using supersymmetry techniques [Opp94, Opp95].

Nongeneric properties, such as bouncing balls, again give rise to deviations from the universal behaviour. For the stadium billiard [Tak92] and the Sinai billiard [Sie95], excesses at small curvatures have been found, caused by the bouncing-ball structures as shown in Fig 5.2.

Another quantity related to avoided crossings is the distribution of closest

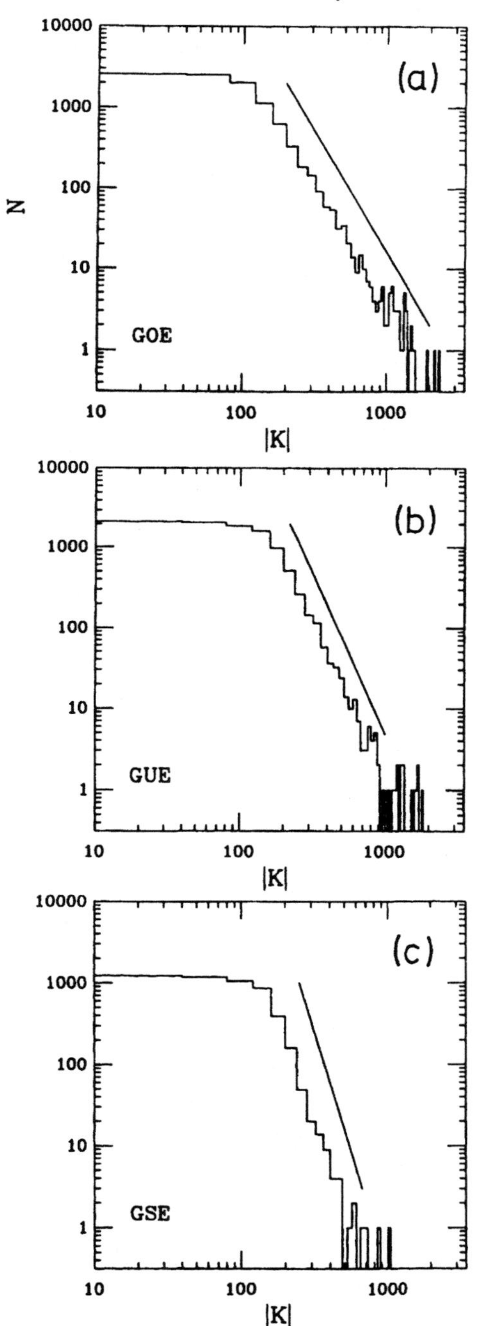

Figure 5.5. Curvature distributions in the spectra of random matrices $\mathcal{H} = \mathcal{H}_0 + Vt$ taken from the orthogonal (a), the unitary (b), and the symplectic (c) ensemble [Gas90]. The slopes of the straight lines are -3, -4, -6 for the GOE, the GUE, and the GSE, respectively, see Eq. (5.2.23) (Copyright 1990 by the American Physical Society).

approach distances $x_{\min}$ between neighbouring eigenvalues [Zak91, Zak93b]. For this quantity we may again restrict the discussion to two-body collisions, and represent the Hamiltonian $\mathscr{H} = \mathscr{H}_0 + Vt$ by a 2×2 matrix. In the present

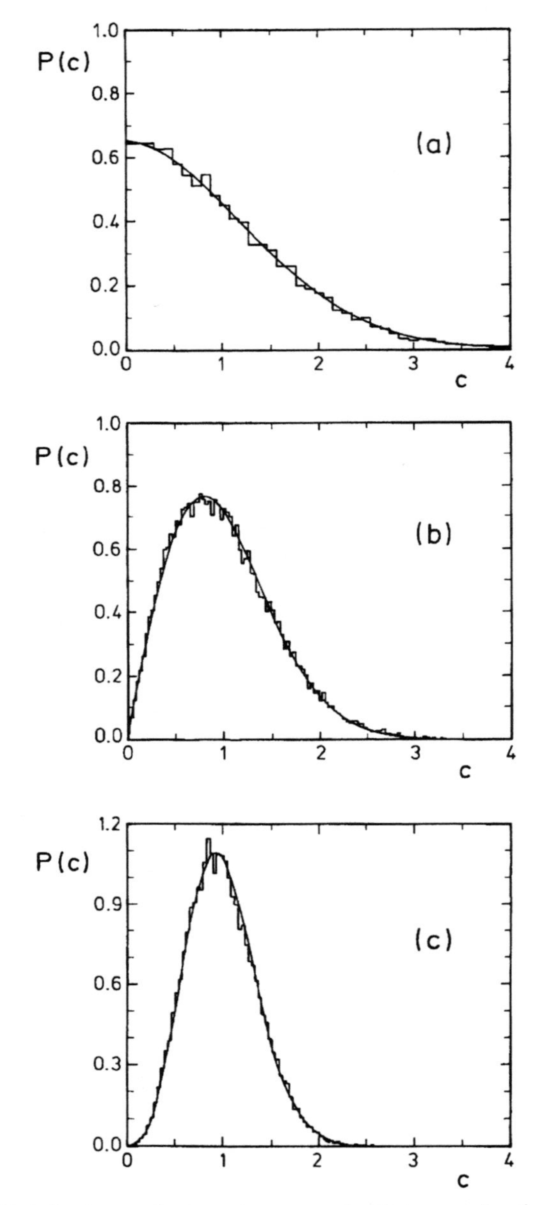

Figure 5.6. Distribution of closest approach distances in the spectra of periodically kicked tops of the orthogonal (a), the unitary (b), and the symplectic (c) universality class [Zak93b] (Copyright 1993 by the American Physical Society).

context it is suitable to adopt a basis system in which V is diagonal. For the GOE case $\mathcal{H}$ can then be written as

$$\mathcal{H} = \begin{pmatrix} a & c \\ c & d \end{pmatrix} + t \begin{pmatrix} v & \cdot \\ \cdot & w \end{pmatrix}, \tag{5.2.26}$$

whence follows for the distance between the two eigenvalues in the moment of closest approach

$$x_{\rm min} = 2|c|. \tag{5.2.27}$$

For random matrices c is Gaussian distributed,

$$P(c) \sim \exp(-Ac^2). \tag{5.2.28}$$

The distribution of closest approach distances is therefore Gaussian as well.

$$P_{\rm min}(x) \sim \exp\left(\frac{A}{4}x^2\right). \tag{5.2.29}$$

For other Gaussian ensembles this result can be generalized to

$$P_{\rm min}(x) = a_\nu x^{\nu-1} \exp\left(-b_\nu x^2\right), \tag{5.2.30}$$

where ν is again the universality index. The coefficients a_ν and b_ν can again be fixed by normalization conditions. Expression (5.2.30) for the distribution of closest approach distances is identical to the Wigner distribution for the level spacings, the only difference being that the repulsion exponent has been diminished by one. Figure 5.6 shows the distribution of closest approach distances for spectra of kicked tops for the three universality classes [Zak93b]. Perfect agreement with the predictions from Eq. (5.2.30) is found.

5.3 Geometrical phases

In the dynamical analogy the nondiagonal elements of chaotic Hamiltonians can be interpreted as repulsive forces, suggesting that in chaotic systems all degeneracies are lifted. But this is not strictly true. For illustrative purposes we consider the Hamiltonian represented by the symmetric 2×2 matrix

$$\mathcal{H} = \begin{pmatrix} x & y \\ y & -x \end{pmatrix}, \tag{5.3.1}$$

depending on the two parameters x and y. Its eigenvalues are given by

$$E_{1/2} = \pm\sqrt{x^2 + y^2}. \tag{5.3.2}$$

In the two-dimensional parameter space all eigenvalues can be found on the mantle of a double cone, also called a 'diabolo' after a toy of this shape. If one of the parameters is fixed, say y, the eigenvalues of $\mathcal{H}$ as a function of x lie on the two branches of a hyperbola. Here we again have the eigenvalue repulsion behaviour studied in the last sections. If, however, y is zero, the two eigenvalues meet at the contact point between the two cones, the so-called diabolic point, leading to an 'accidental' degeneracy.

From this discussion it becomes clear that in one-parameter systems the probability to find degeneracies is zero. For two-parameter families of Hamiltonians the situation is different. By adjusting two parameters, degeneracies can be enforced at diabolic points. Conceptually simple two-parameter systems are triangular billiards. Here the two coordinates of one corner with the base side fixed may be taken as the parameters. The first systematic studies on diabolic points in such systems were performed by Berry and Wilkinson [Ber84b], although there are some earlier works treating these questions (for references see the cited paper).

In two-parameter systems a qualitatively new type of level dynamics can be realized. For illustrative purposes we consider again the above example, but we now vary the parameters according to $x = r\cos\phi$, $y = r\sin\phi$, where r is fixed. Thus the diabolic point is encircled. After a change of ϕ by 2π we again obtain the original point in the parameter space. The eigenvalues $E_{1/2} = \pm r$ do not change during the rotation, but this does not apply to the corresponding eigenvectors. For a given ϕ the eigenvectors corresponding to E_1 and E_2 are given by

$$\psi_1(\phi) = \begin{pmatrix} \cos\dfrac{\phi}{2} \\ \sin\dfrac{\phi}{2} \end{pmatrix}, \qquad \psi_2(\phi) = \begin{pmatrix} -\sin\dfrac{\phi}{2} \\ \cos\dfrac{\phi}{2} \end{pmatrix}. \tag{5.3.3}$$

By means of the spinor rotation matrices introduced in Section 4.2.3 (see Eq. (4.2.57)) this may be written as

$$\psi_1(\phi) = \exp\left(-i\frac{\phi}{2}\sigma_y\right)\begin{pmatrix} 1 \\ 0 \end{pmatrix}, \qquad \psi_2(\phi) = \exp\left(-i\frac{\phi}{2}\sigma_y\right)\begin{pmatrix} 0 \\ 1 \end{pmatrix}. \tag{5.3.4}$$

After a rotation by 2π both ψ_1 and ψ_2 have changed their signs,

$$\psi_1(2\pi) = -\psi_1(0), \qquad \psi_2(2\pi) = -\psi_2(0). \tag{5.3.5}$$

This is a manifestation of the well-known fact that the period of spinor rotations is 4π. This change of sign of spinors after a 2π rotation is not only of academic interest. Readers familiar with nuclear magnetic resonance know that this fact is used to change the sign of the nuclear orientation by the so-called adiabatic fast passage technique (see Section 2.4 of Ref. [Sli80]).

This example shows that the phase of a wave function during a parameter change must not be ignored. Though the absolute phase of a wave function is irrelevant, this is not true for *changes* of the phase, if it is possible to compare the phases prior to and after the parameter change, e.g. by means of an interference experiment. Such changes of the phase of wave functions due to parameter variations are called *Berry's phases* after M. Berry [Ber84a]. We shall follow essentially his presentation.

We start with the time-dependent Schrödinger equation

$$\imath\hbar\frac{\partial\psi}{\partial t} = \mathcal{H}(\lambda)\psi, \tag{5.3.6}$$

where λ is a parameter assumed to vary slowly with time. Let $E_n(\lambda)$ be an eigenvalue of $\mathcal{H}(\lambda)$ to the eigenfunction $|n\rangle$,

$$\mathcal{H}(\lambda)|n\rangle = E_n(\lambda)|n\rangle. \tag{5.3.7}$$

We assume that at the beginning the system is in the state $|k\rangle$. For the time evolution of the wave function we make the ansatz

$$\psi_k(t) = \sum_n a_{kn}(t)\exp\left(-\frac{\imath}{\hbar}\int_0^t E_n(\lambda)\,dt\right)|n\rangle. \tag{5.3.8}$$

Inserting this expression into the Schrödinger equation, we get a differential equation system for the a_n,

$$\dot{a}_{kn} = -\sum_m \exp\left(\frac{\imath}{\hbar}\int_0^t [E_n(\lambda) - E_m(\lambda)]\,dt\right)\left\langle n\middle|\dot{\lambda}\frac{\partial}{\partial\lambda}\middle|m\right\rangle a_{km} \tag{5.3.9}$$

which has to be solved with the initial condition $a_{kn}(0) = \delta_{kn}$. In the adiabatic approximation we assume that λ varies slowly as compared to all other time dependences in the system. We then have $(1/\hbar)\int_0^t(E_n(\lambda) - E_m(\lambda))\,dt \gg 1$ for $n \neq m$. Neglecting the rapidly oscillating terms on the right hand side of Eq. (5.3.9) and restricting the sum to the term with $m = n$, we get

$$\dot{a}_{kn} = -\left\langle n\middle|\dot{\lambda}\frac{\partial}{\partial\lambda}\middle|n\right\rangle a_{kn}. \tag{5.3.10}$$

The integration is trivial and yields

$$a_{kn}(t) = \exp\left(-\int_{\lambda(0)}^{\lambda(t)}\left\langle n\middle|\frac{\partial}{\partial\lambda}\middle|n\right\rangle d\lambda\right)\delta_{kn}. \tag{5.3.11}$$

From the orthogonality relation $\langle n|n\rangle = 1$ we obtain

$$\left\langle n\middle|\frac{\partial n}{\partial\lambda}\right\rangle = -\left\langle\frac{\partial n}{\partial\lambda}\middle|n\right\rangle = -\left\langle n\middle|\frac{\partial n}{\partial\lambda}\right\rangle^*. \tag{5.3.12}$$

Thus the argument of the exponential function in Eq. (5.3.11) is purely imaginary, and we may write

$$\gamma_n = \imath\int_{\lambda(0)}^{\lambda(t)}\left\langle n\middle|\frac{\partial}{\partial\lambda}\middle|n\right\rangle d\lambda, \tag{5.3.13}$$

where γ_n is real. We thus obtain

$$\psi_k(t) = \exp\left(-\frac{\imath}{\hbar}\int_0^t E_k(\lambda)\,dt\right)\exp(\imath\gamma_k)|k\rangle. \tag{5.3.14}$$

This is the main result of this section. There are two different contributions to the phase. The first exponential on the right hand side of Eq. (5.3.14) is the dynamical phase factor. Only the familiar term $E_k t$ has been replaced by an integral of E_k over t to account for the adiabatic change of E_k. The second factor is new. It has been called *geometrical phase* by Berry but is now generally known as Berry's phase. It is termed 'geometrical' as its value depends on the geometry of the path taken in the parameter space.

For a closed circle the integral for γ_n in Eq. (5.3.13) can be transformed into an integral over the enclosed area by means of Stoke's theorem. This integral no longer depends on the parameter variations of the wave functions but depends exclusively on the matrix elements of $\partial\mathscr{H}/\partial\lambda$ which are much easier to calculate. Details can be found in Berry's original work [Ber84a].

The spinor rotation matrices in Eq. (5.3.4) describing the change of the wavefunctions upon circling around a diabolic point, may be considered as generalizations of geometrical phase factors to spinor wavefunctions. Thus even the change of sign observed for spinor wavefunctions after a complete orbit in the parameter space may be interpreted as a manifestation of Berry's phase. Another example is the *Aharonov–Bohm effect*: upon encircling a magnetic flux line a wave function acquires a phase $\oint \boldsymbol{A}\, \boldsymbol{dr}$ from a vector potential [Aha59]. Geometrical phases are reviewed in the monograph by Shapere and Wilczek [Sha89] which also contains reprints of a number of important contributions to the subject.

Berry's phase has been experimentally realized by Lauber *et al.* [Lau94] in a triangular microwave billiard. The authors measured the wave functions for a pair of eigenvalues while varying the position of one corner on a circle. The central part of Fig 5.7 shows the geometry of the studied triangle, and the 12 different positions taken by the corner C. In the upper and the lower panels the wave functions for the two involved eigenvalues are shown. Neighbouring regions separated by node lines are shaded differently to mark the different signs of the wave functions, which have not been determined explicitly in the experiment. We follow the figures clockwise starting at position 1. The appearance of both wave functions changes slowly. It is easy to follow the shaded regions from figure to figure. After a half circle at position 7 the wave functions have changed their identities: the wave function of state 13 at position 7 is identical with the wave function of state 14 at position 1 and vice versa, in the latter case up to a change of the sign. This demonstrates that the two

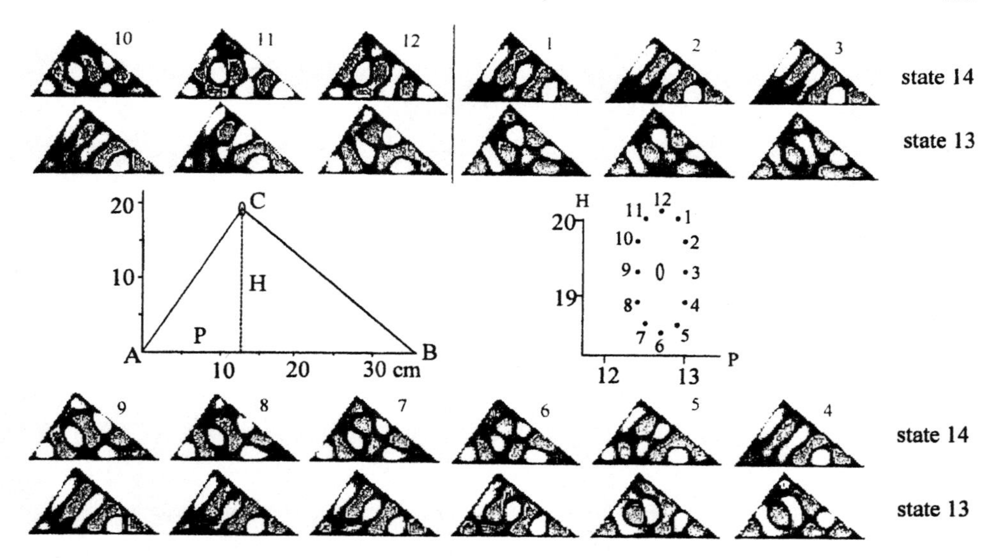

Figure 5.7 Eigenfunctions for a pair of neighbouring resonances in a triangular microwave billiard. The position of corner C has been varied on a circle around a diabolic point. After a complete cycle the signs of the wave functions have changed [Lau94] (Copyright 1994 by the American Physical Society).

adjacent states form a pair. Only their linear combination changes during the displacement of the corner. After a complete circle the original wave functions are restored again, but in both cases with a change of sign as is evident by comparing the patterns at positions 12 and 1.

6
Scattering systems

It is impossible to study a system without disturbing it by the measuring process. To determine the spectrum of a microwave billiard, for example, we have to drill a hole into its wall, introduce a wire, and radiate a microwave field. We learnt in Section 3.2.2 that the spectrum obtained for a rectangular microwave cavity is no longer integrable, but has become pseudointegrable by the presence of the antenna. The measurement thus unavoidably yields an unwanted combination of the system's own properties and those of the measuring apparatus. The mathematical tool to treat the coupling between the system and the environment is provided by scattering theory, which was originally developed in nuclear physics [Mah69, Lew91]. This theory has been successfully applied to mesoscopic systems and microwave billiards as well.

In this chapter scattering theory will be introduced with special emphasis on billiard systems. We shall also discuss the amplitude distributions of wave functions in chaotic systems, since they enter into the calculation of the distribution of scattering matrix elements. The perturbing bead method, used to get information on field distributions in microwave resonators, can also be described by scattering theory. In the last section we shall touch on the discussion of mesoscopic systems, which can be linked to scattering theory via the *Landauer formula* [Lan57] expressing conduction through mesoscopic devices in terms of transmission probabilities. The latter aspects are treated in the proceedings of the Les Houches summer school 1994 on mesoscopic physics [Akk94] addressing in particular questions of quantum chaos and random matrix theory.

6.1 Billiards as scattering systems

6.1.1 Scattering matrix

Scattering theory is treated in most textbooks on quantum mechanics. Therefore only the fundamental concepts indispensable for the understanding of chaotic scattering will be introduced.

A scattering process can be described quantum mechanically by the Schrödinger equation

$$\left(-\frac{\hbar^2}{2m}\Delta + V(r)\right)\psi = E\psi, \tag{6.1.1}$$

where $V(r)$ specifies the potential of the scattering system. In contrast to the situations discussed hitherto we are now interested in the scattering solutions of the equation. Quantum mechanically a beam of incoming particles can be described by a plane wave, whereas the scattered particles correspond to outgoing circular waves. We have therefore to look for solutions of the Schrödinger equation which can be written as a superposition of an incoming plane wave, and a number of outgoing circular waves, far from the nucleus.

Equation (6.1.1) can be rewritten as

$$(\Delta + k^2)\psi = v(r)\psi, \tag{6.1.2}$$

where $k = \sqrt{2mE}/\hbar$, and $v(r) = (2m/\hbar^2)V(r)$. We now interpret this equation as an inhomogeneous differential equation, where the left hand side is the Schrödinger equation of the free particle, and the potential on the right hand side is treated as an inhomogeneous term. The inhomogeneous Schrödinger equation can be solved by means of the Green function $G(r, r', k)$ which is obtained as the solution of the inhomogeneous Helmholtz equation

$$(\Delta + k^2)G(r, r', k) = \delta(r - r'). \tag{6.1.3}$$

In two dimensions $G(r, r', k)$ is given by [Mor53]

$$G(r, r', k) = -\frac{\imath}{4}H_0^{(1)}(k|r - r'|), \tag{6.1.4}$$

where $H_0^{(1)}(x)$ is a Hankel function. It has the asymptotic behaviour

$$H_0^{(1)}(x) \simeq \sqrt{\frac{2}{\pi x}}\exp\left[\imath\left(x - \frac{\pi}{4}\right)\right]. \tag{6.1.5}$$

This shows that $G(r, r', k)$ corresponds asymptotically to an outgoing circular wave. The complementary solution, obtained by replacing $H_0^{(1)}(x)$ by $H_0^{(2)}(x)$ and corresponding to an incoming circular wave, is not of interest here.

In many textbooks, especially those on electrodynamics (see Chapter 7 of Ref. [Mor53]), the inhomogeneous term on the right hand side of Eq. (6.1.3) contains an additional multiplicative factor of -4π. This definition would be in conflict with our former definition of the Green function in Section 3.1.5 and will therefore not be adopted.

By means of $G(r, r', k)$ the Schrödinger equation (6.1.2) can be transformed to

$$\psi(r) = \psi_0(r) + \int G(r, r')v(r')\psi(r')\,dr', \tag{6.1.6}$$

where $\psi_0(r)$ is an arbitrary solution of the homogeneous Schrödinger equation. We have thus obtained the *Lippmann–Schwinger* equation. It is not really a solution of the Schrödinger equation, as the unknown wave function $\psi(r)$ still enters on the right hand side of the equation. We have only converted the differential equation (6.1.2) into the equivalent integral equation (6.1.6).

Nevertheless we have already achieved a good deal. The second term on the right hand side of Eq. (6.1.6) corresponds asymptotically to an outgoing circular wave. This follows immediately from the asymptotic property (6.1.5) of the Green function, provided, of course, that the potential approaches zero rapidly enough. This is the case for nuclei and hard sphere potentials, but not for the Coulomb potential which needs extra treatment. The Lippmann–Schwinger equation therefore has the pleasant property that it automatically takes into account the asymptotic behaviour desired.

We now expand the Green function as well as the incoming and outgoing waves into a suitable set of orthogonal functions. In two dimensions we may use the ordinary Bessel functions. For the Green function the expansion reads

$$G(r, \theta, r', \theta') = -\frac{\imath}{4}\sum_{l=-\infty}^{\infty} e^{\imath l(\theta-\theta')}J_l(kr')H_l^{(1)}(kr), \tag{6.1.7}$$

where $r > r'$. For the incoming wave we assume, for simplicity's sake, that it contains only one component with a well-defined angular momentum,

$$\psi_0(r, \theta) = e^{\imath l\theta}H_l^{(2)}(kr). \tag{6.1.8}$$

Entering the ansatz (6.1.8) and the expansion (6.1.7) into the Lippmann–Schwinger equation, we obtain for $\psi(r, \theta)$ (now denoted $\psi_l(r, \theta)$ to indicate that it is the scattering solution for an incoming wave with angular momentum l):

$$\psi_l(r, \theta) = e^{\imath l\theta}H_l^{(2)}(kr) + \sum_{l'} S_{ll'}e^{\imath l'\theta}H_{l'}^{(1)}(kr), \tag{6.1.9}$$

where

$$S_{ll'} = -\frac{\imath}{4}\int e^{-\imath l'\theta'}J_{l'}(kr')v(r')\psi_l(r', \theta')\,d\theta' dr'. \tag{6.1.10}$$

The $S_{ll'}$ are the components of the scattering matrix. They give the quantum mechanical probability amplitude that a wave entering with an angular momentum l is scattered into an outgoing wave with an angular momentum l'. By means of the scattering matrix Eq. (6.1.9) can be compactly written as

$$\psi = \psi_{\text{in}} + S\psi_{\text{out}}. \tag{6.1.11}$$

The scattering matrix is unitary, $SS^{\dagger} = 1$, although this is not immediately evident from Eq. (6.1.10). The unitarity is a consequence of the time-reversal invariance of the Hamiltonian (6.1.1) and can be proved as follows. From ψ we can construct another solution $T\psi$ obtained from ψ by changing the sign of time. $T\psi$ obeys the equation

$$T\psi = \psi_{\text{out}} + S^{\dagger}\psi_{\text{in}}. \tag{6.1.12}$$

This holds since the time-reversal process (i) interchanges the roles of incoming and outgoing waves, and (ii) replaces all matrix elements by their complex conjugates. But if $T\psi$ is a solution of the Schrödinger equation then

$$\overline{\psi} = ST\psi = S\psi_{\text{out}} + SS^{\dagger}\psi_{\text{in}} \tag{6.1.13}$$

must be a solution as well. Comparing Eqs. (6.1.11) and (6.1.13) we see that the outgoing parts of ψ and $\overline{\psi}$ are identical. But then the two solutions must be identical as a whole, whence follows $SS^{\dagger} = 1$.

Replacing ψ_l on the right hand side of Eq. (6.1.10) by expression (6.1.9) we obtain an equation system for the components of the scattering matrix. Unfortunately the resulting integrals are divergent in most cases because of the singularity of the Hankel function at the origin. Therefore this direct approach is usually not accessible. The problem does not exist for scattering at obstacles with infinitely high walls. As an example we consider a chaotic billiard, but now seen from the outside. In this case we can determine the scattering matrix directly from Eq. (6.1.9), combined with the boundary condition that the wave function must be zero at the surface of the scatterer. The situation is particularly simple for a circular scatterer with the radius R. Here we get from Eq. (6.1.9)

$$e^{\imath l\theta} H_l^{(2)}(kR) + \sum_{l'} S_{ll'} e^{\imath l'\theta} H_{l'}^{(1)}(kR) = 0. \tag{6.1.14}$$

This equation must hold for all values of θ. Therefore all elements of the scattering matrix must vanish with the exception only of the component S_{ll}, for which we get

$$S_{ll} = -\frac{H_l^{(2)}(kR)}{H_l^{(1)}(kR)} = -e^{-2\imath\delta_l}. \tag{6.1.15}$$

The phase angle δ_l is obtained from

$$\tan \delta_l = \frac{N_l(kR)}{J_l(kR)}, \tag{6.1.16}$$

where we have used the relation $H_l^{(1,2)}(x) = J_l(x) \pm \imath N_l(x)$. The expansion in terms of incoming and outgoing circular waves is the only appropriate choice for circular scatterers, but for chaotic scatterers any set of orthogonal functions will do as well. This motivated Blümel and Smilansky to try whether the Bohigas–Giannoni–Schmit conjecture, which proved to be successful in the description of the spectra of chaotic systems (see Section 3.2.1), can be generalized to the eigenphases of scattering matrices [Blü88, Blü90].

The conjecture has been tested in a number of different scattering arrangements [Blü89, Blü92b], and has been verified in all cases. As an example Fig. 6.1 shows the nearest neighbour distance distribution of the eigenphases of the scattering matrix for the 'frying pan', a circle with an attached channel through which the waves can enter and exit [Blü92b]. The variable d is the distance between the centre of the circle and the chord common to the circle and the channel, measured in units of the radius of the circle. The system is regular for

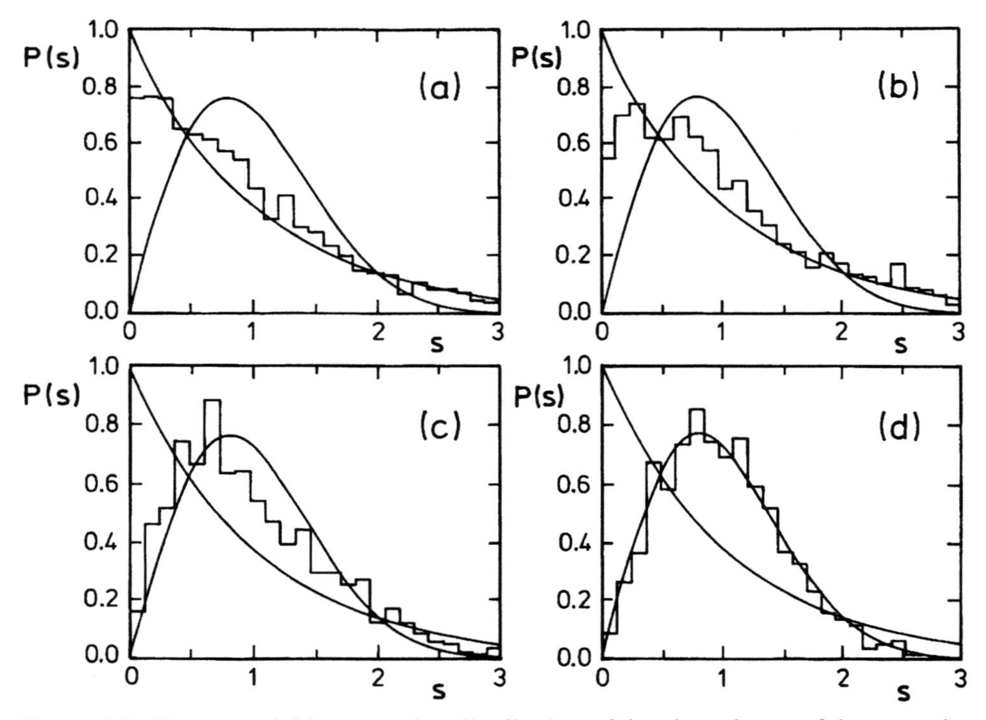

Figure 6.1. Nearest neighbour spacing distribution of the eigenphases of the scattering matrix for the 'frying pan', a circle with an attached channel for $d = 0$ (a), $d = 0.035$ (b), $d = 0.1$ (c), $d = 0.3$ (d). For the definition of d see text [Blü92b].

$d = 0$ and chaotic for all values of d between 0 and 1. This is reflected by the nearest neighbour spacing distribution of the eigenphases displayed in Fig. 6.1. By a variation of d from 0 to 0.3 a gradual change from Poisson to Wigner behaviour is observed.

It is remarkable that the scattering matrix describing the scattering at the outer boundary of a billiard can be used to calculate the eigenvalues of the billiard itself [Dor92]. This can be seen in Eq. (6.1.9). Because of the singularity of the Hankel function at the origin, this equation generally yields a solution of the Schrödinger equation only valid *outside* the billiard. However, for the exceptional situation that all divergent terms cancel, the outside solution is also a solution *inside* the billiard. This leads to the condition

$$-e^{\imath l\theta} N_l(kr) + \sum_{l'} S_{ll'} e^{\imath l'\theta} N_{l'}(kr) = 0. \tag{6.1.17}$$

This relation can only be obeyed if the determinant relation

$$|1 - S| = 0 \tag{6.1.18}$$

holds. Therefore all k values for which one of the eigenphases of S is 0 belong to the spectrum of the billiard. For the circular billiard this can be verified immediately. From Eq. (6.1.15) we learn that relation (6.1.18) holds for $J_l(kR) = 0$. But this is exactly the quantization condition for the circle. This *inside-outside duality* has been used by Smilansky and coworkers to calculate the spectra of a number of chaotic billiards [Dor92, Die93, Sch95a].

6.1.2 Billiard Breit–Wigner formula

A microwave billiard is a special example of a scattering system. Here the measuring antennas take the role of the scattering channels. We are now going to discuss the consequences of the external coupling for resonance positions and widths. In the presentation we shall follow essentially Ref. [Ste95]. The technique is a more or less direct adoption from nuclear physics where analogous questions were studied in the fifties and sixties (see Ref. [Bla52]).

Let us take a two-dimensional billiard of arbitrary shape. To be concise, we restrict the discussion to a coupling via waveguides, but the technique can be applied as well to the coupling via antennas with only minor modifications [Ste95]. Let $\psi(r, k)$ be the amplitude of the field within in the billiard while microwaves enter and exit through the different coupling ports. $\psi(r, k)$ obeys the Helmholtz equation

$$(\Delta + k^2)\psi = 0. \tag{6.1.19}$$

Assuming that the widths of the waveguides are small compared to the wavelengths, the field within the ith waveguide is given by

$$\psi(r, k) = a_i e^{\imath k_i |r - r_i|} - b_i e^{-\imath k_i |r - r_i|}, \tag{6.1.20}$$

where r_i is the point of entrance of the ith waveguide. a_i and b_i are the amplitudes of the waves entering and leaving the billiard through the waveguide. For circular antennas Eq. (6.1.20) has to be replaced by a superposition of incoming and outgoing circular waves [Ste95]. From Eq. (6.1.20) we obtain for the field and its normal derivative at the entrance position

$$\psi(r_i, k) = a_i - b_i, \tag{6.1.21}$$

$$\nabla_\perp \psi(r_i, k) = -\imath k(a_i + b_i). \tag{6.1.22}$$

Here $\nabla_\perp$ is the normal derivative pointing *into* the waveguide.

Since, by assumption, the width of the waveguide is small compared to the wavelength, each waveguide corresponds to exactly one scattering channel, and the scattering matrix S reduces to an $N \times N$ matrix, where N is the number of waveguides. We denote the amplitude vectors of the waves entering or leaving the billiard by $a = (a_1, a_2, \ldots, a_N)$ and $b = (b_1, b_2, \ldots, b_N)$, respectively. The scattering matrix for the billiard system is now defined by the relation

$$b = Sa. \tag{6.1.23}$$

The diagonal elements S_{ii} of S correspond to the reflection amplitudes for the ith channel, whereas the nondiagonal elements S_{ij} constitute the transmission amplitudes between channels i and j, respectively.

To get a relation between the amplitudes of the incoming and outgoing waves, we again apply the Green function technique. The billiard Green function $G(r, r', k)$ has to be modified for this purpose, since it vanishes anywhere on the billiard boundary for both r and r'. Coupling to the waveguides is therefore impossible. The problem can be overcome with a modified Green function $\overline{G}(r, r', k)$, obeying Dirichlet boundary conditions except for the waveguide entrances, where Neumann boundary conditions are applied. Due to this modification a flow through the billiard walls becomes possible. The modified Green function can be expanded as

$$\overline{G}(r, r', k) = \sum_n \frac{\overline{\psi}_n(r)\overline{\psi}_n(r')}{k^2 - \overline{k}_n^2}, \tag{6.1.24}$$

(see Section 3.1.5), where $\overline{k}_n$ and $\overline{\psi}_n(r)$ are eigenvalues and eigenfunctions, respectively, of the billiard with the modified boundary conditions. We now multiply the Helmholtz equation (6.1.19) for ψ by $\overline{G}(r, r')$, and the inhomogeneous Helmholtz equation

$$(\Delta + k^2)\overline{G}(r, r', k) = \delta(r - r') \tag{6.1.25}$$

for $\overline{G}(r, r', k)$ by $\psi(r)$, and take the difference. We then obtain

$$\delta(r - r')\psi(r) = \psi(r)\Delta\overline{G}(r, r', k) - \overline{G}(r, r', k)\Delta\psi(r). \tag{6.1.26}$$

Integrating both sides over r and applying Green's theorem we end with

$$\psi(r') = \sum_i \int_{S_i} \left[\psi(r)\nabla_\perp\overline{G}(r, r', k) - \overline{G}(r, r', k)\nabla_\perp\psi(r)\right] ds$$

$$= -L\sum_i \overline{G}(r_i, r', k)\nabla_\perp\psi(r_i), \tag{6.1.27}$$

where the sum is over all waveguides, and L is their width which has been assumed to be equal for all guides. In the second step $\nabla_\perp\overline{G}(r, r', k)$ vanishes at the entrance positions of the waveguides due to the modified boundary conditions. The integral over the remaining part of the boundary vanishes, since both $\psi(r)$ and $\overline{G}(r, r', k)$ are zero here. In normally conducting microwave billiards the field is not exactly zero because of partial penetration into the walls due to the finite conductivity. In superconducting cavities the influence of the wall is negligible.

Now we insert for r' the attachment position r_j of the jth waveguide. Applying Eqs. (6.1.21) and (6.1.22) we get from Eq. (6.1.27)

$$a_j - b_j = \imath\gamma \sum_i \overline{G}(r_i, r_j, k)(a_i + b_i), \tag{6.1.28}$$

with $\gamma = kL$. In compact matrix notation this may be written as

$$a - b = \imath\gamma\overline{G}(a + b), \tag{6.1.29}$$

where $\overline{G}$ is the matrix with the elements $\overline{G}_{ij} = \overline{G}_{ji} = \overline{G}(r_i, r_j, k)$. Comparing Eq. (6.1.29) with definition (6.1.23) of the scattering matrix, we find that $\overline{G}$ and S are related via

$$S = \frac{1 - \imath\gamma\overline{G}}{1 + \imath\gamma\overline{G}}. \tag{6.1.30}$$

The unitarity of S is immediately evident from this expression. For the case of nonoverlapping resonances the expression can be further simplified. In this case, at most one term n contributes to the sum (6.1.24) for a given k. In this approximation the matrix elements of the denominator on the right hand side of Eq. (6.1.30) are given by

$$M_{ij} = \delta_{ij} + cx_ix_j, \tag{6.1.31}$$

where $x_i = \overline{\psi}_n(r_i)$ and $c = \imath\gamma(k^2 - \overline{k}_n^2)^{-1}$. It can be easily verified that the matrix elements of M^{-1} are given by

$$(M^{-1})_{ij} = \delta_{ij} - \frac{c}{1 + c\sum_k x_k^2} x_ix_j. \tag{6.1.32}$$

Entering this result into Eq. (6.1.30) we get for the matrix elements of the scattering matrix in the neighbourhood of the nth resonance

$$S_{ij} = \delta_{ij} - 2\imath\gamma \frac{\overline{\psi}_n(r_i)\overline{\psi}_n(r_j)}{k^2 - \overline{k}_n^2 + \frac{\imath}{2}\Gamma_n}, \tag{6.1.33}$$

where

$$\Gamma_n = 2\gamma \sum_k |\overline{\psi}_n(r_k)|^2. \tag{6.1.34}$$

The factor 2 has been introduced in accordance with the usual conventions. As by assumption for a given k only one resonance at a time contributes to the Green function, we may take the sum over n on the right hand side of Eq. (6.1.33) and obtain

$$S_{ij} = \delta_{ij} - 2\imath\gamma \hat{G}(r_i, r_j, k), \tag{6.1.35}$$

with

$$\hat{G}(r_i, r_j, k) = \sum_n \frac{\overline{\psi}_n(r_i)\overline{\psi}_n(r_j)}{k^2 - \overline{k}_n^2 + \frac{\imath}{2}\Gamma_n}, \tag{6.1.36}$$

holding for all k values in the nonoverlapping resonance approximation. For the case of coupling via circular antennas the same result is obtained with the only difference that γ may now take complex values [Ste95].

The last two equations are the billiard equivalent of the Breit–Wigner formula well-known from nuclear physics [Bla52]. Equation (6.1.35) shows that the complete modified Green function $\hat{G}(r_i, r_j, k)$ may be obtained if the transmission between two antennas of variable position is measured [Ste95]. A reflection measurement at one antenna only yields the spectrum and the modulus of the wave function [Stö90, Ste92]. We never obtain the 'true' wave function $G(r_i, r_j, k)$: the resonances are shifted from the true k_n^2 to $\overline{k}_n^2$ and are broadened by Γ_n. If γ is complex, somewhat modified expressions for shift and broadening are obtained, see Ref. [Ste95]. Moreover, the wave functions are distorted in the neighbourhood of the coupling points because of the modified boundary conditions which had to be assumed. But for the case of circular antennas the difference between $\psi(r)$ and $\overline{\psi}(r)$ is only of $\mathcal{O}\left[(kR)^2\right]$ as has been shown in Ref. [Ste95] by a slightly different approach.

Due to the negligible loss in the walls superconducting cavities are especially suited to checking these predictions. Figure 6.2 shows part of a spectrum in a superconducting stadium-shaped cavity measured by Alt *et al.* [Alt93] together with a fit. In the lower part of the figure the difference between fit and experimental data is shown, demonstrating convincingly the

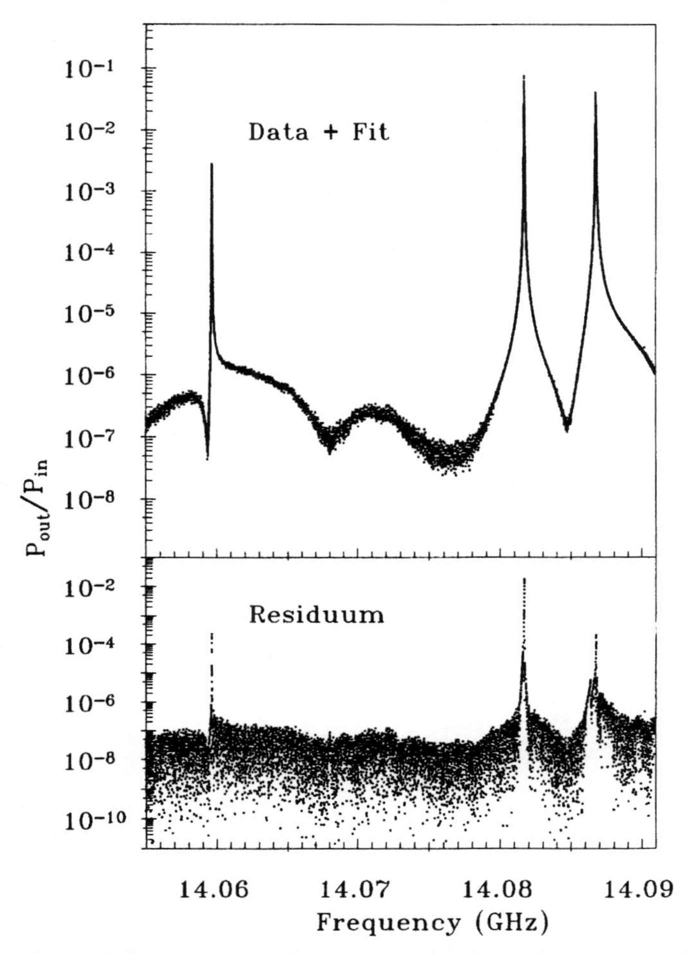

Figure 6.2. Part of the spectrum of a superconducting microwave cavity in the shape of a quarter stadium. The solid line was obtained from a fit using a modified Breit–Wigner function which also considers the case of overlapping resonances [Alt93] (Copyright 1993 by the American Physical Society).

applicability of scattering theory to microwave billiards. For the two resonances at the right the assumption of isolated resonances no longer holds. Here the authors had to go beyond expression (6.1.36) to obtain agreement with the data. The regime of overlapping resonances will be treated in detail in the next section.

6.1.3 Coupled-channel Hamiltonian

As we have seen, the spectra of microwave billiards are modified by the presence of the coupling antennas. For the case of nonoverlapping resonances explicit expressions for shifting and broadening of the resonances have been obtained. Rather frequently, however, the resonance widths are large compared to the mean level spacing. In such cases the approximations applied in Section 6.1.2 are no longer justified, and the single resonance Breit–Wigner formula (6.1.35) becomes invalid. Extending the discussion to overlapping resonances we start with an expression for the scattering matrix,

$$S = \frac{1 - \imath W^\dagger G W}{1 + \imath W^\dagger G W}, \tag{6.1.37}$$

which is somewhat more general than Eq. (6.1.30) derived for microwave billiards. The matrix elements of G are given by

$$G_{nm} = \left(\frac{1}{E - \mathscr{H}_0} \right)_{nm} = \frac{\delta_{nm}}{E - E_n^0} \tag{6.1.38}$$

(see Section 3.1.5), where $\mathscr{H}_0$ and E_n^0 are the Hamiltonian and eigenvalues of the undisturbed system. The small energy shifts due to the modified boundary conditions (see Section 6.1.2) are neglected here. We assume again that $\mathscr{H}_0$ is truncated to an $N \times N$ matrix. The matrix elements w_{nk} of W contain information on the coupling strengths of the kth channel to the nth resonance. w_{nk} may be written as an overlap integral

$$w_{nk} = \int \psi_n(r) u_k^*(r)\, dr, \tag{6.1.39}$$

where $u_k(r)$ is the wave function in the kth channel, and $\psi_n(r)$ is the eigenfunction of the billiard with modified boundary conditions. With a total of K channels W is an $N \times K$ matrix. The integral is over the contact area between the channel and the billiard. For a point-like coupling w_{nk} is proportional to $\psi_n(r_k)$ at the coupling point r_k for the kth channel, and expression (6.1.37) reduces to Eq. (6.1.30).

The next step is an elementary mathematical transformation of expression (6.1.37) for S:

$$\begin{aligned} S &= 1 - 2\imath W^\dagger G W \frac{1}{1 + \imath W^\dagger G W} \\ &= 1 - 2\imath W^\dagger G W \sum_{n=0}^{\infty} (-\imath)^n (W^\dagger G W)^n \\ &= 1 - 2\imath W^\dagger G \sum_{n=0}^{\infty} (-\imath)^n (W W^\dagger G)^n W \\ &= 1 - 2\imath W^\dagger G \frac{1}{1 + \imath W W^\dagger G} W. \end{aligned} \tag{6.1.40}$$

With $G = (E - \mathscr{H}_0)^{-1}$ the latter expression reduces to

$$S = 1 - 2\imath W^\dagger \frac{1}{E - \mathscr{H}_0 + \imath W W^\dagger} W. \tag{6.1.41}$$

The poles of the scattering matrix are thus the eigenvalues of the effective Hamiltonian

$$\mathscr{H} = \mathscr{H}_0 - \imath W W^\dagger. \tag{6.1.42}$$

This is the main result of this section. The information on widths and shifts induced by the presence of the antennas is completely contained in the eigenvalues of the new Hamiltonian $\mathscr{H}$. Effective Hamiltonians of this type have been extensively used especially in nuclear physics [Mah69, Lew91]. Expressions for width distributions, decay behaviour etc. can be found in Refs. [Fyo97, Per96]. For the one-channel case there are explicit formulas for the correlated distribution function of the real and the imaginary parts of the eigenvalues of $\mathscr{H}$ [Sok88, Stö98].

The limiting cases of small and large coupling strengths are particularly simple. In the first case the eigenvalues of $\mathscr{H}$ are given in first order perturbation theory by

$$\begin{aligned} E_n &= E_n^0 - \imath (W W^\dagger)_{nn} \\ &= E_n^0 - \imath \sum_k |w_{nk}|^2. \end{aligned} \tag{6.1.43}$$

For a point-like coupling this is equivalent to the expression already derived in Section 6.1.2 (see Eq. (6.1.34)).

In the other limiting case of large coupling strengths $\mathscr{H}$ is dominated by the term $-\imath W W^\dagger$, suggesting the selection of a basis system where this term is diagonal, and $\mathscr{H}_0$ is treated as a perturbation. Without loss of generality we may assume that the K vectors w_k with the components w_{nk} are mutually orthogonal,

$$w^\dagger_k w_l = \sum_n w^*_{nk} w_{nl} = |w_k|^2 \delta_{kl}. \tag{6.1.44}$$

If this does not apply, we first have to orthogonalize the w_k. We now take as basis vectors the K vectors $v_k = w_k/|w_k|$, obtained from the w_k by normalization, and $N - K$ further normalized vectors u_α being mutually orthogonal to each other and orthogonal to all v_k. Using these basis vectors we obtain for the matrix elements of $\mathscr{H}$

$$\mathscr{H} = \begin{pmatrix} v^\dagger_k \mathscr{H}_0 v_l - \imath |w_k|^2 \delta_{kl} & v^\dagger_k \mathscr{H}_0 u_\beta \\ u^\dagger_\alpha \mathscr{H}_0 v_l & u^\dagger_\alpha \mathscr{H}_0 u_\beta \end{pmatrix}. \tag{6.1.45}$$

The u_α are not yet unequivocally defined by the above orthogonality condition. We still have the free choice of an arbitrary orthogonal transformation in the $N-K$ dimensional subspace spanned by the u_α. We may therefore select a basis in which $\mathscr{H}_0$ is diagonal in the subspace,

$$u^\dagger_\alpha \mathscr{H}_0 u_\beta = u^\dagger_\alpha \mathscr{H}_0 u_\alpha \, \delta_{\alpha\beta}. \tag{6.1.46}$$

The eigenvalues of $\mathscr{H}$ are now obtained in first order perturbation theory as

$$E_n = \begin{cases} v^\dagger_n \mathscr{H}_0 v_n - \imath |w_n|^2 & n \leqslant K \\ u^\dagger_n \mathscr{H}_0 u_n & n > K \end{cases}. \tag{6.1.47}$$

The coupling to the K channels does not influence all eigenvalues in the same way as we might assume. Only K eigenvalues obtain imaginary parts. Because of their large widths there is hardly any chance of detecting these resonances experimentally. The remaining $N-K$ eigenvalues, on the other hand, are not damped (second order perturbation theory shows that there is a residual damping of the order of the reciprocal coupling constant). The spectrum of the surviving eigenvalues is completely different from the original spectrum, as it is obtained from the eigenvalues of $\mathscr{H}_0$ in a reduced subspace.

The eigenvalue distributions of Hamiltonians of the type (6.1.42) have been studied numerically and analytically by Haake, Lehmann and coworkers [Haa92, Leh95]. Figure 6.3 shows the distribution of eigenvalues in the complex plain for $K/N = 0.25$ for three different coupling strengths [Leh95]. $\mathscr{H}_0$ has been taken from the GOE, and Gaussian distributed real matrix elements have been assumed for W. The behaviour observed is in complete correspondence with the predictions from Eq. (6.1.47). For small coupling strengths all eigenvalues are concentrated in a small stripe just below the real axis, but with increasing coupling strength the cloud of eigenvalues separates into two parts. One fourth of the resonances, corresponding to the fraction of coupled channels, obtain large negative imaginary parts, whereas the remaining resonances stay close to the real axis (be aware of the logarithmic scale!). The solid contour lines correspond to analytic boundaries obtained in the limit $N \to \infty$ [Haa92, Leh95].

The microwave billiard with an attached unidirectional waveguide [Sto95] (see Section 2.2.3) is an experimental example for the Hamiltonian (6.1.42) with one effective coupled channel [Haa96]. Figure 6.4 shows experimentally obtained level spacing distributions (histogram) for a rectangle and a Sinai microwave billiard with an attached unidirectional waveguide [Haa96]. The solid lines were obtained numerically from the eigenvalues of $N \times N$ matrices of the form (6.1.42) where $\mathscr{H}_0$ has been taken from the Poissonian and the GOE.

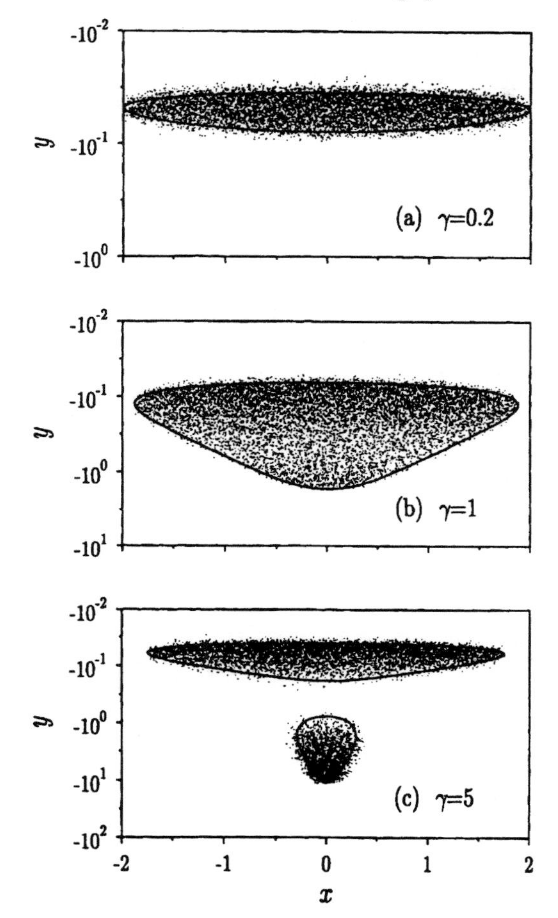

Figure 6.3. Clouds of eigenvalues in the complex plain for the Hamiltonian (6.1.42) with $K/N = 0.25$ for three different coupling strengths γ. The coupling strengths have been assumed to be equal for all channels. The parameter γ corresponds to $|w_k|^2$ of Eq. (6.1.44) [Leh95] (with kind permission from Elsevier Science).

The components of W were assumed to be complex with both the real and imaginary parts being Gaussian distributed. Complete agreement between the experimental results and the simulations is found. Note that the eigenvalue repulsion for small distances is linear for the rectangle and quadratic for the Sinai billiard. As compared to the standard behaviour, the repulsion exponent is increased in both cases by one due to the presence of the unidirectional waveguide.

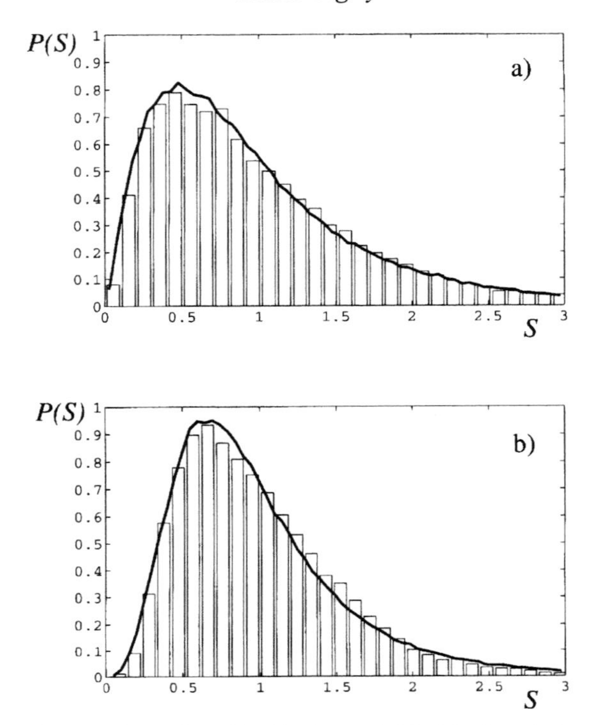

Figure 6.4. Experimental spacing distributions $P(S)$ (bins) and simulations (solid line) for a rectangle (a) and a Sinai (b) microwave billiard with an attached unidirectional waveguide. In the simulations the Hamiltonian (6.1.42) has been used assuming one coupled channel [Haa96].

6.1.4 Perturbing bead method

One method to study field distributions in microwave cavities uses the fact that the eigenfrequencies of microwave billiards are modified if some perturbing object is introduced into the resonator. It is known as the perturbing bead technique. The underlying idea is simple. If a perturbing metallic object, called a 'bead' in the following, is placed just at a node line of the field distribution, it will not influence the eigenfrequency, as long as the bead dimensions are small compared to the wavelengths. If, however, the object is placed close to a field maximum, its detuning effect will be maximal. The technique was applied in the fifties to map field distributions in microwave resonators [Mai52]. In recent years it has been used by Sridhar and coworkers to study the eigenfunctions of two-dimensional billiards [Sri91, Sri92, Kud95]. An experimental example has been given in Section 2.2.2 (see Fig. 2.14). The technique has been extended by Dörr and coworkers [Dör98] to three-dimensional microwave cavities.

The scattering matrix approach could be applied to the perturbing bead method, too, by treating the bead as an additional scattering channel. But in this case Green functions for vector fields would be needed which are somewhat inconvenient to handle [Bal77]. If we are only interested in the shifts induced by the bead, a more direct approach is possible.

In the absence of the bead in the resonator we denote the electric field for an eigenfrequency by E_n, the associate wavenumber by k_n, and the corresponding quantities with the bead placed somewhere in the resonator by E_{1n} and k_{1n}. Both fields obey the Helmholtz equations

$$\left(\Delta + k_n^2\right) E_n = 0, \tag{6.1.48}$$

$$\left(\Delta + k_{1n}^2\right) E_{1n} = 0. \tag{6.1.49}$$

The electromagnetic boundary conditions demand that both fields are perpendicular to the resonator surface. For E_{1n} this boundary condition must hold as well for the surface of the metallic bead.

A combination of Eqs. (6.1.48) and (6.1.49) yields

$$E_{1n}\Delta E_n - E_n \Delta E_{1n} = \left(k_{1n}^2 - k_n^2\right) E_{1n} E_n. \tag{6.1.50}$$

Integrating both sides over the resonator volume, with the exception of the volume of the bead, and applying Green's theorem we obtain

$$\int (E_{1n}\nabla_\perp E_n - E_n \nabla_\perp E_{1n}) dS = \left(k_{1n}^2 - k_n^2\right) \int (E_{1n} E_n) dV, \tag{6.1.51}$$

where the normal derivative points outside. Since the electric field is divergence-free, it follows from the boundary conditions that the normal derivatives of E_n and E_{1n} vanish on the resonator surface. The surface integral on the left hand side is therefore only over the surface of the bead. Let us now assume that the change of the electric field $\epsilon_n = E_{1n} - E_n$ is small due to the presence of the bead. Then we get from Eq. (6.1.51) in linear approximation

$$\begin{aligned} \Delta k_n^2 &= k_{1n}^2 - k_n^2 \\ &= \int (\epsilon_n \nabla_\perp E_n - E_n \nabla_\perp \epsilon_n)\, dS \Big/ \int (E_n)^2\, dV. \end{aligned} \tag{6.1.52}$$

This is the central result. For the frequency shift produced by the bead the field ϵ_n must be known. The ϵ_n are determined by the boundary condition that the total field $E_{1n} = E_n + \epsilon_n$ must be perpendicular to the bead surface. Moreover, the magnetic boundary conditions demand that the magnetic fields are parallel to the surface of the bead.

Details of further calculation depend on the shape of the bead. The situation is particularly simple for quasi-two-dimensional systems with cylindrical beads. In this case the electric field points from the bottom to the top plate of the resonator. We expand E_n at the bead position into cylindrical waves. If the cylinder radius R is small compared to the wavelength λ, we may restrict the expansion to the first term and write

$$E_n \approx E_{n0} J_0(kr), \tag{6.1.53}$$

where E_{n0} is the field value on the cylinder axis. The perturbing field ϵ_n is written as a superposition of outgoing and incoming waves, again considering only s-waves,

$$\begin{aligned} \epsilon_n &= a H_0^{(1)}(kr) + a^* H_0^{(2)}(kr) \\ &= 2|a| [\cos\phi J_0(kr) + \sin\phi N_0(kr)]. \end{aligned} \tag{6.1.54}$$

The incoming wave must be the complex conjugate of the outgoing one, since the resulting field is real. The determination of the phase angle ϕ of the amplitude a is not a trivial procedure, as the reflected wave is built up in a complicated way by reflections from the different parts of the billiard walls. Fortunately ϕ is not needed. For $R \ll \lambda$, corresponding to $kR \ll 1$, the logarithmic singularity of the Neumann function $N_0(x)$ at $x = 0$ dominates the right hand side of Eq. (6.1.54). Then only the second term on the right hand side has to be considered. The prefactor can now be fixed by the condition that $E_n + \epsilon_n$ must be zero on the cylinder mantle, yielding

$$\epsilon_n = -E_{n0} \frac{J_0(kR)}{N_0(kR)} N_0(kr). \tag{6.1.55}$$

We can easily verify that for this case the magnetic boundary condition is automatically obeyed. By means of expressions (6.1.53) and (6.1.55) we obtain

$$\begin{aligned} \int (\epsilon_n \nabla_\perp E_n - E_n \nabla_\perp \epsilon_n) dS &= 2\pi R h \left(-\epsilon_n \frac{\partial E_n}{\partial r} + E_n \frac{\partial \epsilon_n}{\partial r} \right) \Bigg|_{r=R} \\ &= 2\pi k R h (E_{n0})^2 \left(J_0 J_0' - J_0 \frac{J_0 N_0'}{N_0} \right) \\ &= -2\pi k R h (E_{n0})^2 \frac{J_0}{N_0} (J_0 N_0' - N_0 J_0') \\ &= -4h(E_{n0})^2 \frac{J_0}{N_0}, \end{aligned} \tag{6.1.56}$$

where h is the height of the resonator. The argument of all Bessel and Neumann functions is kR. In the last step we have used the Wronskian relation

$$J_0 N_0' - N_0 J_0' = \frac{2}{\pi k R} \tag{6.1.57}$$

(see Chapter 3 of Ref. [Mag66]). Inserting the result (6.1.56) into Eq. (6.1.52) we end with

$$\frac{\Delta k_n^2}{k_n^2} = 2\frac{\Delta \nu_n}{\nu_n} = \gamma(k_n R)\frac{(E_{n0})^2 V_B}{\int (E_n)^2\, dV}, \tag{6.1.58}$$

where $V_B = \pi R^2 h$ is the bead volume and $\gamma(x)$ is given by

$$\gamma(x) = -\frac{4}{\pi}\frac{J_0(x)}{x^2 N_0(x)} \xrightarrow{x\to 0} -\frac{2}{x^2 \ln x}. \tag{6.1.59}$$

Thus the frequency shift due to the presence of the bead is positive. This could already have been concluded from the fact that the bead reduces the billiard area and therefore the mean density of states as well.

Let us now consider the three-dimensional case, using a bead of spherical shape for simplicity's sake. The procedure is essentially the same as in the two-dimensional case. Both the electric and the magnetic field in the unperturbed cavity as well as the bead-induced additional fields are expanded into multipole fields at the bead position. Subsequently the coefficients of the expansion are adjusted to account for the boundary conditions on the surface.

The calculation is a more or less direct application of formulas which can be found in Chapter 16 of Ref. [Jac62]. Here only the result will be given. In the limit $kR \ll 1$ we obtain for the frequency shift induced by the spherical perturber

$$\frac{\Delta k_n^2}{k_n^2} = 2\frac{\Delta \nu_n}{\nu_n} = -\frac{3}{2}\left(2|\boldsymbol{E}_{n0}|^2 - |\boldsymbol{B}_{n0}|^2\right)\frac{V_B}{\int |\boldsymbol{E}_n|^2\, dV}, \tag{6.1.60}$$

where $V_B = (4\pi/3)R^3$ is the bead volume. This expression can be found in Ref. [Mai52]. For spherical beads thus only the field combination $-2\boldsymbol{E}^2 + \boldsymbol{B}^2$ is obtained. However, when using other shapes, such as small sticks or coils, all field components can be determined independently.

Figure 6.5 shows three typical field distributions in a microwave cavity of the shape of a three-dimensional Sinai billiard [Dör98]. The shaded surfaces correspond to surfaces of constant frequency shift, i.e. a constant value of $-2\boldsymbol{E}^2 + \boldsymbol{B}^2$. Although this combination of negative and positive shifts due to electric and magnetic fields is difficult to imagine, the field distributions shown in the figure nevertheless allow a simple interpretation. The eigenmode shown in Fig. 6.5(a) can be associated with a bouncing-ball orbit, whereas the pattern displayed in Fig. 6.5(b) shows a scar associated with a diamond-shaped periodic orbit. The field distribution shown in Fig. 6.5(c) is chaotic with no obvious relation to a periodic orbit. We shall return to this example in Section 6.2.2 where the frequency shift distributions shown in Fig. 6.6 are analyzed within the scope of random matrix theory.

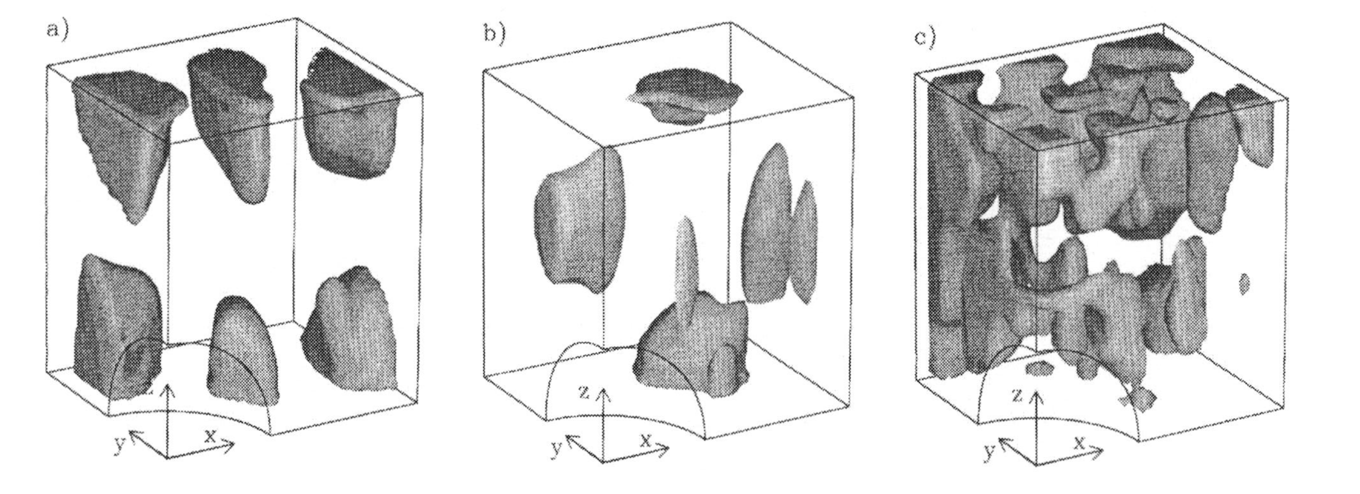

Figure 6.5. Electromagnetic field distributions in a three-dimensional Sinai microwave billiard ($a = 96$ mm, $b = 82$ mm, $c = 106$ mm, $r = 39$ mm) for three eigenfrequencies showing bouncing-ball (a), scarred (b), and chaotic (c) field distributions. The shaded surfaces correspond to surfaces of constant frequency shift $\Delta\nu \sim -2\boldsymbol{E}^2 + \boldsymbol{B}^2$. The unshifted eigenfrequencies for the three modes are 5.208, 2.897, 8.293 GHz, respectively [Dör98] (Copyright 1998 by the American Physical Society).

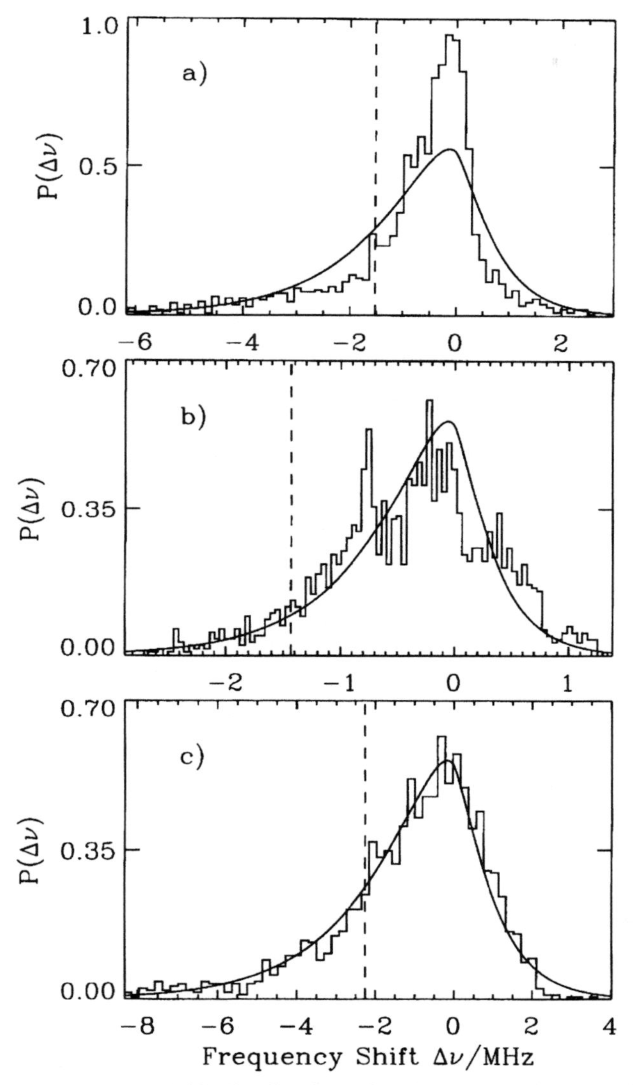

Figure 6.6. Frequency shift distributions for three eigenfrequencies of a three-dimensional Sinai billiard, whose field distributions are shown in Fig. 6.5. The solid lines correspond to the theoretical prediction (6.2.17) for chaotic field distributions. The frequency shifts corresponding to the shaded surfaces in Fig. 6.5 are marked by vertical dashed lines [Dör98] (Copyright 1998 by the American Physical Society).

6.2 Amplitude distribution functions

6.2.1 Random superpositions of plane waves

On the basis of the explicit relation between the scattering matrix and the billiard Green function derived in Section 6.1.2 we are able to analyze the experimental microwave spectra quantitatively. Up to now we have only discussed the positions of resonances, spectral statistics, etc., but ignored the amplitudes and widths of the resonances. From the Breit–Wigner formula derived in Section 6.1.2 we learn that these quantities depend on the squares $|\psi(r_k)|^2$ of the wave functions at the antenna positions. Let us therefore first consider the distribution of wave function amplitudes in billiards.

We have seen that wave functions of chaotic billiards often show extra-high amplitudes in the vicinity of classical periodic orbits. These structures are not generic, and gradually disappear with increasing energy. The wave function displayed in Fig. 6.7(a) for the quarter stadium is of a more typical appearance. The figure is taken from the pathfinding papers of McDonald and Kaufman [McD79, McD88]. The corresponding distribution function for the amplitudes, shown in Fig. 6.7(b), is perfectly well described by a Gaussian function. For scarred wave functions, on the other hand, significant deviations from Gaussian behaviour have been found [McD88]. These findings can be explained on the assumption that chaotic wave functions can be described by a random superposition of plane waves,

$$\psi(r) = \sum_n a_n \cos(k_n r - \phi_n). \qquad (6.2.1)$$

The modulus of the wave vector is the same for all partial waves, i.e. $|k_n| = k$. The amplitudes a_n, the phases ϕ_n, and the directions of the incoming waves, however, are considered as random variables [Ber77]. This ansatz is motivated

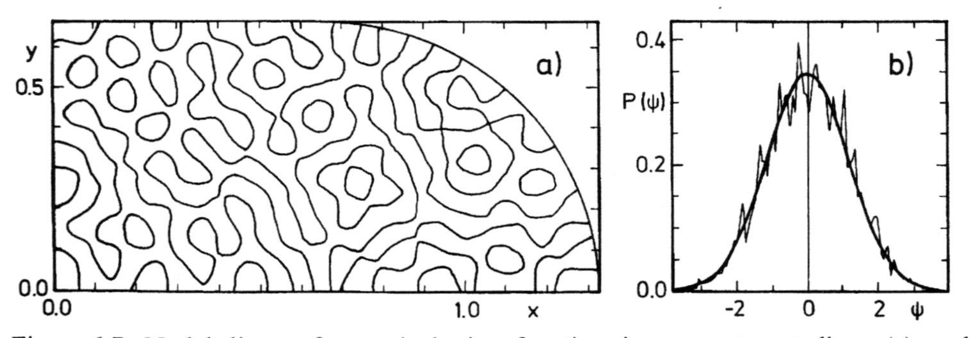

Figure 6.7. Nodal lines of a typical eigenfunction in a quarter stadium (a) and amplitude distribution function for the same eigenfunction (b) [McD88] (Copyright 1988 by the American Physical Society).

by the fact that the waves propagating through the billiard are reflected from all parts of the boundary, and are randomly superimposed in chaotic billards.

It can be easily shown that the amplitude distribution function resulting from the ansatz (6.2.1) is a Gaussian. From the normalization of the wave function we obtain

$$\begin{aligned} 1 &= \int |\psi(r)|^2 \, dA \\ &= \sum_{n,m} a_n a_m \int \cos(k_n r - \phi_n) \cos(k_m r - \phi_m) \, dA \\ &= \frac{A}{2} \sum_n a_n^2 . \end{aligned} \tag{6.2.2}$$

In the second step we have used the fact that for short wavelengths all oscillatory contributions are averaged to zero by the integration. It follows that the squared average of the amplitudes a_n is given by

$$\langle a_n^2 \rangle = \frac{2}{AN}, \tag{6.2.3}$$

where N is the number of partial waves contributing to the sum in Eq. (6.2.1). The amplitude distribution can be written as

$$P(\psi) = \left\langle \delta \left[\psi - \sum_n a_n \cos(k_n r - \phi_n) \right] \right\rangle, \tag{6.2.4}$$

where the bracket denotes averaging over amplitudes, phases, and directions of the incoming waves. Expressing the delta function by its Fourier representation we get

$$P(\psi) = \frac{1}{2\pi} \int_{-\infty}^{\infty} dt \, e^{\imath \psi t} \prod_{n=1}^{N} \langle e^{-\imath t a_n \cos(k_n r - \phi_n)} \rangle . \tag{6.2.5}$$

We have thus obtained a factorization of the averaging over the different contributions. Now we expand the exponential up to the quadratic term,

$$\begin{aligned} \langle e^{-\imath t a_n \cos(k_n r - \phi_n)} \rangle &= 1 - \imath t \langle a_n \cos(k_n r - \phi_n) \rangle \\ &\quad - \frac{t^2}{2} \langle a_n^2 \cos^2(k_n r - \phi_n) \rangle + \cdots \\ &= 1 - \frac{t^2}{2AN} + \cdots . \end{aligned} \tag{6.2.6}$$

In the last step we have used Eq. (6.2.3). Inserting this result into Eq. (6.2.5) we obtain in the limit $N \to \infty$

$$
\begin{aligned}
P(\psi) &= \frac{1}{2\pi}\int_{-\infty}^{\infty} dt\, e^{\imath\psi t} \lim_{N\to\infty}\left(1 - \frac{t^2}{2AN} + \cdots\right)^N \\
&= \frac{1}{2\pi}\int_{-\infty}^{\infty} dt\, e^{\imath\psi t} \exp\left(-\frac{t^2}{2A}\right) \\
&= \sqrt{\frac{A}{2\pi}} \exp\left(-\frac{A\psi^2}{2}\right).
\end{aligned} \tag{6.2.7}
$$

Thus the randon superposition of plane waves leads to a Gaussian amplitude distribution. This is once more a manifestation of the central limit theorem.

With gradually increasing numbers of superimposed plane waves, the appearance of the wave function changes dramatically. Figure 6.8 shows a pattern obtained by the superposition of 10 000 plane waves with random amplitudes and directions [O'Con87]. The more or less smoothly varying amplitudes observed for low-lying wave functions have disappeared. Instead a random pattern of ridge-like structures is found. Such patterns have been experimentally found for water surface waves [Blü92a] and sound waves in water-filled vessels [Chi96], see Sections 2.1.2 and 2.1.4.

The random behaviour of the wave functions can also be found in the spatial autocorrelation function, defined by

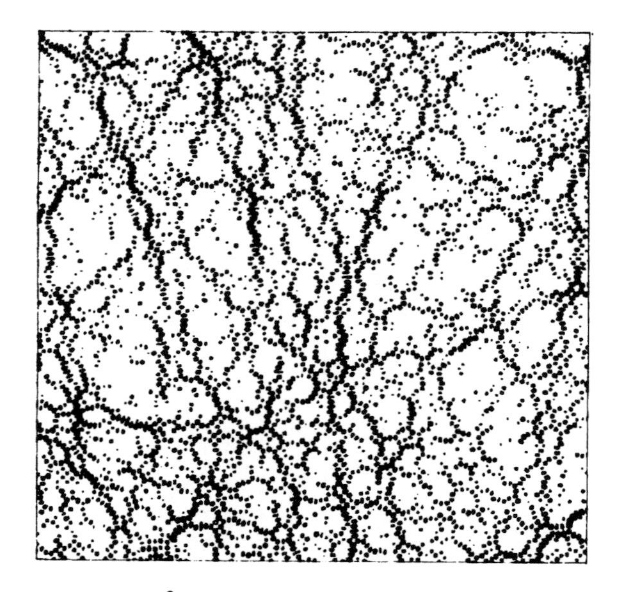

Figure 6.8. Pattern for $|\psi(r)|^2$ obtained by a random superposition of 10 000 plane waves [O'Con87] (Copyright 1987 by the American Physical Society).

$$C(r) = \langle \psi(\overline{r} + r)\psi(\overline{r}) \rangle. \tag{6.2.8}$$

The average is again over directions, amplitudes etc. of the components of the incoming waves, and additionally over $\overline{r}$. Using again expression (6.2.1) for the wave function, we get for the autocorrelation function

$$\begin{aligned} C(r) &= \sum_{n,m} \langle a_n a_m \cos(k_n(\overline{r} + r) - \phi_n) \cos(k_m \overline{r} - \phi_m) \rangle \\ &= \sum_n \langle a_n^2 \rangle \langle \cos(k_n(\overline{r} + r) - \phi_n) \cos(k_n \overline{r} - \phi_n) \rangle \\ &= \frac{2}{AN} \sum_n \Big[\cos k_n r \langle \cos^2(k_n \overline{r} - \phi_n) \rangle. \\ &\quad - \sin k_n r \langle \sin(k_n \overline{r} - \phi_n) \cos(k_n \overline{r} - \phi_n) \rangle \Big] \\ &= \frac{1}{A} \langle \cos k_n r \rangle. \end{aligned} \tag{6.2.9}$$

In the first step the amplitudes have been assumed to be uncorrelated, $\langle a_n a_m \rangle = 0$ for $n \neq m$, and in the second step the oscillating terms do not survive the averaging process. The average over the cosine function can be performed under the assumption that all directions are equally probable:

$$\begin{aligned} C(r) &= \frac{1}{2\pi A} \int_0^{2\pi} \cos(|k||r|) \cos\phi) \, d\phi \\ &= \frac{1}{A} J_0(|k||r|). \end{aligned} \tag{6.2.10}$$

Various other spatial correlation functions can be calculated similarly [Sre96a, Sre96b]. Experimental results from microwave billiards are in complete agreement with the model of a random superposition of plane waves [Kud95, Pri95].

Spatial correlations of wave functions, and also correlation functions with respect to some external parameter, are a very active field of research which cannot be treated in an adequate way in an introductory presentation. Details can be found in the review article by Guhr *et al.* [Guh98].

6.2.2 Porter–Thomas distributions

According to the billiard Breit–Wigner formula (6.1.35) the depth of a resonance is proportional to the square $|\psi_n(r_k)|^2$ of the wave function at the position of the antenna, and its width is proportional to $\sum_k |\psi_n(r_k)|^2$ where the sum is over the positions of all antennas. The distribution of widths is given by the probability $P_\nu(x)$ that this sum takes the value x, where ν is the number of channels. The depth distribution is obtained from the special case $\nu = 1$.

Proceeding in a similar way as in the last section we obtain for the distribution function required

$$
\begin{aligned}
P_\nu(x) &= \left\langle \delta\left(x - \sum_{k=1}^{\nu} |\psi_n(r_k)|^2 \right) \right\rangle \\
&= \left(\frac{A}{2\pi} \right)^{\nu/2} \int \delta\left(x - \sum_{k=1}^{\nu} |\psi_k|^2 \right) \prod_k \exp\left(-\frac{A}{2} |\psi_k|^2 \right) d\psi_k, \qquad (6.2.11)
\end{aligned}
$$

where Eq. (6.2.7) has been used. Introducing ν-dimensional polar coordinates, Eq. (6.2.11) reads

$$
P_\nu(x) = \left(\frac{A}{2\pi} \right)^{\nu/2} \int \delta(x - r^2) \exp\left(-\frac{A}{2} r^2 \right) r^{\nu-1} \, dr \, d\Omega_\nu, \qquad (6.2.12)
$$

where $d\Omega_\nu$ is the surface element of the ν-dimensional unit sphere (see Section 3.3.1). The integrations are easy and yield

$$
P_\nu(x) = \left(\frac{A}{2} \right)^{\nu/2} \frac{1}{\Gamma\left(\frac{\nu}{2} \right)} x^{\nu/2-1} \exp\left(-\frac{A}{2} x \right). \qquad (6.2.13)
$$

We have obtained the χ^2 distribution, which plays a central role in the analysis of statistical errors. For the distribution of resonance depths we get the Porter–Thomas distribution

$$
P_1(x) = \sqrt{\frac{A}{2\pi x}} \exp\left(-\frac{A}{2} x \right), \qquad (6.2.14)
$$

which was originally derived to explain resonance depth distributions in nuclear cross-sections [Por56].

Figure 6.9 shows an experimental example obtained by Schardt [Sch95b] in a superconducting cavity in the shape of a quarter stadium billiard with three antennas (see also Refs. [Alt95, Ric98]). In the upper part of the figure the distribution of the resonance depths at one antenna is plotted. In nuclear physics this quantity is known as the partial width. The solid line corresponds to a Porter–Thomas distribution. In the centre part of the figure the sum of the partial widths from two antennas is shown. The solid line corresponds to a χ^2 distribution with $\nu = 2$, which is a single exponential. The bottom part shows the distribution of the widths of resonances. As there are three antennas, a χ^2 distribution with $\nu = 3$ is expected. In all three cases we find perfect agreement between experiment and theory.

Up to now we have looked for the resonance depth distributions of different

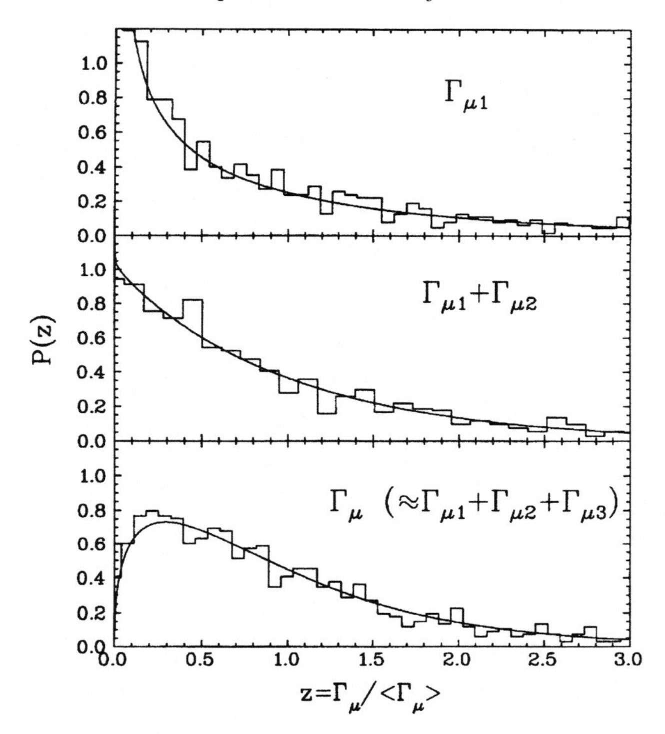

Figure 6.9. Distribution of the partial widths at one antenna (top), of the sum of the partial widths from two antennas (middle), and of the total width (bottom) in a superconducting quarter stadium billiard with three antennas. The solid lines correspond to χ^2 distributions with $\nu = 1, 2, 3$, respectively [Sch95b].

wave functions at a fixed position r. Alternatively we may look for the spatial distribution of $|\psi_n(r)|^2$ for single wave functions. Here again we find Porter–Thomas distributions. Only for scarred and localized wave functions have significant deviations been observed [Kud95, Mül97]. Similar deviations from the usual Gaussian behaviour have been observed in the amplitude distribution for scarred wave functions (see Section 6.2.1). This is not a coincidence: the Gaussian distributions for the wave function amplitudes, and the Porter–Thomas distributions for the resonance depths only differ by a change of variables from ψ to $|\psi|^2$. Porter–Thomas distributions are not restricted to quantum mechanical systems. As we have learnt in Section 2.1.3, the amplitude distributions of the eigenmodes of vibrating plates also obey Porter–Thomas distributions (see Fig. 2.7).

A variant of the Porter–Thomas distribution has been observed in the perturbing bead experiments in three-dimensional microwave cavities [Dör98] presented in Section 6.1.4. For spherical beads a resonance frequency shift of

$\Delta\nu$ proportional to $-2\boldsymbol{E}^2+\boldsymbol{B}^2$ is observed, where $\boldsymbol{E}$ and $\boldsymbol{B}$ are electric and magnetic fields at the bead position. On the assumption that all six field components are randomly Gaussian distributed [Dör98], we may calculate the distribution $P(\Delta\nu)$ of frequency shifts. Both $\boldsymbol{E}^2$ and $\boldsymbol{B}^2$ are χ^2-distributed with $\nu=3$. We thus obtain for $P(\Delta\nu)$

$$P(\Delta\nu) \sim \int \delta(\Delta\nu+2x-y)\sqrt{xy}e^{-\alpha(x+y)}\,dx\,dy. \qquad (6.2.15)$$

The parameter α describes the width of the distributions and has to be adjusted to the experiment. It follows that

$$\begin{aligned} P(\Delta\nu) &\sim \int_{\max(-\Delta\nu/2,0)}^{\infty} \sqrt{x(2x+\Delta\nu)}e^{-\alpha(3x+\Delta\nu)}\,dx \\ &= \exp\left(-\alpha\frac{\Delta\nu}{4}\right)\sqrt{2}\int_{|\Delta\nu|/4}^{\infty}\sqrt{t^2-\left(\frac{\Delta\nu}{4}\right)^2}\,e^{-3\alpha t}\,dt, \end{aligned} \qquad (6.2.16)$$

where the substitution $x=t-\Delta\nu/4$ has been applied in the second step. The remaining integral can be expressed in terms of a modified Bessel function (see Ref. [Mag66]):

$$P(\Delta\nu) = \frac{\sqrt{2}\alpha^2}{3\pi}|\Delta\nu|\exp\left(-\alpha\frac{\Delta\nu}{4}\right)K_1\left(\frac{3}{4}\alpha|\Delta\nu|\right), \qquad (6.2.17)$$

where the prefactor has been chosen in such a manner that the distribution is normalized to one.

In Fig. 6.6 the predictions of Eq. (6.2.17) are compared with the experimental results. The figure shows the frequency shift distributions for three eigenfrequencies of a three-dimensional Sinai microwave billiard, whose field distributions have been displayed in Fig. 6.5. For the chaotic looking field pattern displayed in Fig. 6.5(c) the frequency shift distribution is correctly described by Eq. (6.2.17), but for the bouncing ball and the scarred pattern shown in Figs 6.5(a) and (b) significant deviations from the theoretical curve are found.

These findings can be understood by assuming that, just as in two-dimensional billiards, in three-dimensional chaotic microwave resonators, too, the fields are composed by a random superposition of plane waves. A generalization of the technique described in Section 6.2.1 to electromagnetic fields exactly yields the conjectured uncorrelated Gaussian distributions for the six field components.

6.3 Fluctuation properties of the scattering matrix

6.3.1 Ericson fluctuations

The concept of the spectral autocorrelation function, introduced in Section 3.2.5 for the analysis of chaotic spectra, is now generalized to the elements of the scattering matrix. The energy autocorrelation function for the scattering matrix element S_{ij} is defined by

$$C_{ij}(E) = \left\langle S_{ij}^{*}\left(\overline{E} - \frac{E}{2}\right) S_{ij}\left(\overline{E} + \frac{E}{2}\right)\right\rangle - |\langle S_{ij}(\overline{E})\rangle|^2. \tag{6.3.1}$$

The brackets denote an average over $\overline{E}$. In the pioneering article by Verbaarschot *et al.* [Ver85] on the supersymmetry technique (see Section 3.3) this average has been determined on the assumption that the scattering matrix elements are Gaussian distributed.

Here the simpler case will be considered, where the S_{ij} are given by the Breit–Wigner expression (6.1.35). For the calculation of the average over $\overline{E}$ we apply the technique introduced in Section 3.2.5 for the calculation of the spectral autocorrelation function, and truncate the scattering matrix element by means of a window function

$$\left(S_{ij}\right)_w(E) = w(E) S_{ij}(E), \tag{6.3.2}$$

where $w(E)$ is normalized to one and has a width of ΔE. The average on the right hand side of Eq. (6.3.1) can be written in terms of the truncated scattering matrix elements as

$$C_{ij}(E) = \int_{-\infty}^{\infty} (\Delta S_{ij})_w^{*}\left(\overline{E} - \frac{E}{2}\right)(\Delta S_{ij})_w\left(\overline{E} + \frac{E}{2}\right) d\overline{E}, \tag{6.3.3}$$

where $\left(\Delta S_{ij}\right)_w(E) = (S_{ij})_w(E) - \langle (S_{ij})_w(\overline{E})\rangle$. Using the Fourier transform convolution theorem we see that the Fourier transform

$$\hat{C}_{ij}(t) = \int C_{ij}(E) \exp\left(\frac{\imath}{\hbar} Et\right) dE \tag{6.3.4}$$

of the autocorrelation function can be written as

$$\hat{C}_{ij}(t) = |(\Delta \hat{S}_{ij})_w(t)|^2, \tag{6.3.5}$$

where

$$(\Delta \hat{S}_{ij})_w(t) = \int \left(\Delta S_{ij}\right)_w(E) \exp\left(\frac{\imath}{\hbar} Et\right) dE \tag{6.3.6}$$

is the Fourier transform of $\left(\Delta S_{ij}\right)_w(E)$. Inserting the Breit–Wigner expression (6.1.35) for the scattering matrix element we are able to perform the integration and obtain

$$(\Delta\hat{S}_{ij})_w(t) = 4\gamma\pi \sum_n w(E_n)\psi_n(r_i)\psi_n(r_j)\exp\left[\frac{t}{\hbar}\left(\imath E_n - \frac{\Gamma_n}{2}\right)\right]. \quad (6.3.7)$$

The constant part of $(\Delta S_{ij})_w(E)$ gives rise to an additional delta contribution at $t = 0$ on the right hand side, which has been omitted. Inserting expression (6.3.7) into Eq. (6.3.5) and discarding the oscillating terms we end with

$$\hat{C}_{ij}(t) \sim \left\langle |\psi_n(r_i)|^2 |\psi_n(r_j)|^2 \exp\left(-\frac{\Gamma_n}{\hbar}t\right)\right\rangle. \quad (6.3.8)$$

If there were only one resonance, the Fourier transform of the autocorrelation function would decay exponentially, resulting in a Lorentzian shape of the autocorrelation function. But we have to average over many resonances. If there are ν coupled channels, the line width is given by $\Gamma_n = 2\gamma\sum_{n=1}^{\nu}|\psi_n|^2$ (see Eq. (6.1.34)). Assuming a Gaussian distribution of the wave function amplitudes, the average yields

$$\hat{C}_{ij}(t) \sim \int_{-\infty}^{\infty} d\psi_1 \ldots d\psi_\nu |\psi_i|^2|\psi_j|^2 \exp\left(-\frac{2\gamma t}{\hbar}\sum_{n=1}^{\nu}|\psi_n|^2\right)$$

$$\times \exp\left(-\frac{A}{2}\sum_{n=1}^{\nu}|\psi_n|^2\right). \quad (6.3.9)$$

Substituting $x_n = (1 + 4\gamma t/\hbar A)|\psi_n|^2$ we obtain

$$\hat{C}_{ij}(t) \sim \left(1 + \frac{4\gamma t}{\hbar A}\right)^{-(2+\nu/2)}. \quad (6.3.10)$$

Due to the averaging process the decay of $\hat{C}_{ij}(t)$ has become algebraic. This is the very behaviour found in the experiment [Alt95]. Figure 6.10 shows the autocorrelation function $C_{ii}(E)$ and its Fourier transform $\hat{C}_{ii}(t)$, obtained from the reflection spectrum in the superconducting quarter stadium billiard already described in Section 6.2.2. The dashed lines correspond to the predictions of the theory. In the calculations the authors assumed $\nu = 3$, corresponding to the number of antennas. There is complete correspondence with the experimental results, the errors of which are indicated by the shaded bands.

If the widths of the individual resonances are large compared to the mean distance between neighbouring resonances, we can no longer expect sharp resonances, but only chaotic fluctuations. As an example Fig. 6.11 shows the transition probability $|S_{12}|^2$ between two scattering channels for a system consisting of a periodic array of disks [Blü89].

Similar fluctuations have been observed for many years in the cross-sections of nuclear reactions and are known as *Ericson fluctuations* [Eri63]. In nuclear scattering the number of scattering channels is usually large [Lew91]. In this

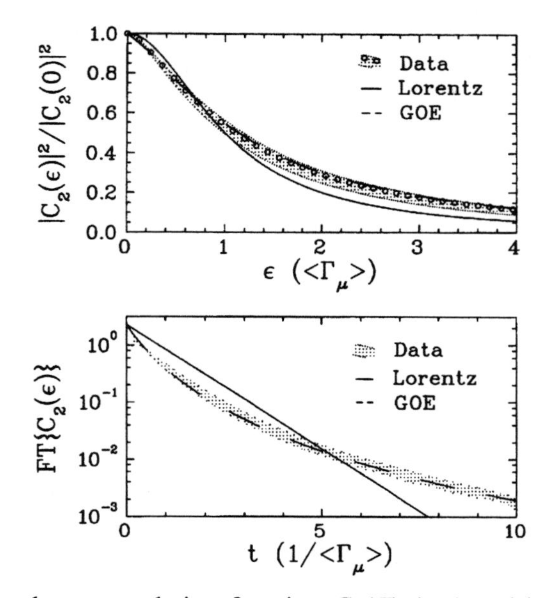

Figure 6.10. Spectral autocorrelation function $C_{ii}(E)$ (top) and its Fourier transform $\hat{C}_{ii}(t)$ (bottom) obtained from a reflection measurement in a superconducting quarter stadium billiard. The experimental results, indicated by the shaded bands, closely follow the prediction from Eq. (6.3.10). The Lorentzian and the exponential behaviour expected for $C_{ii}(E)$ and $\hat{C}_{ii}(t)$, respectively, for the single resonance case, do not accord with the data [Alt95] (Copyright 1995 by the American Physical Society).

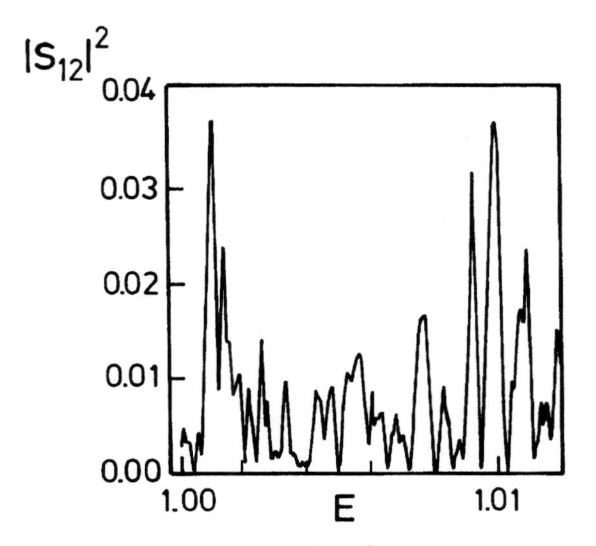

Figure 6.11. Transmission probability $|S_{12}|^2$ through a periodic array of disks [Blü89] (with kind permission from Elsevier Science).

limit the χ^2 distribution (6.2.13) becomes more and more peaked and can eventually be replaced by a delta function. This is another consequence of the central limit theorem: as Γ is a sum over ν independent contributions, its variance decreases for large ν values with $\nu^{-1/2}$. We may then replace all Γ_n in Eq. (6.3.8) by their average values $\overline{\Gamma}$. The Fourier transform of the autocorrelation function now decays exponentially,

$$\hat{C}_{ij}(t) \sim \exp\left(-\frac{\overline{\Gamma}}{\hbar}t\right), \tag{6.3.11}$$

and the autocorrelation function is Lorentzian,

$$C_{ij}(E) \sim \frac{1}{\overline{\Gamma} + \imath E}. \tag{6.3.12}$$

Equation (6.3.5) shows that $\hat{C}_{ij}(t)$ may be interpreted as the probability for a particle, leaving channel i at time $t = 0$, to arrive at channel j at a time t. $T = \hbar/\overline{\Gamma}$ can hence be associated with the mean staying time of a classical particle in the system. Simulations support this interpretation [Dor91, Eck93]. Figure 6.12 shows $\hat{C}_{11}(t)$ taken from an elbow-shaped microwave resonator [Dor90]. The histogram shows the classical staying time distribution, which is in perfect agreement with the overall decay of $\hat{C}_{11}(t)$. In the inset the

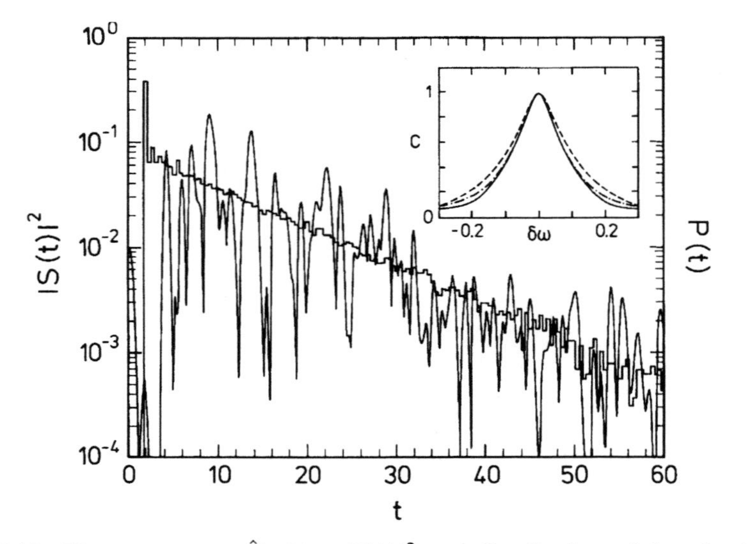

Figure 6.12. Time spectrum $\hat{C}_{11}(t) = |S(t)|^2$ and distribution of the classical staying times in an elbow-shaped microwave resonator. The inset shows the corresponding spectral autocorrelation function $C_{11}(E)$. The dashed and the dashed-dotted lines correspond to a Lorentzian curve and to a simulation, respectively [Dor90] (Copyright 1990 by the American Physical Society).

corresponding $C_{11}(E)$ is displayed, deviating significantly from Lorentzian behaviour. This is not surprising, since the number of channels in the experiment was only one.

6.3.2 Conductance fluctuations in mesoscopic systems

Scattering theory was originally developed in nuclear physics, but has been repeatedly applied to other disciplines as well. We have already learnt that microwave billiards can be properly described within the formalism of scattering theory. Another topic is the transport of electrons through mesoscopic devices, which will be discussed in this section.

The transport of electrons at low temperatures implies a conceptual difficulty. At room temperature the electronic resistance is mainly limited by the inelastic scattering of electrons at phonons, transferring energy from the electrons to the lattice and causing an ohmic heating of the conductor. At low temperatures, however, the lattice vibrations are frozen out, and the omnipresent impurities only scatter the electrons elastically. Under these conditions we may wonder whether there is any resistance at all. Landauer [Lan57, Lan87] has given an answer to this question. He proposed that 'conduction is transmission' between two reservoirs which are maintained at different potentials. The Landauer approach has been discussed in a review article by Baranger and Westerveld [Bar99] where all relevant references can be found. Details, in particular on the random matrix aspects of mesoscopic systems, are given in the proceedings of the 1994 Les Houches summer school [Akk94], a well as in a recent report by Beenakker [Bee97].

In Landauer's theory the electrical conductance through a mesoscopic device with two attached leads is given by

$$G = \frac{2e^2}{h} \sum_{n,m=1}^{N} |t_{nm}|^2, \tag{6.3.13}$$

where N is the number of channels supported by the entrance and the exit leads, and where t_{nm} is the transmission amplitude between the nth incoming and the mth outgoing channel. Up to a universal factor the conductance is hence the total transmission probability between the incoming and the outgoing leads.

The situation is completely analogous to the scattering problems discussed in the preceding sections. Thus it is not surprising that the resistance of mesoscopic devices exhibits similar fluctuations as found for the scattering matrix elements in billiards and atomic nuclei. Figure 6.13 represents a calculation of the Hall resistance in a four-disk junction and the transmission

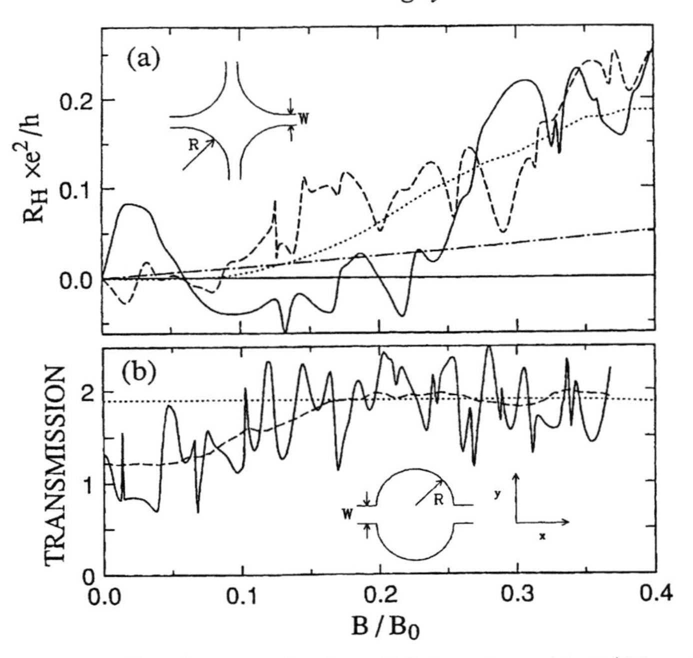

Figure 6.13. (a) Hall resistance of a four-disk junction with $R/W = 4$. The solid and the broken lines represent two quantum mechanical calculations with slightly different energies. For the dotted and the dashed-dotted lines see the original work. (b) Transmission through an open stadium (solid line) with $R/W = 2$ [Jal90] (Copyright 1990 by the American Physical Society).

through a stadium with two coupled channels as a function of the magnetic field applied [Jal90]. The similarity to the Ericson fluctuations discussed in Section 6.3.1 is unmistakable. Note that the four-disk junction is obtained by a two-fold reflection from the elbow-shaped microwave cavity introduced in Section 6.3.1. The measurements by Marcus and coworkers [Mar92] in differently shaped mesoscopic billiards (see Section 2.3.2) represent an experimental example. These fluctuations are known as *universal conductance fluctuations* [Lee85].

Since the conductance is proportional to the total transmission $T = \sum |t_{nm}|^2$, the fluctuations depend on the channel number in a characteristic manner. Figure 6.14 shows the intensity distributions of transmission through a chaotic billard with two attached leads depending on the channel number [Bar94]. The analytical results were obtained by applying random matrix theory to the scattering matrix. Note the very distinct line shapes for $N = 1$ and $N = 2$, and the clear difference between the results with and without a magnetic field. For $N = 3$ the distributions again approach a Gaussian.

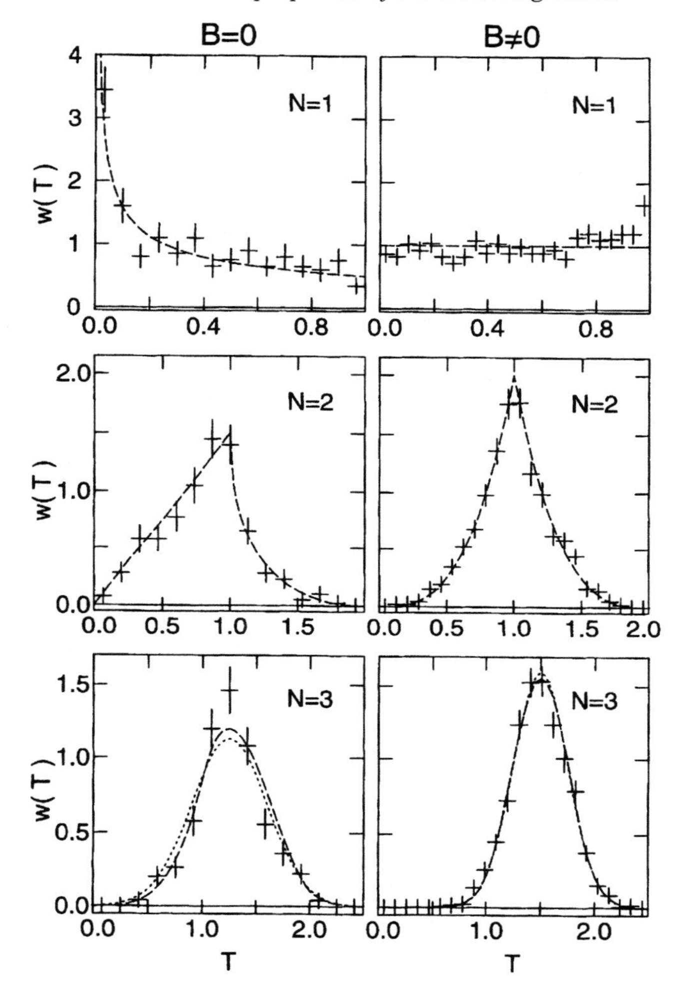

Figure 6.14. Distribution of transmission intensities through a chaotic billiard with two attached leads for different channel numbers N, both without (left column) and with (right column) application of a magnetic field B. The numerical results (crosses) are in good agreement with predictions from circular ensembles [Bar94] (Copyright 1994 by the American Physical Society).

The magnetic field dependence at small fields requires special treatment. Figure 6.15 shows the resistances of a stadium and a circular mesoscopic billiard. The measurements were performed by connecting 48 cavities to improve the signal-to-noise ratio and to average over the individual fluctuations. The figure shows a conspicuous enhancement of the resistance at $B = 0$ [Cha94]. This effect has already been observed in the measurements by Marcus *et al.* [Mar92] mentioned above (see Fig. 2.21). This is a manifestation of the

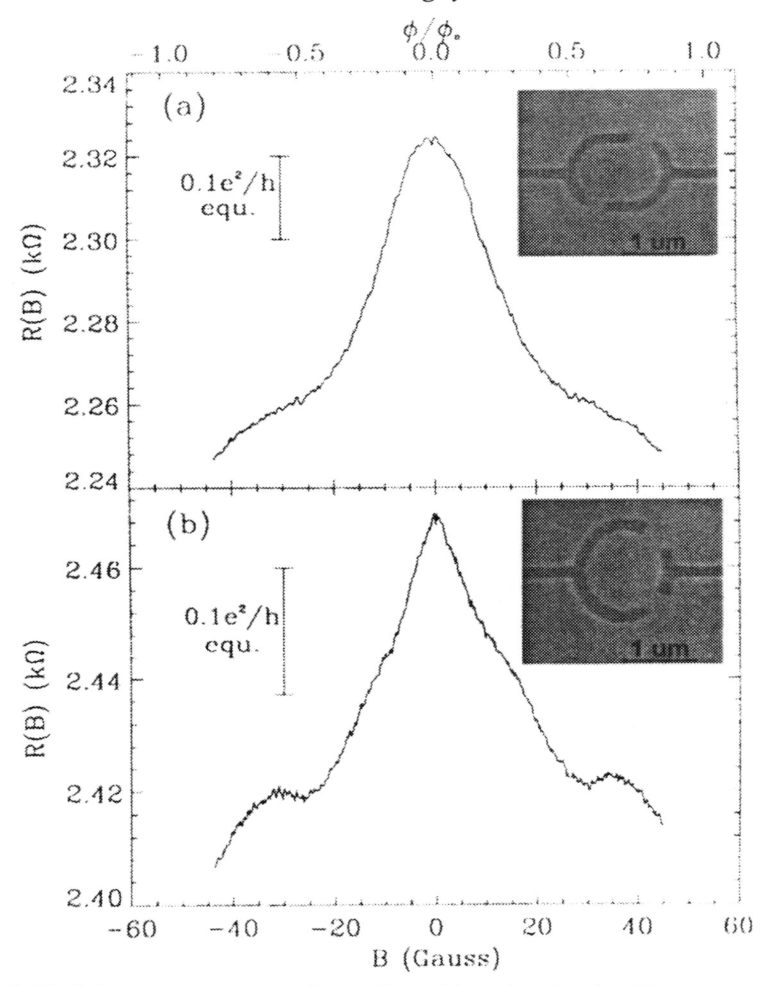

Figure 6.15. Magnetoresistance of a stadium (a) and a circular (b) mesoscopic billiard obtained by adding the resistances of 48 billiards. The weak localization peak shows a Lorentzian shape for the stadium billiard, whereas for the circular billiard a triangular shape is found [Cha94] (Copyright 1994 by the American Physical Society).

weak localization phenomenon. At $B = 0$ any path through the billiard constructively interferes with its time-reversed equivalent, thus leading to enhanced backscattering. The coherence is destroyed if the magnetic field is turned on, since now the time-reversal symmetry is broken. Figure 6.15 shows characteristic differences in the shape of the weak localization peak for chaotic and nonchaotic billiards. For the stadium the peak has a Lorentzian shape, but for the circle the shape is triangular (see also Ref. [Lee97]). To understand this difference it should be remembered that in the presence of a magnetic field the electronic wave function acquires a phase shift. For closed orbits the phase shift is proportional to the encircled area. The distribution of these areas enters into the calculation of the line shape. In this respect chaotic and nonchaotic billiards differ qualitatively, thus causing the shape differences of the weak localization peak [Bar93].

In an alternative approach Pluhař *et al.* [Plu95] studied the weak localization peak within the framework of random matrix theory by replacing the billiard by the Hamiltonian

$$\mathscr{H} = \mathscr{H}_1 + \imath B \mathscr{H}_2 - \imath W W^\dagger, \tag{6.3.14}$$

where both $\mathscr{H}_1$ and $\mathscr{H}_2$ belong to the GOE. The elements of W describing the attachment to the leads (see Section 6.1.3) were assumed to be real and Gaussian distributed. The ensemble averages were performed by means of supersymmetry techniques. The authors calculated the transmission as a function of B and found that the shape of the weak localization peak is close to, but not exactly, a Lorentzian in agreement with the experimentally found behaviour for chaotic billiards.

As we have seen in this chapter, completely different systems, such as atomic nuclei, mesoscopic devices, and microwave billiards, can be equally well described by scattering theory. This example convincingly demonstrates the path of scientific progress: a technique developed in one field of application is transferred to another apparently unrelated discipline. This monograph contains a number of other examples of this type. Universalists are needed to discover such analogies between different objects. Students in particular should take the following advice: although this is not always easy, do *not* specialize, but try to maintain a universal point of view!

7
Semiclassical quantum mechanics

In the preceding chapters we have learnt that random matrix theory is perfectly able to explain the universal properties of the spectra of chaotic systems, and this in spite of the oversimplifying assumptions applied. On the one hand it is very satisfactory that one single theory can cope with such a variety of systems as nuclei, mesoscopic structures, or microwave billiards, on the other hand this is a bit disappointing. If there is no possibility of discriminating between the spectra of a nucleus and a quantum dot, then there is little hope of learning anything of relevance about it.

Fortunately, random matrix theory is only one side of the coin. We have already come across some examples demonstrating its limits of validity. Remember the spectral level dynamics where bouncing balls disturbed the otherwise universal Gaussian velocity distribution (see Section 5.2.2). Another example is the scarring phenomenon observed in many wave functions. Here obviously closed classical orbits have left their fingerprints in the amplitude patterns. We cannot expect that the *universal* random matrix theory can correctly account for *individual* features such as periodic orbits.

We now come to an alternative approach to analysing the spectra. As we know from the correspondence principle, in the semiclassical limit quantum mechanics eventually turns into classical mechanics. That is why classical dynamics must be hidden somewhere in the spectra, at least in the limit of high quantum numbers. In the introduction we have already discussed this connection for a particle in a one-dimensional box. The general case has been treated by M. Gutzwiller in a series of pioneering publications [Gut67, Gut69, Gut70, Gut71]. He has given a survey of his results in *Chaos in Classical and Quantum Mechanics* [Gut90].

In the present monograph the development of semiclassical quantum mechanics is divided into two parts. In this chapter the theory will be developed step by step up to the Gutzwiller trace formula, establishing a correspondence between the quantum mechanical spectrum and the periodic orbits of a system. In the next chapter the consequences and a number of applications of the trace formula will be examined. Very useful discussions of the trace formula can be found in the review articles by Berry and Mount [Ber72] and Eckhardt [Eck88]. Moreover, the author has repeatedly used private notes of a series of lectures given by B. Eckhardt at the University of Marburg during the summer term 1989. The topics of these last two chapters are covered by the monograph by Brack and Bhaduri on *Semiclassical Physics* [Bra97], which is recommended to everyone who is interested in more details.

Although an attempt has been made to reduce formalism as much as possible, and to subdivide the presentation of the theory into a number of small, hopefully digestible, portions, readers not familiar with the subject may have some initial problems. In this case they should not hesitate to skip this chapter and proceed with the next one. Most of the remaining part of this book should be understandable even without explicit knowledge of the technical details which enter into the derivation of the trace formula.

7.1 Integrable systems

7.1.1 One-dimensional case

From Bohr's correspondence principle we expect a gradual transition from quantum to classical mechanics in the limit $\hbar \to 0$, corresponding to high quantum numbers. In this transition region we expect to find the signatures of the classical trajectories in the quantum mechanical spectra as well as in the wave functions.

We begin with the one-dimensional Schrödinger equation

$$-\frac{\hbar^2}{2m_e}\frac{d^2\psi}{dx^2} + V(x)\psi = E\psi, \tag{7.1.1}$$

where it is assumed that the potential energy $V(x)$ has one minimum at $x = 0$ and increases monotonically both for positive and negative x values. A classical particle with energy E performs a periodic motion between the two turning points a and b, which are obtained as the two solutions of the equation $V(x) = E$. With the notation

$$k(x) = \sqrt{\frac{2m_e}{\hbar^2}[E - V(x)]} \tag{7.1.2}$$

Eq. (7.1.1) can be written as

$$\frac{d^2\psi}{dx^2} + k^2(x)\psi = 0. \tag{7.1.3}$$

$k(x)$ can be interpreted as the local wave number, from which the local de Broglie wavelength is obtained as $\lambda(x) = 2\pi/k(x)$. In the classically allowed region $k(x)$ is real and corresponds to an oscillatory behaviour of the wave function, whereas in the classically forbidden region $k(x)$ becomes imaginary, and the wave function is exponentially damped.

We now decompose the wave function into an amplitude prefactor and a phase factor,

$$\psi(x) = A(x) \exp\left[\frac{i}{\hbar} S(x)\right], \tag{7.1.4}$$

where both A and S are assumed to be real. Entering this ansatz into the Schrödinger equation and separating real and imaginary parts, we obtain the two equations

$$A(S')^2 = Ap^2 + \hbar^2 A'', \tag{7.1.5}$$

$$S''A + 2S'A' = 0, \tag{7.1.6}$$

where $p(x) = \hbar k(x)$ is the local momentum. These equations are still exact and are equivalent to the original Schrödinger equation. In the semiclassical approximation the term $\hbar^2 A''$ is considered as small and thus neglected. Now the equations are immediately integrated. From the first equation we obtain

$$S(x) = \int_{x_0}^{x} p(x)\, dx. \tag{7.1.7}$$

$S(x)$ corresponds to the classical action. The integration of Eq. (7.1.6) yields

$$A = \frac{\psi_0}{\sqrt{|S'|}} = \frac{\psi_0}{\sqrt{|p(x)|}}, \tag{7.1.8}$$

where we have written ψ_0 for the integration constant. Inserting expressions (7.1.7) and (7.1.8) into Eq. (7.1.4) we get the semiclassical approximation

$$\psi(x) = \frac{\psi_0}{\sqrt{|p(x)|}} \exp\left(\frac{i}{\hbar}\int_{x_0}^{x} p(x)\, dx\right) \tag{7.1.9}$$

for the wave function. This is the *WKB approximation* (after Wentzel, Kramer, Brillouin, see Chapter 6 of Ref. [Mes61]). We have already applied the WKB approximation in Section 3.2.3 to derive the asymptotic behaviour of the harmonic oscillator eigenfunctions.

A more thorough examination shows that the approximation works well as long as the relative change of the local momentum within the range of one de Broglie wavelength is small. The approximation becomes invalid in particular for $p = 0$, i.e. at the classical turning points.

Applying an approach developed by Maslov [Mas81, Eck88], the wave function can be extended into this region. If the approximation is poor in position space then it should work well in momentum space, and vice versa. The wave function $\tilde{\psi}(p)$ in momentum space is obtained from $\psi(x)$ by a Fourier transformation,

$$\begin{aligned}\tilde{\psi}(p) &= \frac{1}{\sqrt{2\pi\hbar}}\int dx\psi(x)\exp\left(-\frac{\imath}{\hbar}xp\right)\\ &= \frac{\psi_0}{\sqrt{2\pi\hbar}}\int\frac{dx}{\sqrt{|p(x)|}}\exp\left[\frac{\imath}{\hbar}\left(\int_{x_0}^{x}p(x)\,dx - xp\right)\right].\end{aligned} \quad (7.1.10)$$

$p(x)$ is the local momentum, whereas p is the argument of $\tilde{\psi}$. We should not mix up these two quantities. The integral is of the type

$$I = \int dx\,A(x)\exp\left[\frac{\imath}{\hbar}\Phi(x)\right]. \quad (7.1.11)$$

In the semiclassical limit $\hbar \to 0$ the phase $\frac{1}{\hbar}\Phi(x)$ oscillates rapidly. These oscillations cause a cancellation of nearly all contributions to the integral. Exceptions are the neighbourhoods of the points x_s where the phase is stationary, i.e. where $\Phi'(x_s) = 0$. We expand $\Phi(x)$ at the point x_s into a Taylor series up to the quadratic term

$$\Phi(x) = \Phi(x_s) + \frac{(x-x_s)^2}{2}\Phi''(x_s). \quad (7.1.12)$$

If $A(x)$ is regular at $x = x_s$, it may be considered as constant and taken out of the integral. We thus obtain

$$I \approx A(x_s)\exp\left(\frac{\imath}{\hbar}\Phi(x_s)\right)\int\exp\left[\frac{\imath}{\hbar}\frac{(x-x_s)^2}{2}\Phi''(x_s)\right]dx. \quad (7.1.13)$$

The integral on the right hand side is a *Fresnel integral*

$$\int_{-\infty}^{\infty} e^{\imath\alpha x^2}\,dx = \sqrt{\frac{\pi}{|\alpha|}}\exp\left[\imath\frac{\pi}{4}\operatorname{sgn}(\alpha)\right]. \quad (7.1.14)$$

This relation can be derived from the well-known integral

$$\int_{-\infty}^{\infty} e^{-x^2}\,dx = \sqrt{\pi} \quad (7.1.15)$$

by a shift of the integration path in the complex plane. We thus end with

$$I \approx \sqrt{\frac{2\pi\hbar}{|\Phi''(x_s)|}}A(x_s)\exp\left\{\frac{\imath}{\hbar}\Phi(x_s) + \frac{\imath\pi}{4}\operatorname{sgn}[\Phi''(x_s)]\right\}. \quad (7.1.16)$$

By fixing the sign of the square root of $\imath$ according to

$$\sqrt{\imath} = \exp\left(\frac{\imath\pi}{4}\right), \quad (7.1.17)$$

this may alternatively be written as

$$I \approx \sqrt{\frac{2\pi\imath\hbar}{\Phi''(x_s)}} A(x_s) \exp\left[\frac{\imath}{\hbar}\Phi(x_s)\right]. \tag{7.1.18}$$

This is the *stationary phase approximation*. It is the standard technique for performing integrations in the semiclassical region, and we shall use it repeatedly.

The method is easily extended to higher dimensions. For the d-dimensional variant of the integral (7.1.11) the stationary phase approximation yields

$$I \approx \frac{(2\pi\imath\hbar)^{d/2}}{\sqrt{|\partial^2\Phi/\partial x_n \partial x_m|}} A(x_s) \exp\left[\frac{\imath}{\hbar}\Phi(x_s)\right], \tag{7.1.19}$$

where $|\partial^2\Phi/\partial x_n \partial x_m|$ is the determinant of the matrix of the second derivatives of $\Phi(x)$ at the stationary phase point.

The attentive reader may have noticed a similarity with the saddle-point integration discussed in Section 3.3.4. In fact both techniques are more or less equivalent. They only differ in the direction of the integration path in passing the saddle.

The stationary phase approximation is now applied to the integral in Eq. (7.1.10). Here the phase is given by

$$\Phi(x) = \frac{1}{\hbar}\left[\int_{x_0}^{x} p(x)\, dx - xp\right], \tag{7.1.20}$$

and its first two derivatives are

$$\Phi'(x) = \frac{1}{\hbar}[p(x) - p], \tag{7.1.21}$$

$$\Phi''(x) = \frac{1}{\hbar}p'(x). \tag{7.1.22}$$

The points x_s of the stationary phase are hence the solutions of the equation

$$p(x_s) = p. \tag{7.1.23}$$

From this we have for the wave function in momentum space

$$\tilde{\psi}(p) = \frac{\psi_0}{\sqrt{|p(x_s)p'(x_s)|}} \exp\left\{\frac{\imath}{\hbar}\left[\int_{x_0}^{x_s} p(x)\, dx - px_s\right] + \frac{\imath\pi}{4}\operatorname{sgn}[p'(x_s)]\right\}. \tag{7.1.24}$$

The expression under the square root can be written as

$$p(x_s)p'(x_s) = \frac{1}{2}\frac{d}{dx}p^2\bigg|_{x=x_s} = m_e\frac{d}{dx}[E - V(x)]\bigg|_{x=x_s} = -m_e V'(x_s). \tag{7.1.25}$$

The prefactor in Eq. (7.1.24) is hence nonsingular, provided that the potential derivative $V'(x)$ does not vanish at the turning point. This situation is pathological even in classical mechanics and has to be excluded. A particle on such a trajectory will never arrive at the turning point within a finite time.

We are now going to study the evolution of the phase of the wave function on a complete orbit in phase space (see Fig. 7.1). To this end we divide the trajectory into four parts so that each quadrant contains just one division point. We start in the section between points 1 and 2 with the position space representation of the wave function:

$$\psi(x) = \frac{\psi_0}{\sqrt{|p(x)|}} \exp\left[\frac{i}{\hbar}\int_{x_0}^{x} p(x)\,dx\right]. \tag{7.1.26}$$

At point 2 we transform from position to momentum space and obtain

$$\tilde{\psi}(p) = \psi_0 \sqrt{\left|\frac{x'(p)}{p}\right|} \exp\left\{\frac{i}{\hbar}\left[\int_{x_0}^{x(p)} p(x)\,dx - px(p)\right] - \frac{i\pi}{4}\right\}. \tag{7.1.27}$$

In contrast to Eq. (7.1.24) we have now written $x(p)$ instead of x_s to indicate that via relation (7.1.23) x_s is a function of p. We have further used that $x'(p)$ is negative in the lower left quadrant. This representation becomes invalid at the point where the trajectory crosses the p axis. Therefore we transform the wave function back to position space at point 3:

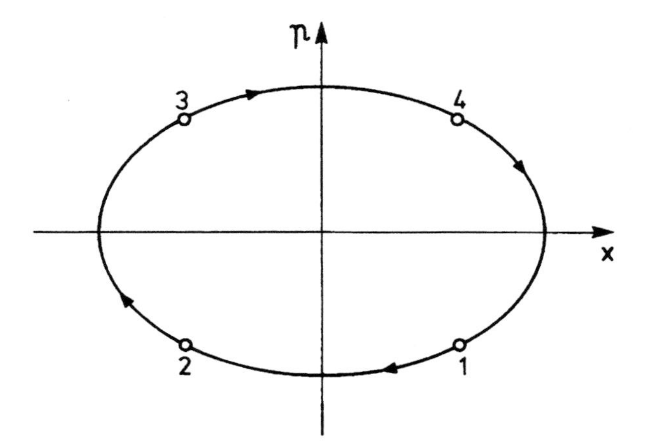

Figure 7.1. Typical phase portrait of a particle in a one-dimensional potential with one minimum.

$$\psi(x) = \frac{1}{\sqrt{2\pi\hbar}} \int dp \tilde{\psi}(p) \exp\left(\frac{\imath}{\hbar} xp\right)$$

$$= \frac{\psi_0}{\sqrt{2\pi\hbar}} \int dp \sqrt{\left|\frac{x'(p)}{p}\right|}$$

$$\times \exp\left\{\frac{\imath}{\hbar}\left[\int_{x_0}^{x(p)} p(x)\,dx - px(p) + px\right] - \frac{\imath\pi}{4}\right\}. \tag{7.1.28}$$

The integral is again evaluated by the stationary phase approximation. The phase and its first two derivatives are now given by

$$\Phi(p) = \frac{1}{\hbar}\left[\int_{x_0}^{x(p)} p(x)\,dx - px(p) + xp\right], \tag{7.1.29}$$

$$\Phi'(p) = x - x(p), \tag{7.1.30}$$

$$\Phi''(p) = x'(p). \tag{7.1.31}$$

Repeating the same steps as above we obtain for the representation of the wave function in the coordinate space

$$\psi(x) = \frac{\psi_0}{\sqrt{|p(x)|}} \exp\left(\frac{\imath}{\hbar}\int_{x_0}^{x} p(x)\,dx - \frac{\imath\pi}{2}\right). \tag{7.1.32}$$

A comparison with expression (7.1.26) shows that whilst passing the turning point a phase loss of $\pi/2$ has occurred. We continue in this way and obtain, after a complete orbit, a phase shift of

$$\Delta\Phi = \frac{1}{\hbar} S - \frac{\nu\pi}{2}, \tag{7.1.33}$$

where $S = \oint p(x)\,dx$ is the action for the complete orbit, and ν, called the *Maslov index*, is the number of turning points on the orbit. In the present case we have $\nu = 2$.

Equation (7.1.33) has to be modified for billiard hard-wall reflections. For Dirichlet boundary conditions each reflection leads to a phase jump of π, i.e. a hard-wall reflection contributes twice to the Maslov index. For Neumann boundary conditions, on the other hand, there is no phase jump.

Equation (7.1.33) suggests that only such orbits are semiclassically allowed for which the phase differences on a complete orbit are multiple integers of 2π,

$$\frac{1}{\hbar}\oint p\,dx - \frac{\nu\pi}{2} = 2\pi n, \qquad n = 1, 2, \dots . \tag{7.1.34}$$

This is essentially the Bohr–Sommerfeld quantization rule of early quantum mechanics. The Maslov index, however, was not yet known at that time.

As an illustrative example we take the harmonic oscillator

$$\mathcal{H} = \frac{p^2}{2m_e} + \frac{m_e\omega^2}{2}x^2. \tag{7.1.35}$$

For a given energy E we have for the action

$$\begin{aligned} S &= \oint p\,dx \\ &= 2\int_{-\sqrt{2E/m_e\omega^2}}^{\sqrt{2E/m_e\omega^2}} \sqrt{2m_e\left(E - \frac{m_e\omega^2}{2}x^2\right)}\,dx \\ &= 2\pi\frac{E}{\omega}. \end{aligned} \tag{7.1.36}$$

Inserting this expression into Eq. (7.1.34) we obtain the energy quantization condition

$$E_n = \hbar\omega\left(n + \frac{1}{2}\right), \tag{7.1.37}$$

which is identical with the exact quantum mechanical result. This correspondence would not have been obtained without the Maslov index, which is responsible for the $\frac{1}{2}$-term on the right hand side.

7.1.2 Multidimensional integrable systems

The semiclassical quantization procedure introduced in Section 7.1.1 can be easily extended to integrable systems with an arbitrary number of variables. For d degrees of freedom the trajectories are on the surface of a d-dimensional torus in the $2d$-dimensional phase space. Every possible closed orbit along the torus gives rise to a separate quantization condition

$$\frac{1}{\hbar}\oint p_i\,dx_i = 2\pi\left(n_i + \frac{\nu_i}{4}\right), \quad i = 1, \ldots, d. \tag{7.1.38}$$

Each eigenvalue is thus classified unequivocally by d quantum numbers n_i, $i = 1, \ldots, d$:

$$E_{n_1,\ldots,n_d} = H(n_1, \ldots, n_d), \tag{7.1.39}$$

where H is the Hamilton function of the system. Take as an example the d-dimensional harmonic oscillator with the eigenfrequencies ω_i ($i = 1, \ldots, d$) and the eigenenergies

$$E_{n_1,\ldots,n_d} = \sum_{i=1}^{d} \hbar\omega_i\left(n_i + \frac{1}{2}\right). \tag{7.1.40}$$

Another example is the particle in a d-dimensional rectangular box with side lengths a_i ($i = 1, \ldots, d$), whose eigenvalues for Dirichlet boundary conditions are given by

$$E_{n_1,\ldots,n_d} = \sum_{i=1}^{d} \left(\frac{\pi n_i}{a_i} \right)^2 . \tag{7.1.41}$$

From Eq. (7.1.39) we obtain for the density of states

$$\rho(E) = \sum_n \delta[E - H(n)], \tag{7.1.42}$$

where we have introduced the vector $n = (n_1, \ldots, n_d)$ of quantum numbers to abbreviate the notation. The mean density of states

$$\rho_0(E) = \int \delta[E - H(n)]\, dn \tag{7.1.43}$$

will not be constant in general. We therefore introduce a new energy variable $\hat{E}$ defined by

$$\hat{E} = n_0(E) = \int_0^E \rho_0(E)\, dE, \tag{7.1.44}$$

where $n_0(E)$ is the integrated mean density of states. If we take on both sides the differentials,

$$d\hat{E} = \rho_0(E)\, dE, \tag{7.1.45}$$

we see that for the rescaled energies the mean density of states is constant and one. The Hamilton function has to be rescaled correspondingly. We thus get for the harmonic oscillator the rescaled Hamilton function

$$\hat{H}(m) = \frac{\left(\sum_{i=1}^{d} \omega_i m_i \right)^d}{d! \prod_{i=1}^{d} \omega_i} . \tag{7.1.46}$$

The rescaled Hamilton function for the particle in a d-dimensional rectangular box is given by

$$\hat{H}(m) = \frac{1}{\Gamma(1 + \frac{d}{2})} \left(\frac{\pi}{4} \right)^{d/2} \prod_{i=1}^{d} a_i \left[\sum_{i=1}^{d} \left(\frac{m_i}{a_i} \right)^2 \right]^{d/2} . \tag{7.1.47}$$

In the following we shall omit the hats and assume that $H(n)$ has already been normalized to a constant density of one. The Hamilton function on the defolded energy scale obeys an interesting scaling relation. Substituting $n_i = E^{1/d} n_i'$ in Eq. (7.1.43) for the mean density of states we get

$$\begin{aligned} \rho_0(E) &= \int \delta\left[E - H\left(E^{1/d} n'\right)\right] E\, dn' \\ &= \int \delta\left[1 - \frac{1}{E} H\left(E^{1/d} n'\right)\right] dn', \end{aligned} \tag{7.1.48}$$

using elementary relations of the delta function. Since $\rho_0(E)$ is constant, the right hand side of the equation, too, must not depend on E, whence follows

$$H(\lambda n) = \lambda^d H(n). \tag{7.1.49}$$

H is hence a homogeneous function of the quantum numbers of degree d. For the two examples given above this is immediately evident from Eqs. (7.1.46) and (7.1.47).

With the scaling relation (7.1.49) as the main ingredient Berry and Tabor [Ber77] could show that in integrable systems the level spacing distribution is Poissonian. By means of the Poisson sum rule (1.19) they transformed the sum (7.1.42) to

$$\rho(E) = \sum_m \int \delta[E - H(n)] e^{2\pi\imath nm}\, dn, \tag{7.1.50}$$

and evaluated the integral in the stationary phase approximation. Unfortunately their calculations are tricky and require some skill in the treatment of multi-dimensional integrals. We therefore confine ourselves to the particularly simple example of a particle in a two-dimensional rectangular box, which nevertheless contains all essential elements of the calculation. The rescaled Hamilton function is obtained from Eq. (7.1.47) as

$$H(n, m) = \frac{\pi}{4} ab \left[\left(\frac{n}{a} \right)^2 + \left(\frac{m}{b} \right)^2 \right], \tag{7.1.51}$$

where we have now written a and b for the sides of the rectangle. Introducing the wave number

$$k_{nm} = \sqrt{\left(\frac{\pi n}{a} \right)^2 + \left(\frac{\pi m}{b} \right)^2}, \tag{7.1.52}$$

this may be rewritten as

$$H(n, m) = \frac{ab}{4\pi} k_{nm}^2. \tag{7.1.53}$$

In Section 7.3.1 we shall see that the mean level density of a particle with mass m_e in a rectangular box is given by

$$\rho_0 = \frac{2m_e}{\hbar^2} \frac{ab}{4\pi}. \tag{7.1.54}$$

Since ρ_0 has been normalized to one, it follows that

$$\frac{ab}{4\pi} = \frac{\hbar^2}{2m_e}. \tag{7.1.55}$$

Thus Eq. (7.1.53) is the usual expression for the kinetic energy.

Entering expression (7.1.53) into Eq. (7.1.42) and applying Poisson's sum rule (1.19), we get for the density of states

$$\rho(E) = \sum_{n,m} \int \delta[E - H(x, y)] e^{2\pi\imath(nx+my)}\, dx\, dy$$

$$= \frac{ab}{\pi^2} \sum_{n,m} \int_0^\infty k'\, dk' \int_0^{\pi/2} d\phi\, \delta\left(E - \frac{ab}{4\pi} k'^2\right) e^{2\imath k'(na\cos\phi + mb\sin\phi)}. \quad (7.1.56)$$

With the substitution $E = (ab/4\pi)k^2$ we finally obtain

$$\rho(E) = \frac{2}{\pi} \sum_{n,m} \int_0^{\pi/2} d\phi e^{2\imath k(na\cos\phi + mb\sin\phi)}$$

$$= \frac{1}{2\pi} \sum_{n,m} \int_0^{2\pi} d\phi e^{2\imath k(na\cos\phi + mb\sin\phi)}$$

$$= \sum_{n,m} J_0(kl_{nm}), \quad (7.1.57)$$

with

$$l_{nm} = 2\sqrt{(na)^2 + (mb)^2}. \quad (7.1.58)$$

In Eq. (7.1.57) we have used in the second step the fact that the sum is running over positive and negative values. We may therefore replace the integral $\int_0^{\pi/2} d\phi$ by $\frac{1}{4}\int_0^{2\pi} d\phi$. The l_{nm} are the lengths of the periodic orbits of the rectangle. This can best be visualized by interpreting the rectangle as the unit cell of a rectangular lattice. Then the l_{nm} are twice the lengths of the segments connecting two lattice points, each line corresponding to a periodic orbit. The factor 2 takes into account that the segment must be retraced to obtain a closed orbit.

In the next step we calculate the spectral autocorrelation function (3.2.84). In addition we replace the Bessel functions by their asymptotic behaviour. The latter step is not essential, but facilitates the calculations. We then have

$$C(E) = \left\langle \left[\rho\left(\overline{E} + \frac{E}{2}\right) - 1\right]\left[\rho\left(\overline{E} - \frac{E}{2}\right) - 1\right]\right\rangle$$

$$= \left\langle \frac{2}{\pi\sqrt{k_1 k_2}} \sum_{n,m,n',m'}{}' \frac{1}{\sqrt{l_{nm} l_{n'm'}}} \cos\left(k_1 l_{nm} - \frac{\pi}{4}\right) \cos\left(k_2 l_{n'm'} - \frac{\pi}{4}\right)\right\rangle, \quad (7.1.59)$$

where the prime denotes that the term $n = n' = m = m' = 0$ is absent in the sum. k_1 and k_2 are given by

$$k_{1/2} = \sqrt{\frac{4\pi}{ab}\left(\overline{E} \pm \frac{E}{2}\right)}$$
$$= \overline{k} \pm \frac{\pi}{ab\overline{k}} E, \tag{7.1.60}$$

where we have assumed $|E| \ll \overline{E}$. In averaging over $\overline{E}$, all rapidly oscillating terms vanish. The only terms remaining are the diagonal ones with $l_{nm} = l_{n'm'}$ whence follows

$$C(E) = \frac{1}{\pi\overline{k}} \sum_{n,m}{}' \frac{\epsilon_{nm}}{l_{nm}} \cos[(k_1 - k_2) l_{nm}]$$
$$= \frac{1}{\pi\overline{k}} \sum_{n,m}{}' \frac{\epsilon_{nm}}{l_{nm}} \cos\left(\frac{2\pi l_{nm}}{ab\overline{k}} E\right), \tag{7.1.61}$$

where

$$\epsilon_{nm} = \begin{cases} 4 & n \neq 0,\ m \neq 0 \\ 2 & n = 0,\ m \neq 0 \text{ or } n \neq 0,\ m = 0 \end{cases}. \tag{7.1.62}$$

The factor ϵ_{nm} accounts for the fact that the sum over n and m runs over positive and negative values, i.e. there are diagonal contributions from the terms $n = \pm n'$ and $m = \pm m'$.

As the diagonal approximation is crucial, a short explanation is appropriate. If the expression on the right hand side of Eq. (7.1.59) is averaged over a wave number window of width Δk, only those terms survive whose length differences $\Delta l = |l_{nm} - l_{n'm'}|$ are of the order of $(\Delta k)^{-1}$ or smaller. The number of periodic orbits with lengths in the range $\langle l_{nm}, l_{nm} + \Delta l\rangle$ is of the order of $l_{nm}\Delta l \sim l_{nm}/\Delta k$, as follows from Eq. (7.1.58). Since the weight of an orbit of length l_{nm} is of $\mathcal{O}(l_{nm}^{-1})$ according to Eq. (7.1.61), the contribution of the nondiagonal terms is of $\mathcal{O}[(\Delta k)^{-1}]$ and becomes negligibly small in the limit of large window widths.

The same argumentation also applies to higher dimensional integrable billiards, but not to chaotic systems, where the number of periodic orbits increases exponentially with the length. In chaotic systems the diagonal approximation therefore generally produces the wrong results. We shall come back to this point in Section 8.2.2.

The spectral form factor is obtained as the Fourier transform of the spectral autocorrelation function (see Section 3.2.5)

$$K(t) = \frac{1}{\pi \bar{k}} \sum_{n,m}{}' \frac{\epsilon_{nm}}{l_{nm}} \int \cos\left(\frac{2\pi l_{nm}}{ab\bar{k}} E\right) \exp\left(-\frac{\imath}{\hbar} Et\right) dE$$

$$= \frac{ab}{2\pi} \sum_{n,m}{}' \frac{\epsilon_{nm}}{l_{nm}} \delta(l_{nm} - vt). \tag{7.1.63}$$

In the second step we have introduced the velocity $v = \hbar\bar{k}/m_e$, and have used once more the normalization (7.1.55).

Since the density of periodic orbits increases with the inverse of the length, we may replace the sum by an integral (in the range of shortest orbits the approximation becomes doubtful; this aspect will be discussed in Section 8.2.1). In addition we may replace ϵ_{nm} by 4, since the fraction of orbits where either n or m is zero becomes negligibly small in the limit of large lengths. We then get

$$K(t) = \frac{2ab}{\pi} \int_{-\infty}^{\infty} dn \int_{-\infty}^{\infty} dm \, \frac{1}{2\sqrt{(an)^2 + (bm)^2}} \delta\left(2\sqrt{(an)^2 + (bm)^2} - vt\right). \tag{7.1.64}$$

The integrations are immediately carried out and yield the surprisingly simple result

$$K(t) = 1. \tag{7.1.65}$$

The spectral form factor is constant and has a value of one! This is equivalent to the statement that the eigenvalues are uncorrelated and Poisson distributed, as we have learnt in Section 3.2.5.

The same is true for nearly all integrable systems as has been shown by Berry and Tabor [Ber77]. The multidimensional harmonic oscillator is exceptional, since here the stationary phase approximation cannot be applied to calculate the integral in Eq. (7.1.50). The difficulties are caused by the fact that the quantum numbers enter linearly into expression (7.1.40) for the Hamilton function. Consequently the surfaces of constant energy are planes in the d-dimensional space of quantum numbers. It is, however, essential for the functioning of the stationary phase approximation that the curvatures do not vanish.

As we already know from elementary quantum mechanics, the eigenvalues of the harmonic oscillator form an equidistant sequence, which obviously does not obey a Poisson level spacing distribution. We might think that the situation changes with incommensurable eigenfrequencies, but this is not true either, as has been shown by Berry and Tabor.

Meanwhile it has become evident that the number of exceptions to the behaviour predicted by Berry and Tabor for integrable systems is much larger than previously believed. The small but significant deviations from the Poisson level spacing distribution observed for rectangular billiards have already been mentioned in Section 3.1.1. It has been proved by Creham [Cre95] that any spectral sequence allowing a certain growth law can be associated with an infinite family of classically integrable Hamiltonians whose quantum spectrum coincides with this sequence.

7.2 Gutzwiller trace formula

7.2.1 Feynman path integral

For nonintegrable systems the motion in phase space is no longer confined to multidimensional tori, and a semiclassical quantization is not possible. Therefore Einstein [Ein17] had rejected the semiclassical approach as a whole in 1917. Further development took another course and led to the formulation of modern quantum mechanics by Heisenberg, Schrödinger, Pauli and others. Nearly half a century later interest in semiclassical quantum mechanics revived, when Gutzwiller found a way to master the difficulties resulting from the fact that torus quantization is not possible in nonintegrable systems.

In the next sections we shall develop his ideas step by step until we arrive at an expression which may be considered as a generalization of the WKB approximation to nonintegrable systems. We shall begin with the Feynman path integral formulation of the quantum mechanical propagator [Fey65]. We shall now rederive the essential relations, since probably many readers are not familiar with them. We start with the quantum mechanical propagator $K(q_A, q_B, t)$. It is zero for $t < 0$, and is given by

$$K(q_A, q_B, t) = \langle q_B | U(t) | q_A \rangle \tag{7.2.1}$$

for $t > 0$, where

$$U(t) = \exp\left(-\frac{\imath}{\hbar} \mathscr{H} t \right) \tag{7.2.2}$$

is the time-evolution operator. Remember that the eigenfunctions $|q_A\rangle$ are delta functions of the position. $|K(q_A, q_B, t)|^2$ is the quantum mechanical probability density of finding a particle at a time t at the position q_B having started at time $t = 0$ at the position q_A. The Fourier transform of the propagator is the Green function

$$G(q_A, q_B, E) = -\frac{\imath}{\hbar}\int_0^\infty dt\, K(q_A, q_B, t) \exp\left(\frac{\imath}{\hbar}Et\right)$$
$$= \left\langle q_B \left| \frac{1}{E - \mathscr{H}} \right| q_A \right\rangle, \tag{7.2.3}$$

which has already been introduced in Section 3.1.5. Remember that E has been assumed to have an infinitesimally small positive imaginary part, guaranteeing that the integral in the above equation is well-defined.

For the calculation of the propagator we divide the time t into N intervals of equal length $\tau = t/N$. As the time-evolution operator obeys the relation

$$U(t) = [U(\tau)]^N, \tag{7.2.4}$$

the propagator may be written as

$$K(q_A, q_B, t) = \int dq_1 \ldots dq_{N-1} \langle q_B|U(\tau)|q_{N-1}\rangle \ldots \langle q_1|U(\tau)|q_A\rangle. \tag{7.2.5}$$

In the further calculation we take

$$\mathscr{H} = \frac{p^2}{2m_e} + V(q) \tag{7.2.6}$$

for the Hamiltonian. For small τ values we may expand the time-evolution operator as

$$U(\tau) = \exp\left[-\frac{\imath\tau}{\hbar}\left(\frac{p^2}{2m_e} + V(q)\right)\right]$$
$$\approx \exp\left(-\frac{\imath\tau}{\hbar}\frac{p^2}{2m_e}\right)\exp\left(-\frac{\imath\tau}{\hbar}V(q)\right). \tag{7.2.7}$$

The errors introduced by the approximation are of $\mathscr{O}(\tau^2)$ (see Eq. (4.3.48)), and can be neglected in the limit of large N. The matrix elements of $U(\tau)$ now read

$$\langle q_B|U(\tau)|q_A\rangle = \left\langle q_B \left| \exp\left(-\frac{\imath\tau}{\hbar}\frac{p^2}{2m_e}\right)\right| q_A \right\rangle \exp\left(-\frac{\imath\tau}{\hbar}V(q_A)\right). \tag{7.2.8}$$

The remaining matrix element can be calculated by replacing $|q_A\rangle$ by its Fourier representation

$$|q_A\rangle = \delta(q_A - q) = \frac{1}{(2\pi\hbar)^d}\int \exp\left[\frac{\imath}{\hbar}(q_A - q)p\right] dp. \tag{7.2.9}$$

We then obtain

$$\left\langle q_B \left| \exp\left(-\frac{\imath\tau}{\hbar}\frac{p^2}{2m_e}\right) \right| q_A \right\rangle$$

$$= \int \delta(q_B - q) \exp\left(-\frac{\imath\tau}{\hbar}\frac{p^2}{2m_e}\right) \times \frac{1}{(2\pi\hbar)^d} \exp\left[\frac{\imath}{\hbar}(q_A - q)p\right] dq\, dp$$

$$= \frac{1}{(2\pi\hbar)^d} \int \exp\left\{\frac{\imath}{\hbar}\left[-\frac{\tau p^2}{2m_e} + (q_A - q_B)p\right]\right\} dp. \quad (7.2.10)$$

After a shift of variables the integral reduces to a Fresnel integral and yields

$$\left\langle q_B \left| \exp\left(-\frac{\imath\tau}{\hbar}\frac{p^2}{2m_e}\right) \right| q_A \right\rangle = \left(\frac{m_e}{2\pi\imath\hbar\tau}\right)^{d/2} \exp\left[\frac{\imath}{\hbar}\frac{m_e}{2\tau}(q_A - q_B)^2\right]. \quad (7.2.11)$$

Inserting this expression into Eq. (7.2.8) we obtain for the propagator in the limit of small τ values

$$\langle q_B | U(\tau) | q_A \rangle = \left(\frac{m_e}{2\pi\imath\hbar\tau}\right)^{d/2} \exp\left\{\frac{\imath}{\hbar}\left[\frac{m_e}{2\tau}(q_A - q_B)^2 - \tau V\left(\frac{q_A + q_B}{2}\right)\right]\right\}. \quad (7.2.12)$$

We have substituted q_A by $(q_A + q_B)/2$ in the argument of V, which is allowed in the limit of small τ. Any position between q_A and q_B would do equally well. If the Hamiltonian contains a vector potential term, however, it is mandatory to choose the position halfway between q_A and q_B, as discussed in Ref. [Gut89]. For the expression entering into the exponential we introduce the notation

$$W(q_A, q_B, \tau) = W_{BA}(\tau) = \frac{m_e}{2\tau}(q_A - q_B)^2 - \tau V\left(\frac{q_A + q_B}{2}\right). \quad (7.2.13)$$

$W_{BA}(\tau)$ allows a simple classical interpretation. For small τ we may write

$$\frac{q_A - q_B}{\tau} \approx \dot{q}_A \approx -\dot{q}_B, \quad (7.2.14)$$

whence follows

$$W_{BA}(\tau) = \tau\left[\frac{m_e}{2}(\dot{q}_A)^2 - V\left(\frac{q_A + q_B}{2}\right)\right]. \quad (7.2.15)$$

This can be written as

$$W_{BA}(\tau) = \int_0^\tau L(q, \dot{q})\, dt, \quad (7.2.16)$$

where

$$L(q, \dot{q}) = \frac{m_e}{2}\dot{q}^2 - V(q) \quad (7.2.17)$$

is the classical *Lagrange function*. Thus $W_{BA}(t)$ is exactly *Hamilton's principal function* known from classical mechanics.

The prefactor entering into the right hand side of Eq. (7.2.12) may be expressed in terms of the second derivatives of the principal function,

$$\frac{m_e}{\tau} = -\frac{\partial^2 W_{BA}(\tau)}{\partial q_{A_i} \partial q_{B_i}}, \tag{7.2.18}$$

where q_{A_i} and q_{B_i} are arbitrary components of q_A and q_B, respectively. For $i \neq j$ the second derivatives $\partial^2 W_{BA}(\tau)/\partial q_{A_i} \partial q_{B_j}$ vanish for small τ values. We may therefore write

$$\left(\frac{m_e}{\tau}\right)^d = \left| -\frac{\partial^2 W_{BA}(\tau)}{\partial q_{A_i} \partial q_{B_i}} \right| = |D_{BA}|, \tag{7.2.19}$$

where we have introduced D_{BA} as the negative of the matrix of the second derivatives of W_{BA}. We have thus obtained the following formulation for the propagator, holding for small times,

$$\langle q_B | U(\tau) | q_A \rangle = \left(\frac{1}{2\pi \imath \hbar}\right)^{d/2} |D_{BA}|^{1/2} \exp\left(\frac{\imath}{\hbar} W_{BA}\right). \tag{7.2.20}$$

Inserting expression (7.2.20) into Eq. (7.2.5) for the propagator we end with Feynman's path integral formulation for the propagator,

$$K(q_A, q_B, t) = \lim_{N \to \infty} \left(\frac{1}{2\pi \imath \hbar}\right)^{Nd/2} \int dq_1 \ldots dq_{N-1} \left| \prod_{i=0}^{N-1} D_{i,i+1} \right|^{1/2} \exp\left(\frac{\imath}{\hbar} \sum_{i=0}^{N-1} W_{i,i+1}\right), \tag{7.2.21}$$

where $q_0 = q_A$ and $q_N = q_B$. Equation (7.2.21) may be written in a symbolic short-hand notation as

$$K(q_A, q_B, t) = \int \mathscr{D}(q) \exp\left[\frac{\imath}{\hbar} \int_0^t L(q, \dot{q})\, dt\right], \tag{7.2.22}$$

where the differential $\mathscr{D}(q)$ denotes that the integration is over all paths connecting q_A and q_B. It is understood that the weight factors are incorporated into the differential.

It is remarkable that only classical quantities enter into the right hand side of the equation. Quantum mechanics comes in only via the factor $\hbar^{-1}$ in the exponential. Nevertheless in the limit $N \to \infty$ the integral reproduces the quantum mechanical properties correctly. The path integral is definitely not an effective way to calculate the propagator, but it is, as we shall see, the ideal starting point for semiclassical calculations.

7.2.2 A short excursion in classical mechanics

Before we evaluate the Feynman path integral by means of the stationary phase approximation, let us first recapitulate a number of definitions and relations from classical mechanics, which will be frequently used.

The symbolic expression (7.2.22) for the quantum mechanical propagator shows that the phases are stationary for all paths for which the relation

$$\delta \int_0^t L(q, \dot{q})\, dt = 0 \tag{7.2.23}$$

holds. This is exactly the Hamilton principle of classical mechanics, from which the classical equations of motion can be derived. Consequently in the stationary phase approximation only the classically allowed paths contribute to the Feynman path integral. But expression (7.2.22) is only symbolic, and therefore this conclusion must be considered as preliminary.

Carrying out the variation, we get

$$\begin{aligned} 0 &= \int_0^t \left(\frac{\partial L}{\partial q} \delta q + \frac{\partial L}{\partial \dot{q}} \delta \dot{q} \right) dt \\ &= \int_0^t \left[\frac{\partial L}{\partial q} - \frac{d}{dt} \left(\frac{\partial L}{\partial \dot{q}} \right) \right] \delta q\, dt. \end{aligned} \tag{7.2.24}$$

In the second step a partial integration has been performed using the fact that the variation δq must vanish at the limits of integration. Apart from this restriction the variation is arbitrary, hence the integrand must vanish as a whole. In this way we obtain the Lagrange differential equation

$$\frac{\partial L}{\partial q} = \frac{d}{dt} \left(\frac{\partial L}{\partial \dot{q}} \right) = \dot{p}, \tag{7.2.25}$$

where

$$p = \frac{\partial L}{\partial \dot{q}} \tag{7.2.26}$$

is the momentum. From the Lagrange differential equation a first integral of the equations of motion is immediately obtained. To this end we differentiate the Lagrange function with respect to time,

$$\begin{aligned}\dot{L} &= \frac{\partial L}{\partial q}\frac{dq}{dt} + \frac{\partial L}{\partial \dot{q}}\frac{d\dot{q}}{dt} \\ &= \dot{p}\frac{dq}{dt} + p\frac{d\dot{q}}{dt} \\ &= \frac{d}{dt}(p\dot{q}). \end{aligned} \tag{7.2.27}$$

In the second step Eqs. (7.2.25) and (7.2.26) have been used. We have thus found one constant of motion

$$E = p\dot{q} - L, \tag{7.2.28}$$

corresponding to the total energy. Equation (7.2.28) allows an alternative notation of the principal function $W(q_A, q_B, t)$ as

$$\begin{aligned}W(q_A, q_B, t) &= \int_0^t L(q, \dot{q})dt \\ &= \int_0^t (p\dot{q} - E)dt \\ &= \int_{q_A}^{q_B} p\,dq - Et. \end{aligned} \tag{7.2.29}$$

This expression has the advantage that the arguments q_A, q_B, and t of W directly enter into the expression on the right hand side. The partial derivatives of W are thus calculated in a straightforward manner and yield

$$\frac{\partial W}{\partial q_B} = p_B, \quad \frac{\partial W}{\partial q_A} = -p_A, \quad \frac{\partial W}{\partial t} = -E. \tag{7.2.30}$$

The action $S(q_A, q_B, E)$, which has already been introduced in Section 7.1.1, is closely related to the principal function. It is defined by

$$S(q_A, q_B, E) = \int_{q_A}^{q_B} p\,dq, \tag{7.2.31}$$

whence follows

$$S(q_A, q_B, E) = W(q_A, q_B, t) + Et. \tag{7.2.32}$$

The replacement of W by S corresponds to a change of variables from t to E. Therefore on the right hand side t has now to be interpreted as a function of E. Bearing this in mind we get for the partial derivative of S with respect to E

$$\frac{\partial S}{\partial E} = \frac{\partial W}{\partial t}\frac{dt}{dE} + t + E\frac{dt}{dE} = t, \tag{7.2.33}$$

where $\partial W/\partial t = -E$ has been used. The partial derivatives of S with respect to q_A and q_B are the same as for W. We thus have

$$\frac{\partial S}{\partial q_B} = p_B, \quad \frac{\partial S}{\partial q_A} = -p_A, \quad \frac{\partial S}{\partial E} = t. \tag{7.2.34}$$

These relations as well as the corresponding ones for W are vital for the semiclassical theory. It has been the main intention of the present section to give a straightforward derivation of these equations, since in most standard textbooks they are mentioned only cursorily.

The following equations, however, can be found in any introduction on classical mechanics. They are repeated here for the sake of completeness. From the Lagrange function we obtain the Hamilton function as

$$H(p, q) = p\dot{q} - L(q, \dot{q}). \tag{7.2.35}$$

While L is a function of q and $\dot{q}$, H is a function of p and q. Consequently $\dot{q}$ on the right hand side is a function of p and q. We obtain for the partial derivatives of H with respect to p and q the canonical equations

$$\frac{\partial H}{\partial p} = \dot{q} + p\frac{\partial \dot{q}}{\partial p} - \frac{\partial L}{\partial \dot{q}}\frac{\partial \dot{q}}{\partial p} = \dot{q}, \tag{7.2.36}$$

$$\frac{\partial H}{\partial q} = p\frac{\partial \dot{q}}{\partial q} - \frac{\partial L}{\partial q} - \frac{\partial L}{\partial \dot{q}}\frac{\partial \dot{q}}{\partial q} = -\dot{p}, \tag{7.2.37}$$

where the Lagrange differential equation (7.2.25) and definition (7.2.26) of p have been used.

The Hamilton function is a constant of motion,

$$H(p, q) = E, \tag{7.2.38}$$

which follows from Eq. (7.2.28) and definition (7.2.35). Since p_A and p_B can be expressed via Eqs. (7.2.34) in terms of the partial derivatives of S, Eq. (7.2.38) may be alternatively written as

$$H\left(\frac{\partial S}{\partial q_B}, q_B\right) = H\left(-\frac{\partial S}{\partial q_A}, q_A\right) = E. \tag{7.2.39}$$

These are the Hamilton–Jacobi differential equations, which are equivalent to the canonical equations.

7.2.3 Semiclassical propagator

We are now going to continue the discussion of Section 7.2.1. Our starting point is the expression (7.2.20) for the propagator,

$$\begin{aligned} K(q_A, q_B, \tau) &= \langle q_B|U(\tau)|q_A\rangle \\ &= \left(\frac{1}{2\pi\imath\hbar}\right)^{d/2} |D_{BA}|^{1/2} \exp\left[\frac{\imath}{\hbar} W_{BA}(\tau)\right], \end{aligned} \tag{7.2.40}$$

valid in the limit of infinitely small τ. We shall now show that in the semiclassical approximation this expression holds for *all* τ values, apart from the fact that classical turning points, focal points etc. will give rise to additional phases, just as in the one-dimensional case.

From Eq. (7.2.40) we obtain for the propagator after the time $t = 2\tau$

$$K(q_A, q_B, 2\tau) = \int \langle q_B|U(\tau)|q_C\rangle\langle q_C|U(\tau)|q_A\rangle dq_C$$
$$= \left(\frac{1}{2\pi\imath\hbar}\right)^d \int dq_C |D_{BC} D_{CA}|^{1/2} \exp\left\{\frac{\imath}{\hbar}[W_{BC}(\tau) + W_{CA}(\tau)]\right\}. \tag{7.2.41}$$

The integral is calculated by means of the stationary phase approximation. The phase of the exponential function as well as its first and second derivatives are given by

$$\Phi(q_C) = \frac{1}{\hbar}[W_{BC}(\tau) + W_{CA}(\tau)], \tag{7.2.42}$$

$$\frac{\partial \Phi}{\partial q_C} = \frac{1}{\hbar}\left(\frac{\partial W_{BC}}{\partial q_C} + \frac{\partial W_{CA}}{\partial q_C}\right), \tag{7.2.43}$$

$$\frac{\partial^2 \Phi}{\partial q_C^2} = \frac{1}{\hbar}\left(\frac{\partial^2 W_{BC}}{\partial q_C^2} + \frac{\partial^2 W_{CA}}{\partial q_C^2}\right). \tag{7.2.44}$$

The stationary phase point q_C is obtained from the condition

$$\frac{\partial W_{BC}}{\partial q_C} + \frac{\partial W_{CA}}{\partial q_C} = -p_C^{(BC)} + p_C^{(CA)} = 0, \tag{7.2.45}$$

where we have used Eq. (7.2.30) for the partial derivatives of the principal function. $p_C^{(CA)}$ is the momentum at the end of the trajectory leading from q_A to q_C, and $p_C^{(BC)}$ is the momentum at the beginning of the trajectory from q_C to q_B. The stationary phase condition demands that both momenta are equal, i.e. that the momentum is continuous in the point q_C. Expressed in other words: the trajectory passing from q_A via q_C to q_B is classically allowed. This is one of the central results of semiclassical quantum mechanics: only classically allowed trajectories from q_A to q_B survive the stationary phase approximation!

But if the point q_C is on a classically allowed trajectory from q_A to q_B, the principal function is additive,

$$W_{BC}(\tau) + W_{CA}(\tau) = W_{BA}(2\tau). \tag{7.2.46}$$

The stationary phase approximation (7.1.19) now yields

$$K(q_A, q_B, 2\tau) = \left(\frac{1}{2\pi\imath\hbar}\right)^{d/2} \left\{ \left| -\frac{\partial^2 W_{BC}}{\partial q_B \partial q_C} \right| \left| -\frac{\partial^2 W_{CA}}{\partial q_C \partial q_A} \right| \Big/ \left| \frac{\partial^2 W_{BC}}{\partial q_C^2} + \frac{\partial^2 W_{CA}}{\partial q_C^2} \right| \right\}^{1/2} \times \exp\left[\frac{\imath}{\hbar} W_{BA}(2\tau)\right]. \quad (7.2.47)$$

The determinant prefactor can be simplified considerably. To this end we differentiate Eq. (7.2.45) with respect to q_C keeping the starting position q_A fixed. The end position q_B has then to be considered as a function of q_C. We thus get

$$\frac{\partial^2 W_{BC}}{\partial q_B \partial q_C}\frac{\partial q_B}{\partial q_C} + \frac{\partial^2 W_{BC}}{\partial q_C^2} + \frac{\partial^2 W_{CA}}{\partial q_C^2} = 0. \quad (7.2.48)$$

Multiplying the equation by $-\partial^2 W_{BA}/\partial q_B \partial q_A$, shifting the first term to the right hand side, and taking the determinant on both sides, we obtain

$$\begin{aligned}
\left| \frac{\partial^2 W_{BC}}{\partial q_C^2} + \frac{\partial^2 W_{CA}}{\partial q_C^2} \right| \left| -\frac{\partial^2 W_{BA}}{\partial q_B \partial q_A} \right| &= \left| -\frac{\partial^2 W_{BC}}{\partial q_B \partial q_C} \right| \left| \frac{\partial q_B}{\partial q_C} \right| \left| -\frac{\partial^2 W_{BA}}{\partial q_B \partial q_A} \right| \\
&= \left| -\frac{\partial^2 W_{BC}}{\partial q_B \partial q_C} \right| \left| -\frac{\partial^2 W_{BA}}{\partial q_B \partial q_A}\frac{\partial q_B}{\partial q_C} \right| \\
&= \left| -\frac{\partial^2 W_{BC}}{\partial q_B \partial q_C} \right| \left| -\frac{\partial^2 W_{BA}}{\partial q_C \partial q_A} \right| \\
&= \left| -\frac{\partial^2 W_{BC}}{\partial q_B \partial q_C} \right| \left| -\frac{\partial^2 W_{CA}}{\partial q_C \partial q_A} \right|. \quad (7.2.49)
\end{aligned}$$

In the last step we have used the identity $\partial W_{BA}/\partial q_A = \partial W_{CA}/\partial q_A$ following from the additivity relation (7.2.46). The determinant factor entering into Eq. (7.2.47) thus yields

$$\left| -\frac{\partial^2 W_{BC}}{\partial q_B \partial q_C} \right| \left| -\frac{\partial^2 W_{CA}}{\partial q_C \partial q_A} \right| \Big/ \left| \frac{\partial^2 W_{BC}}{\partial q_C^2} + \frac{\partial^2 W_{CA}}{\partial q_C^2} \right| = \left| -\frac{\partial^2 W_{BA}}{\partial q_B \partial q_A} \right| = |D_{BA}|. \quad (7.2.50)$$

We have hence obtained for the propagator

$$K(q_A, q_B, 2\tau) = \langle q_B | U(2\tau) | q_A \rangle = \left(\frac{1}{2\pi\imath\hbar}\right)^{d/2} |D_{BA}|^{1/2} \exp\left[\frac{\imath}{\hbar} W_{BA}(2\tau)\right]. \quad (7.2.51)$$

This solution is identical with the initial expression (7.2.40), with the only exception that τ is replaced by 2τ.

Nowhere in the derivation have we assumed that τ is small. We can therefore conclude that the relation holds in the semiclassical approximation for arbitrary values of τ, as has been stated at the beginning of this section.

Two addenda, however, are necessary. If the determinant $|\partial^2 W_{BA}/\partial q_B \partial q_A|$ becomes singular, the stationary phase approximation breaks down. This happens at the so-called *conjugated points*. Here we adopt the technique described in Section 7.1.1 for one dimension, and bypass the conjugated point via the momentum space. In the multidimensional case this produces an additional phase of $-\frac{\nu\pi}{2}$, where ν is the number of vanishing eigenvalues of the determinant.

Furthermore it is possible that there are several classically allowed trajectories leading from q_A to q_B in time t. In this case we have

$$K(q_A, q_B, t) = \left(\frac{1}{2\pi\imath\hbar}\right)^{d/2} \sum_r |D_{BA,r}|^{1/2} \exp\left(\frac{\imath}{\hbar} W_{BA,r}(t) - \imath\frac{\nu_r \pi}{2}\right), \quad (7.2.52)$$

where the sum is over all classically allowed trajectories leading from q_A to q_B, and where ν_r is the number of conjugated points on the rth trajectory, including possible multiplicities.

Equation (7.2.52) allows a simple semiclassical interpretation [Ber72]. Take a swarm of trajectories leaving a point q_A isotropically. Quantum mechanically the number of distinguishable initial conditions is limited by Heisenberg's uncertainty relation. The density $w(q_A, p_A)$ of trajectories for a particle to leave point q_A with momentum p_A, is therefore given by

$$w(q_A, p_A) = \left(\frac{1}{2\pi\hbar}\right)^d. \quad (7.2.53)$$

Let us denote the probability density that a trajectory will arrive at point q_B by $p(q_A, q_B)$. It is obtained from $w(q_A, p_B)$ by means of the relation

$$p(q_A, q_B)\, dq_B = w(q_A, p_A) \left|\frac{\partial p_A}{\partial q_B}\right| dq_B. \quad (7.2.54)$$

Expressing p_A in terms of a partial derivative of the action integral (see Eq. (7.2.30)) $p(q_A, q_B)$ can be written as

$$p(q_A, q_B) = \left(\frac{1}{2\pi\hbar}\right)^d \left| -\frac{\partial^2 W_{BA}}{\partial q_A \partial q_B}\right| = \left(\frac{1}{2\pi\hbar}\right)^d |D_{BA}|. \quad (7.2.55)$$

The right hand side of the equation is identical with the square of the modulus of the quantum mechanical propagator (7.2.51). If there is only one classically allowed trajectory connecting q_A and q_B, the classical and the semiclassical calculations thus yield exactly the same result. If there are different paths contributing to the propagator, the situation changes. In the quantum mechanical calculation we have to add the amplitudes and take the square of the modulus of the sum, whereas classically the probabilities of the different paths add up incoherently.

The conjugated points can be interpreted classically as well. These are the points where the determinant $|-\partial^2 W_{BA}/\partial q_B \partial q_A| = |\partial p_A/\partial q_B|$ becomes singular, or, alternatively, the points where the inverse matrix $\partial q_B/\partial p_A$ has vanishing eigenvalues. This situation is illustrated in Fig. 7.2(a). The direction of any trajectory emitted at q_A can be varied without changing the component of q_B along the dashed line, showing that the determinant $|\partial q_B/\partial p_A|$ is zero. The dashed line is a so-called *caustic*, the number of vanishing eigenvalues of $\partial q_B/\partial p_A$ is called the *order* of the caustic. If the order is identical to the dimension of the space, we have a *focal point*. This situation is shown in Fig. 7.2(b): now *all* components of q_B are unchanged, if the direction of the trajectories emitted from q_A is varied.

The studies of Tomsovic and Heller [Tom91, Tom93] on the pulse propagation in a stadium billiard nicely demonstrate the results of this section. Figure 7.3 shows a sequence of snapshots of a wave packet. It starts travelling to the right and is destroyed after a small number of reflections. For a quantitative analysis the authors studied the autocorrelation function

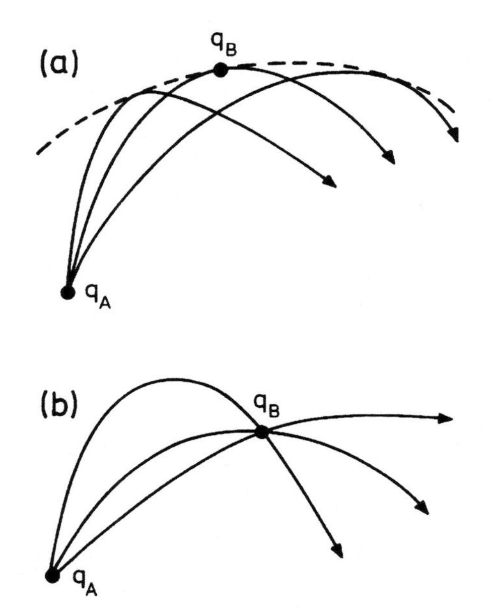

Figure 7.2. Caustic (a) and focal point (b).

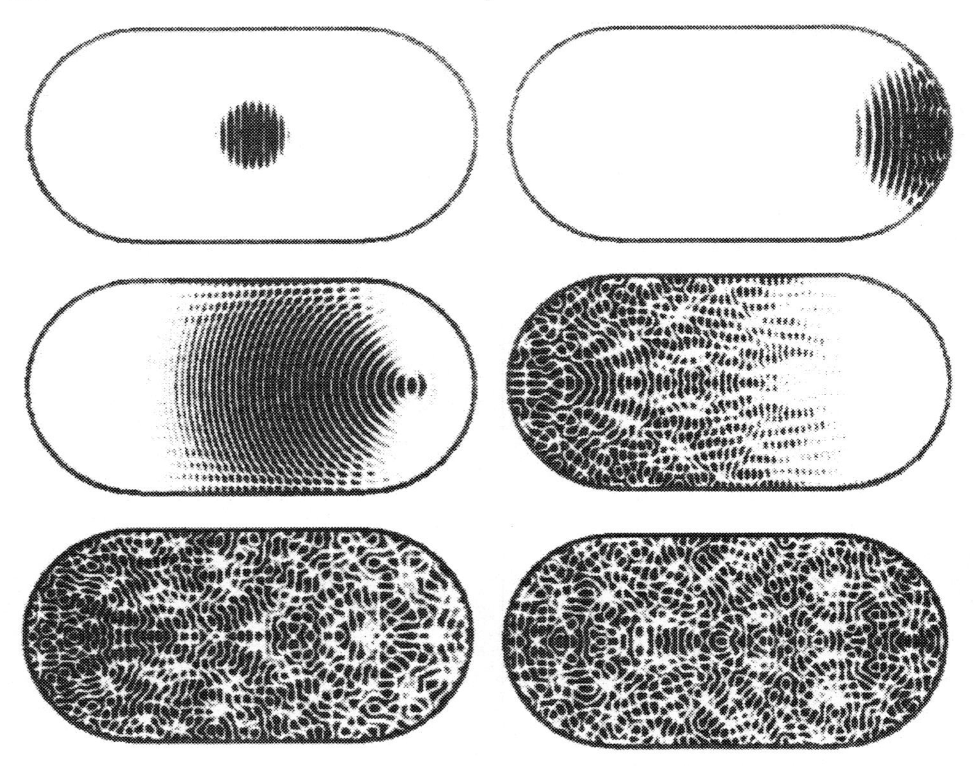

Figure 7.3. Propagation of a Gaussian wave packet in a stadium billiard. At time $t = 0$ the packet starts travelling to the right. After several reflections the packet is uniformly distributed over the complete stadium. The times are $t/T = 0, 0.5, 1.5, 2, 3, 6$, where T is the time needed for a particle to propagate along the stadium axis [Tom93] (Copyright 1993 by the American Physical Society).

$$C(t) = \int \psi(q, 0)\psi(q, t)\, dq, \tag{7.2.56}$$

where $\psi(q, 0)$ and $\psi(q, t)$ describe the packet at the time $t = 0$ and at a later time t, respectively. $C(t)$ may be written in terms of the quantum mechanical propagator as

$$C(t) = \int |\psi(q, 0)|^2 K(q, q, t)\, dq. \tag{7.2.57}$$

The authors calculated the propagator first by solving the Schrödinger equation for the stadium, and second by using the semiclassical approximation (7.2.52). Figure 7.4 shows that for periods corresponding to a six-fold propagation of a particle along the stadium axis both calculations yield essentially the same result! The limits of the semiclassical calculation are

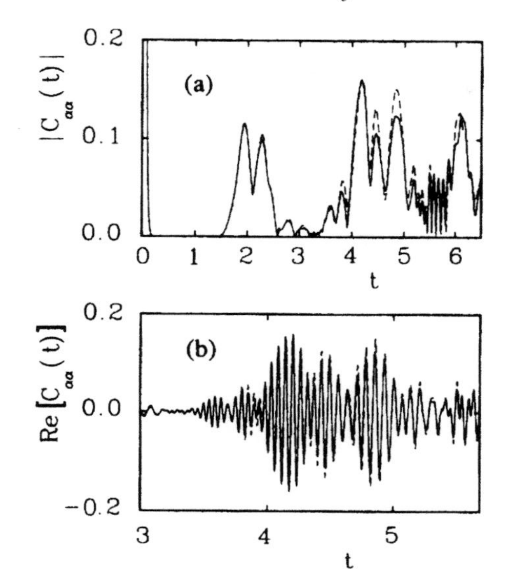

Figure 7.4. Quantum mechanical (solid line) and semiclassical (dashed line) autocorrelation function $C(t)$ describing the wave packet propagation displayed in Fig. 7.3. The upper part shows $|C(t)|$, the lower one $\mathrm{Re}[C(t)]$ [Tom93] (Copyright 1993 by the American Physical Society).

purely technical. To get the result 30 000 trajectories had to be considered. To extend the calculation to eight stadium lengths 10^6 trajectories would have been needed. Here the exponential increase in the number of trajectories with the length, typical for all chaotic systems, becomes manifest. An introduction to the propagation of wave packets has been given by Heller in the Les Houches lectures [Hel89].

The pulse propagation in a stadium-shaped microwave billiard discussed in the introduction [Ste95] constitutes an experimental equivalent to the studies of Tomsovic and Heller. In Fig. 1.1 the electromagnetic propagator $K(q_A, q_B, t)$ is plotted for a sequence of times. In the experiment the position q_A of the entrance antenna was fixed, whereas the position of the exit antenna q_B was varied. The electromagnetic and the quantum mechanical propagator only differ in their dispersion behaviour, as has been discussed in the introduction. The most conspicuous result of the experiment is pulse reconstruction, which is caused by the focusing properties of the concave mirror formed by the circular part of the stadium boundary. This is a direct demonstration of the close correspondence between classical and semiclassical dynamics. Alternatively we may look upon pulse reconstruction as an experimental manifestation of a conjugate point.

7.2.4 Semiclassical Green function

In the second step towards the trace formula we are going to derive a semiclassical expression for the Green function. $G(q_A, q_B, E)$ is obtained as the Fourier transform of the propagator,

$$G(q_A, q_B, E) = -\frac{\imath}{\hbar}\int_0^\infty dt \exp\left(\frac{\imath}{\hbar}Et\right) K(q_A, q_B, t) \tag{7.2.58}$$

(see Eq. (7.2.3)). Inserting the semiclassical approximation (7.2.52) for the propagator we get

$$G(q_A, q_B, E) =$$
$$-\frac{\imath}{\hbar}\left(\frac{1}{2\pi\imath\hbar}\right)^{d/2}\sum_r \int_0^\infty dt\, |D_{BA,r}|^{1/2} \exp\left\{\frac{\imath}{\hbar}[W_r(q_A, q_B, t) + Et] - \imath\frac{\nu_r\pi}{2}\right\}. \tag{7.2.59}$$

The integral is once more evaluated by means of the stationary phase approximation. The phase of the exponential function and its derivatives are now given by

$$\Phi(t) = \frac{1}{\hbar}[W_r(q_A, q_B, t) + Et], \tag{7.2.60}$$

$$\dot{\Phi}(t) = \frac{1}{\hbar}\left(\frac{\partial W_r}{\partial t} + E\right), \tag{7.2.61}$$

$$\ddot{\Phi}(t) = \frac{1}{\hbar}\frac{\partial^2 W_r}{\partial t^2}. \tag{7.2.62}$$

The stationary phase time t_0 is obtained from the solution of the equation

$$\left.\frac{\partial W_r}{\partial t}\right|_{t=t_0} = -E. \tag{7.2.63}$$

Due to this relation t_0 becomes a function of q_A, q_B, and E. The phase at time t_0 is given by

$$\Phi(t_0) = \frac{1}{\hbar}[W_r(q_A, q_B, t_0) + Et_0] = \frac{1}{\hbar}S_r(q_A, q_B, E), \tag{7.2.64}$$

where the relation (7.2.32) between action and principal function has been used. Performing the stationary phase approximation we get for the Green function

$$G(q_A, q_B, E) = -\frac{\imath}{\hbar}\left(\frac{1}{2\pi\imath\hbar}\right)^{d/2}\sqrt{2\pi\hbar}\sum_r\left\{|D_{BA,r}|\bigg/\left|\frac{\partial^2 W_{BA,r}}{\partial t^2}\right|\right\}^{1/2}$$
$$\times\exp\left[\frac{\imath}{\hbar}S_r(q_A, q_B, E) - \imath\frac{\nu_r\pi}{2} + \frac{\imath\pi}{4}\mathrm{sgn}\left(\frac{\partial^2 W_{BA,r}}{\partial t^2}\right)\right]. \tag{7.2.65}$$

In Eq. (7.2.65) the argument of the exponential no longer depends on the principal function but on the action. We are therefore going to express the determinant prefactor in terms of the action as well. The calculation is somewhat technical, and anybody not interested in the details may proceed directly to Eq. (7.2.78).

We start by differentiating Eq. (7.2.63) with respect to E,

$$\frac{\partial^2 W}{\partial t^2}\frac{\partial t_0}{\partial E} = -1. \tag{7.2.66}$$

The partial derivatives of W have to be taken at the stationary phase point t_0. Expressing t_0 in terms of a partial derivative of the action, $t_0 = \partial S/\partial E$ (see Eq. (7.2.34)), we obtain

$$\frac{\partial^2 W}{\partial t^2} = -\left(\frac{\partial^2 S}{\partial E^2}\right)^{-1}. \tag{7.2.67}$$

From definition (7.2.31) of the action

$$S(q_A, q_B, E) = \int_{q_A}^{q_B} p\,dq$$
$$= \int_{q_A}^{q_B}\sqrt{2m_e[E - V(q)]}\,|dq|, \tag{7.2.68}$$

where $|dq|$ is the modulus of the path differential along the trajectory, we see that the second derivative of S with respect to E is negative:

$$\frac{\partial^2 S}{\partial E^2} = -\frac{\sqrt{2m_e}}{4}\int_{q_A}^{q_B}[E - V(q)]^{-3/2}\,|dq|. \tag{7.2.69}$$

$\partial^2 W/\partial t^2$ is therefore positive, resulting in an additional phase of $+\frac{\pi}{4}$ from the stationary phase approximation (see Eq. (7.2.65)).

In the next step we differentiate Eq. (7.2.64) with respect to q_A,

$$\frac{\partial S}{\partial q_A} = \frac{\partial W}{\partial q_A} + \frac{\partial W}{\partial t}\frac{\partial t_0}{\partial q_A} + E\frac{\partial t_0}{\partial q_A} = \frac{\partial W}{\partial q_A}. \tag{7.2.70}$$

A further differentiation with respect to q_B yields

$$\frac{\partial^2 S}{\partial q_A \partial q_B} = \frac{\partial^2 W}{\partial q_A \partial q_B} + \frac{\partial^2 W}{\partial q_A \partial t}\frac{\partial t_0}{\partial q_B}$$
$$= \frac{\partial^2 W}{\partial q_A \partial q_B} + \frac{\partial^2 W}{\partial q_A \partial t}\frac{\partial^2 S}{\partial q_B \partial E}. \tag{7.2.71}$$

From Eq. (7.2.63), we obtain, by a differentiation with respect to q_A,

$$\frac{\partial^2 W}{\partial q_A \partial t} + \frac{\partial^2 W}{\partial t^2}\frac{\partial t_0}{\partial q_A} = 0, \tag{7.2.72}$$

or

$$\frac{\partial^2 W}{\partial q_A \partial t} = \frac{\partial^2 S}{\partial q_A \partial E} \Big/ \frac{\partial^2 S}{\partial E^2}. \tag{7.2.73}$$

Entering this result into Eq. (7.2.71) we get

$$\frac{\partial^2 W}{\partial q_A \partial q_B} = \frac{\partial^2 S}{\partial q_A \partial q_B} - \frac{\partial^2 S}{\partial q_A \partial E}\frac{\partial^2 S}{\partial q_B \partial E} \Big/ \frac{\partial^2 S}{\partial E^2}. \tag{7.2.74}$$

Collecting the results we obtain for the determinant in Eq. (7.2.65)

$$|\Delta_{BA}| = |D_{BA}| \Big/ \left| \frac{\partial^2 W_{BA}}{\partial t^2} \right|$$
$$= (-1)^{d+1} \frac{\partial^2 S}{\partial E^2} \cdot \left| \frac{\partial^2 S}{\partial q_A \partial q_B} - \frac{\partial^2 S}{\partial q_A \partial E}\frac{\partial^2 S}{\partial q_B \partial E} \Big/ \frac{\partial^2 S}{\partial E^2} \right|. \tag{7.2.75}$$

This can be written somewhat more conveniently by means of the determinant relation

$$\begin{vmatrix} A & B \\ C & D \end{vmatrix} = |A - BD^{-1}C||D|. \tag{7.2.76}$$

Here A, B, and C are $N \times N$, $N \times M$, and $M \times N$ matrices, respectively, and D is a nonsingular $M \times M$ matrix.

The relation is proved by taking the determinant on both sides of the matrix identity

$$\begin{pmatrix} A & B \\ C & D \end{pmatrix} \begin{pmatrix} 1 & 0 \\ -D^{-1}C & 1 \end{pmatrix} = \begin{pmatrix} A - BD^{-1}C & B \\ 0 & D \end{pmatrix}.$$

Application to Eq. (7.2.75) yields for the determinant

$$|\Delta_{BA}| = (-1)^{d+1} \begin{vmatrix} \dfrac{\partial^2 S}{\partial q_A \partial q_B} & \dfrac{\partial^2 S}{\partial q_A \partial E} \\ \dfrac{\partial^2 S}{\partial q_B \partial E} & \dfrac{\partial^2 S}{\partial E^2} \end{vmatrix} = \begin{vmatrix} -\dfrac{\partial^2 S}{\partial q_A \partial q_B} & -\dfrac{\partial^2 S}{\partial q_A \partial E} \\ -\dfrac{\partial^2 S}{\partial q_B \partial E} & -\dfrac{\partial^2 S}{\partial E^2} \end{vmatrix}, \tag{7.2.77}$$

whence follows for the semiclassical Green function

$$G(q_A, q_B, E) = -\frac{\imath}{\hbar}\left(\frac{1}{2\pi\imath\hbar}\right)^{(d-1)/2} \sum_r |\Delta_{BA,r}|^{1/2} \exp\left[\frac{\imath}{\hbar} S_r(q_A, q_B, E) - \imath\frac{\nu_r \pi}{2}\right]. \tag{7.2.78}$$

In microwave billiards the Green function can be directly obtained from the experiment, as we have learnt in Section 6.1.2. This allows an easy test of the semiclassical theory. The action for a trajectory of length l in billiards is given by the simple expression

$$S = \hbar k l, \tag{7.2.79}$$

whence follows for the semiclassical Green function

$$\overline{G}(q_A, q_B, k) = -\frac{\imath}{\hbar}\left(\frac{1}{2\pi\imath\hbar}\right)^{(d-1)/2} \sum_r |\Delta_{BA,r}|^{1/2} \exp\left[\imath\left(k l_r - \frac{\nu_r \pi}{2}\right)\right]. \tag{7.2.80}$$

We have taken the wave number k as variable, which is more appropriate for billiards than E, and have denoted the corresponding Green function by $\overline{G}(q_A, q_B, k)$. The contributions of the different trajectories are obtained from the Fourier transform

$$\hat{G}(q_A, q_B, l) = \int \overline{G}(q_A, q_B, k) e^{-\imath k l}\, dk \tag{7.2.81}$$

of $\overline{G}(q_A, q_B, k)$. $\hat{G}(q_A, q_B, l)$ shows a maximum for each trajectory of length l running from q_A to q_B.

The predictions from the semiclassical approximation have been tested in a quarter stadium microwave billiard. The reflection probability has been measured as a function of k for different positions q of the antenna [Ste92]. This measurement directly yields the diagonal part $\overline{G}(q, q, k)$ of the Green function (see Section 6.1.2). Its Fourier transform $\hat{G}(q, q, l)$ should thus exhibit maxima at all l values corresponding to lengths of *closed* trajectories starting and ending at q. A closed trajectory should not be mistaken for a periodic one. In contrast to periodic trajectories, the directions of closed ones may differ at the starting and the end points. On the left column of Fig. 7.5 the Fourier transform $\hat{G}(q, q, l)$ is shown for a sequence of different l values. On the right column the results of a classical calculation are given. They have been obtained by colouring each pixel q being the starting and the end point of a trajectory of length l. The figure demonstrates the nearly perfect correspondence between experimental data and simulation.

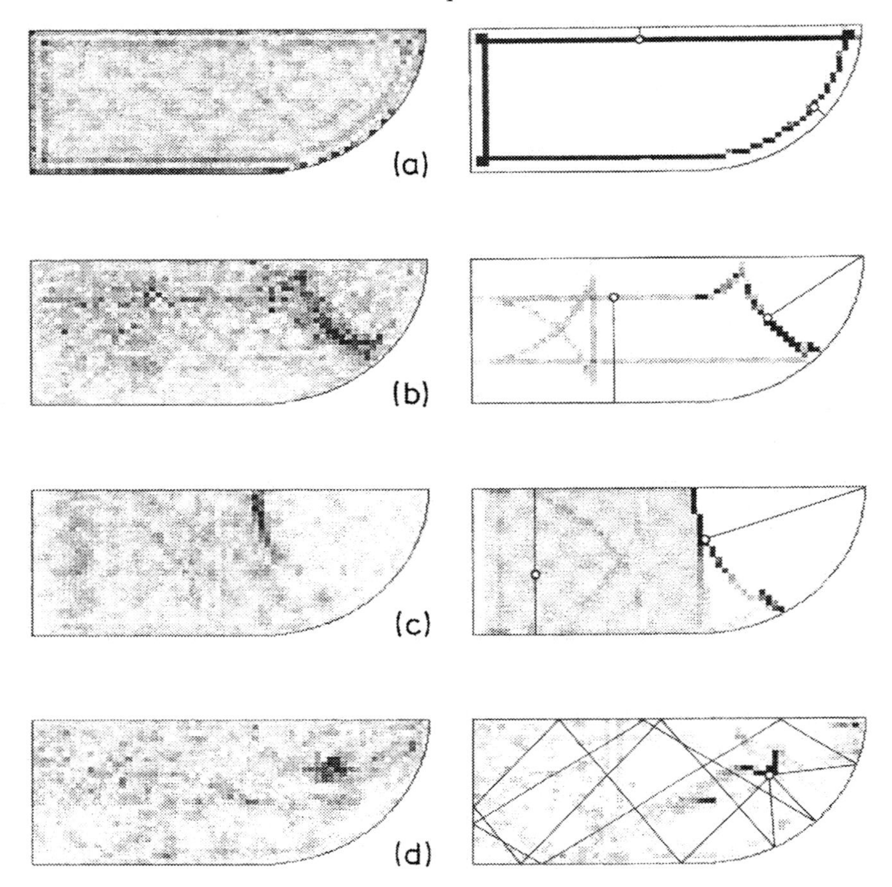

Figure 7.5. Fourier transform $\hat{G}(q, q, l)$ of the reflection spectrum of a quarter stadium microwave billiard ($L = 18$ cm, $R = 13.5$ cm) for different lengths of closed trajectories $l = 2.4$ cm (a), 19.5 cm (b), 27.4 cm (c), 84.1 cm (d) (left column). For the presentation $|\hat{G}(q, q, l)|$ has been turned into a grey scale. The right part of the figure shows the result of a classical calculation. For details see text [Ste92] (Copyright 1992 by the American Physical Society).

7.2.5 Monodromy matrix

In the last section we derived a semiclassical expression for the quantum mechanical Green function, depending only on properties of the classical trajectories. In the next section we shall calculate the spectral density from the trace of the Green function, but some preparatory steps are necessary.

Let as assume that a trajectory starts from a point q_A with momentum p_A and ends at a point q_B with momentum p_B. We then take another trajectory starting with the slightly different initial conditions $q_A + \delta q_{A\perp}$, $p_A + \delta p_{A\perp}$.

The index '$\perp$' denotes that the deviations are taken perpendicular to the initial trajectory. At q_B the new trajectory deviates from the original one by $\delta q_{B\perp}$ and $\delta p_{B\perp}$. The deviations at the end point are obtained from the deviations at the starting point in linear approximation as

$$\begin{pmatrix} \delta q_{B\perp} \\ \delta p_{B\perp} \end{pmatrix} = M_{BA} \begin{pmatrix} \delta q_{A\perp} \\ \delta p_{A\perp} \end{pmatrix}. \tag{7.2.82}$$

M_{BA} is called the *monodromy matrix*. Since both $\delta q_{A\perp}$ and $\delta p_{A\perp}$ are $(d-1)$-dimensional vectors, the rank of the monodromy matrix is $2(d-1)$. The components of M_{BA} can be calculated from the second derivatives of the action. To this end we start from the relations

$$p_A = -\frac{\partial S}{\partial q_A}, \qquad p_B = \frac{\partial S}{\partial q_B}, \tag{7.2.83}$$

for the partial derivatives of S (see Eq. (7.2.34)). Replacing q_A by $q_A + \delta q_A$, p_A by $p_A + \delta p_A$, etc., and expanding the right hand side of the equations up to the linear terms, we get

$$\begin{aligned} \delta p_{A\perp} &= -S_{AA}\delta q_{A\perp} - S_{AB}\delta q_{B\perp}, \\ \delta p_{B\perp} &= S_{AB}\delta q_{A\perp} + S_{BB}\delta q_{B\perp}. \end{aligned} \tag{7.2.84}$$

S_{AA}, S_{AB}, etc., are the matrices of the second derivatives of S with respect to the perpendicular components of the starting and the end point. Rearranging the equations we get

$$\begin{aligned} \delta q_{B\perp} &= -S_{AB}^{-1}S_{AA}\delta q_{A\perp} - S_{AB}^{-1}\delta p_{A\perp}, \\ \delta p_{B\perp} &= \left(S_{AB} - S_{BB}S_{AB}^{-1}S_{AA}\right)\delta q_{A\perp} - S_{BB}S_{AB}^{-1}\delta p_{A\perp}. \end{aligned} \tag{7.2.85}$$

Comparing this expression with definition (7.2.82), we see that the monodromy matrix is given by

$$M_{BA} = \begin{pmatrix} -S_{AB}^{-1}S_{AA} & -S_{AB}^{-1} \\ S_{AB} - S_{BB}S_{AB}^{-1}S_{AA} & -S_{BB}S_{AB}^{-1} \end{pmatrix}. \tag{7.2.86}$$

From the matrix identity

$$\begin{aligned} &\begin{pmatrix} \lambda\cdot 1 - S_{AB}^{-1}S_{AA} & -S_{AB}^{-1} \\ S_{AB} - S_{BB}S_{AB}^{-1}S_{AA} & \lambda\cdot 1 - S_{BB}S_{AB}^{-1} \end{pmatrix} \\ &\qquad = \begin{pmatrix} \lambda\cdot 1 & -1 \\ S_{AB} - \lambda S_{AA} & \lambda S_{AB} - S_{BB} \end{pmatrix} \begin{pmatrix} 1 & \cdot \\ S_{AB}^{-1}S_{AA} & S_{AB}^{-1} \end{pmatrix} \end{aligned} \tag{7.2.87}$$

we obtain, by taking the determinant on both sides,

$$|M_{BA} + \lambda\cdot 1| = |(\lambda^2 + 1)S_{AB} - \lambda(S_{AA} + S_{BB})| / |S_{AB}|. \tag{7.2.88}$$

Taking in particular $\lambda = 0$ we see that the determinant of the monodromy matrix is one:

$$|M_{BA}| = 1. \tag{7.2.89}$$

Another special case of importance is obtained from relation (7.2.88) by taking $\lambda = -1$:

$$|M_{BA} - 1| = |S_{AA} + 2S_{AB} + S_{BB}|/|S_{AB}|. \tag{7.2.90}$$

The monodromy matrix obeys the multiplicative property

$$M_{BA} = M_{BC} M_{CA}, \tag{7.2.91}$$

where q_C is an arbitrary point on the trajectory between q_A and q_B. This follows from definition (7.2.82). For a periodic orbit the determinant $|M_{AA} - 1|$ is consequently independent of the choice of the starting position q_A:

$$\begin{aligned} |M_{AA} - 1| &= |M_{AC}(M_{CA} - M_{AC}^{-1})| \\ &= |(M_{CA} - M_{AC}^{-1}) M_{AC}| \\ &= |M_{CC} - 1|. \end{aligned} \tag{7.2.92}$$

In two dimensions the rank of the monodromy matrix is two. As the determinant of the monodromy matrix is one, its two eigenvalues occur in pairs of λ and λ^{-1}. Furthermore, since the elements of the monodromy matrix are real, the eigenvalues are either real, or they are complex conjugates of each other. There are only three possible situations:

In the *elliptic* case the two eigenvalues are given by $\lambda_{1/2} = e^{\pm \imath \alpha}$, which corresponds to a stable orbit. A trajectory which starts with initial conditions slightly deviating from a periodic orbit will oscillate about the orbit, but will always remain close to it. For the *hyperbolic* and the *inverse-hyperbolic* case the eigenvalues are given by $\lambda_{1/2} = e^{\pm \alpha}$ and $\lambda_{1/2} = -e^{\pm \alpha}$, respectively. In the two latter cases the orbits are unstable, and any deviation from the orbit will increase exponentially with the length. α is called the *instability exponent.*

As the trace of a matrix is given by the sum of its eigenvalues, we may, as an alternative, discriminate between the three cases by means of the trace:

(i) elliptic case: $\mathrm{Tr}(M) < 2$
(ii) hyperbolic case: $\mathrm{Tr}(M) > 2$
(iii) inverse-hyperbolic case: $\mathrm{Tr}(M) < -2$.

Again billiard systems are best suited to illustrate the concept of the monodromy matrix. Here every periodic orbit consists of a sequence of straight lines and reflections. It is thus sufficient to consider the monodromy matrix for the case of a reflection at a boundary with a radius of curvature of R. The monodromy matrix for the complete orbit is then obtained by multiplying the monodromy matrices for all reflections.

The action of a trajectory passing from q_A to q_B and being reflected at the boundary at a point q_R is given by

$$S = \hbar k(|q_A - q_R| + |q_B - q_R|) \tag{7.2.93}$$

(see Eq. (7.2.79)). q_R has to be chosen in such a way that the reflection law is obeyed (see Fig. 7.6). To determine the monodromy matrix we need the second derivatives of S with respect to the perpendicular components of the starting and the end positions. After a somewhat tedious calculation, which is omitted here, we get

$$S_{AA} = \frac{q}{q_A}\left(\frac{2}{R} - \frac{\cos\gamma}{q_B}\right), \tag{7.2.94}$$

$$S_{AB} = -\frac{q\cos\gamma}{q_A q_B}, \tag{7.2.95}$$

$$S_{BB} = \frac{q}{q_B}\left(\frac{2}{R} - \frac{\cos\gamma}{q_A}\right), \tag{7.2.96}$$

where

$$q = \left(\frac{2}{R} - \frac{\cos\gamma}{q_A} - \frac{\cos\gamma}{q_B}\right)^{-1}. \tag{7.2.97}$$

Using Eq. (7.2.86) we obtain for the monodromy matrix

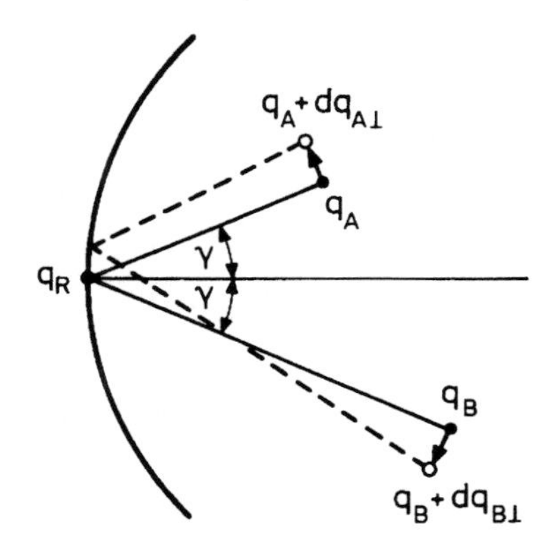

Figure 7.6. Sketch illustrating the monodromy matrix for reflection at a circular boundary. For details see text.

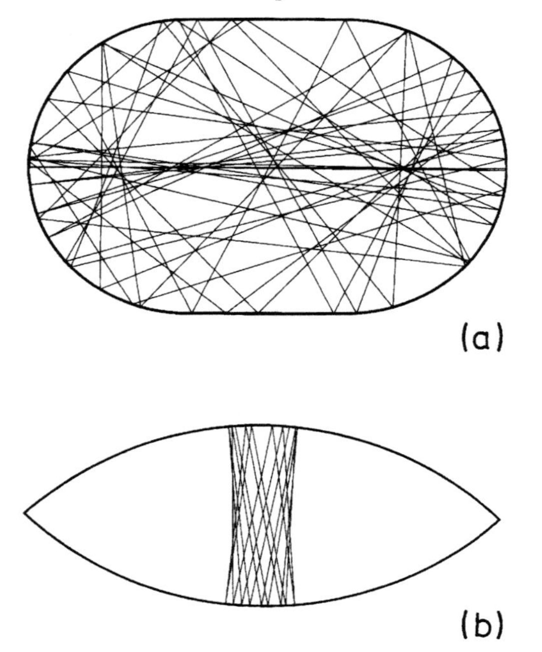

Figure 7.7. Classical trajectories in a stadium (a) and a lemon (b) billiard. In both cases the trajectories start along the symmetry axis with a slight deviation from the ideal line. For the stadium the trajectory becomes chaotic after a short time, but for the lemon billiard the trajectory remains always close to the symmetry axis.

$$M_{BA} = \begin{pmatrix} -1 + \dfrac{2q_B}{2\cos\gamma} & \dfrac{q_A q_B}{q\cos\gamma} \\ \dfrac{2}{R\cos\gamma} & -1 + \dfrac{2q_A}{R\cos\gamma} \end{pmatrix}. \tag{7.2.98}$$

In the limiting case $q_A, q_B \ll R$, corresponding to the reflection at a plane, M_{BA} reduces to the negative unit matrix, i.e. $\delta q_{A\perp}$ and $\delta p_{A\perp}$ just change their signs. Monodromy matrices for a number of periodic orbits in the stadium billiard have been calculated by Bogomolny [Bog88]. He adopted a somewhat different sign convention with the consequence that his monodromy matrix for the reflection at the plane is the *positive* unit matrix.

As an example we shall calculate the stability of the periodic orbit along the long axis in a stadium billiard. Applying Eq. (7.2.98) twice (for the two reflections at the end caps) we obtain for the trace of the monodromy matrix for the complete orbit

$$\mathrm{Tr}(M) = 2 + 16\frac{A}{R}\left(\frac{A}{R} - 1\right), \tag{7.2.99}$$

independent of the choice of the starting point on the orbit. A is the length of the semimajor axis. Therefore the case $A = R$ corresponds to the circle. The orbit is hyperbolic for all values $A > R$. After a short time a trajectory becomes chaotic, even if it starts with only a minute deviation from the axis (see Fig. 7.7(a)). In fact each orbit in the stadium billiard is hyperbolically unstable, as has been proved by Bunimovich [Bun74].

Expression (7.2.99) also holds for the case $A < R$. This corresponds to a billiard in the shape of an eye or a lemon (see Fig. 7.7(b)). Here the orbit is elliptic. A trajectory starting with a small deviation from the symmetry line always remains in the central region of the billiard. The inverse-hyperbolic case cannot occur, since the expression on the right hand side of Eq. (7.2.99) never assumes a value smaller than -2.

7.2.6 *Trace formula*

In Section 3.1.5 we expressed the density of states in terms of the trace of the Green function:

$$\begin{aligned}\rho(E) &= \sum_n \delta(E - E_n) \\ &= -\frac{1}{\pi}\,\mathrm{Im}\left[\int G(q, q, E)dq\right].\end{aligned} \tag{7.2.100}$$

We are now going to derive a semiclassical approximation for the trace

$$g(E) = \int G(q, q, E)dq. \tag{7.2.101}$$

Inserting the semiclassical approximation (7.2.78) for $G(q_A, q_B, E)$ we get

$$g(E) = -\frac{\imath}{\hbar}\left(\frac{1}{2\pi\imath\hbar}\right)^{(d-1)/2} \sum_r \int dq\, |\Delta_r|^{1/2} \exp\left[\frac{\imath}{\hbar} S_r(q, q, E) - \imath\frac{\nu_r \pi}{2}\right], \tag{7.2.102}$$

where

$$|\Delta_r| = (-1)^{d+1} \left| \begin{matrix} \dfrac{\partial^2 S}{\partial q_A \partial q_B} & \dfrac{\partial^2 S}{\partial q_A \partial E} \\ \dfrac{\partial^2 S}{\partial q_B \partial E} & \dfrac{\partial^2 S}{\partial E^2} \end{matrix} \right|_{q_A = q_B = q}. \tag{7.2.103}$$

The sum is over all closed orbits starting and ending at q. The orbits of length zero which do not leave the point q are included. We shall return to these somewhat pathological trajectories in Section 7.3.1.

For the last time we apply the stationary phase approximation. The phases of the exponential and its derivatives are now given by

$$\Phi(q) = \frac{1}{\hbar} S_r(q, q, E), \tag{7.2.104}$$

$$\Phi'(q) = \frac{1}{\hbar}\left(\frac{\partial S}{\partial q_A} + \frac{\partial S}{\partial q_B}\right)\bigg|_{q_A=q_B=q} = \frac{1}{\hbar}\left(-p_A|_q + p_B|_q\right), \tag{7.2.105}$$

$$\Phi''(q) = \frac{1}{\hbar}\left(\frac{\partial^2 S}{\partial q_A^2} + 2\frac{\partial^2 S}{\partial q_A \partial q_B} + \frac{\partial^2 S}{\partial q_B^2}\right)\bigg|_{q_A=q_B=q}. \tag{7.2.106}$$

The phase is stationary for all trajectories obeying the relation

$$p_A|_q = p_B|_q. \tag{7.2.107}$$

Thus, after a complete traversal, the momentum points in the same direction as at the beginning, i.e. the trajectory is a periodic orbit. This is the central result of the semiclassical theory: it is only the classical periodic orbits that contribute to the density of states!

To carry out the integration, we introduce a coordinate system with one coordinate $q_\parallel$ parallel to the trajectory and the remaining $(d-1)$ coordinates $q_{\perp i} (i = 1, \ldots, d-1)$ perpendicular to it.

In this coordinate system we expand $S(q, q, E)$ in powers of the perpendicular coordinates up to the quadratic term:

$$\begin{aligned} S(q, q, E) = {} & S(q_\parallel, q_\parallel, E) \\ & + \frac{1}{2}\sum_{i,j} q_{\perp i}\left(\frac{\partial^2 S}{\partial q_{A\perp i}\partial q_{A\perp j}} + 2\frac{\partial^2 S}{\partial q_{A\perp i}\partial q_{B\perp j}} + \frac{\partial^2 S}{\partial q_{B\perp i}\partial q_{B\perp j}}\right) q_{\perp j} \\ = {} & S(q_\parallel, q_\parallel, E) + \frac{1}{2} q_\perp^T (S_{AA} + 2S_{AB} + S_{BB}) q_\perp. \end{aligned} \tag{7.2.108}$$

It should be remembered that S_{AA}, S_{AB}, S_{BB} have been introduced as the $(d-1)$-dimensional matrices of the second derivatives of the action with respect to the perpendicular components. As $S(q_\parallel, q_\parallel, E)$ is calculated from an integral over a closed orbit, it is independent of $q_\parallel$. This allows us to omit the $q_\parallel$ in the arguments of S,

$$S(q_\parallel, q_\parallel, E) = S(E) = \oint p\, dq. \tag{7.2.109}$$

For nonisolated periodic orbits an expansion of the action in powers of the perpendicular variables is not possible. As an example take the bouncing-ball orbits in the stadium billiard, whose actions are completely independent of the perpendicular components. Therefore bouncing-ball orbits need a separate treatment, which will be given in Section 7.3.3.

We start with the calculation of the determinant prefactor (7.2.103) which is now written as

$$|\Delta_r| = (-1)^{d+1} \begin{vmatrix} \dfrac{\partial^2 S}{\partial q_{A\|}\partial q_{B\|}} & \dfrac{\partial^2 S}{\partial q_{A\|}\partial q_{B\perp}} & \dfrac{\partial^2 S}{\partial q_{A\|}\partial E} \\ \dfrac{\partial^2 S}{\partial q_{A\perp}\partial q_{B\|}} & \dfrac{\partial^2 S}{\partial q_{A\perp}\partial q_{B\perp}} & \dfrac{\partial^2 S}{\partial q_{A\perp}\partial E} \\ \dfrac{\partial^2 S}{\partial q_{B\|}\partial E} & \dfrac{\partial^2 S}{\partial q_{B\perp}\partial E} & \dfrac{\partial^2 S}{\partial E^2} \end{vmatrix}. \tag{7.2.110}$$

The second derivatives of the action involving the parallel coordinate $q_\|$ can be explicitly calculated. To this end we differentiate the Hamilton–Jacobi equation

$$H\left(\frac{\partial S}{\partial q_B}, q_B\right) = E \tag{7.2.111}$$

(see Eq. (7.2.39)) with respect to E and get

$$\frac{\partial^2 S}{\partial q_B \partial E}\frac{\partial H}{\partial p_B} = 1. \tag{7.2.112}$$

Using the canonical equation $\partial H/\partial p_B = \dot{q}_B$ this may be written as

$$1 = \frac{\partial^2 S}{\partial q_B \partial E}\dot{q}_B = \frac{\partial^2 S}{\partial q_{B\|}\partial E}\dot{q}_{B\|}. \tag{7.2.113}$$

The second equality holds, since the perpendicular components of the velocity $\dot{q}_{B\perp}$ vanish. We have thus obtained

$$\frac{\partial^2 S}{\partial q_{B\|}\partial E} = \frac{1}{\dot{q}_{B\|}}. \tag{7.2.114}$$

In the next step we differentiate the Hamilton–Jacobi equation (7.2.111) with respect to a component q_{Ai} of q_A, which may be either parallel or perpendicular. We now get

$$0 = \frac{\partial^2 S}{\partial q_{Ai}\partial q_B}\dot{q}_B = \frac{\partial^2 S}{\partial q_{Ai}\partial q_{B\|}}\dot{q}_{B\|}, \tag{7.2.115}$$

whence follows

$$\frac{\partial^2 S}{\partial q_{Ai}\partial q_{B\|}} = 0. \tag{7.2.116}$$

Repeating the calculations with the Hamilton–Jacobi equation (7.2.39) for q_A, we get analogously

$$\frac{\partial^2 S}{\partial q_{A\|}\partial E} = -\frac{1}{\dot{q}_{A\|}}, \tag{7.2.117}$$

and

$$\frac{\partial^2 S}{\partial q_{A\|}\partial q_{Bi}} = 0. \tag{7.2.118}$$

Inserting these results into the determinant (7.2.110), we end with

$$|\Delta_r| = \frac{(-1)^{d+1}}{|\dot{q}_\||^2}\left|\frac{\partial^2 S}{\partial q_{A\perp i}\partial q_{B\perp j}}\right| = \frac{1}{|\dot{q}_\||^2}|-S_{AB}|. \tag{7.2.119}$$

Since $|\Delta_r|$ is positive, the determinant $|-S_{AB}|$ must be positive too. The trace of the Green function Eq. (7.2.102) can thus be written as

$$g(E) = -\frac{\imath}{\hbar}\left(\frac{1}{2\pi\imath\hbar}\right)^{(d-1)/2}\sum_r \int dq_\| \, dq_\perp \frac{1}{|\dot{q}_\||}\|S_{AB,r}\|^{1/2}$$
$$\times \exp\left\{\frac{\imath}{\hbar}\left[S_r(E) + \frac{1}{2}(q_\perp)^T(S_{AA,r} + 2S_{AB,r} + S_{BB,r})q_\perp\right] - \imath\frac{\nu_r\pi}{2}\right\}. \tag{7.2.120}$$

The integrals over the perpendicular components are Fresnel integrals and yield

$$g(E) = -\frac{\imath}{\hbar}\sum_r \int \frac{dq_\|}{|\dot{q}_\||}\{\|S_{AB,r}\|/\|S_{AA,r} + 2S_{AB,r} + S_{BB,r}\|\}^{1/2}$$
$$\times \exp\left[\frac{\imath}{\hbar}S_r(E) - \imath\frac{\mu_r\pi}{2}\right], \tag{7.2.121}$$

(see Eq. (7.1.14)). Here $\mu_r = \nu_r + \alpha_r$, where α_r is the number of negative eigenvalues of the matrix $(S_{AA,r} + 2S_{AB,r} + S_{BB,r})$. The determinant prefactor can be expressed in terms of the monodromy matrix (see Eq. (7.2.90)), with the result

$$g(E) = -\frac{\imath}{\hbar}\sum_r \int \frac{dq_\|}{|\dot{q}_\||}\frac{1}{\|M_r - 1\|^{1/2}}\exp\left[\frac{\imath}{\hbar}S_r(E) - \imath\frac{\mu_r\pi}{2}\right], \tag{7.2.122}$$

where M_r is the monodromy matrix for the rth orbit. The determinant $\|M_r - 1\|$ does not depend on $q_\|$, as we have seen in the last section. The same is true for the argument of the exponential. Thus the remaining $q_\|$ integration is trivial. After a substitution of variables we have

$$\int \frac{dq_\parallel}{|\dot{q}_\parallel|} = \oint dt = T_p, \tag{7.2.123}$$

where T_p is the period for the primitive orbit, i.e. the time needed for one passage. We have thus obtained the surprisingly simple expression

$$g(E) = -\frac{\imath}{\hbar} \sum_r \frac{(T_p)_r}{\|M_r - 1\|^{1/2}} \exp\left[\frac{\imath}{\hbar} S_r(E) - \imath \frac{\mu_r \pi}{2}\right] \tag{7.2.124}$$

for the trace of the Green function. Note that, though $(T_p)_r$ is the period of the primitive orbit, the sum in Eq. (7.2.124) is over *all* orbits, including repetitions. From Eq. (7.2.124) we obtain the density of states using Eq. (7.2.100),

$$\rho(E) = \frac{1}{\pi\hbar} \sum_r \frac{(T_p)_r}{\|M_r - 1\|^{1/2}} \cos\left[\frac{1}{\hbar} S_r(E) - \frac{\mu_r \pi}{2}\right]. \tag{7.2.125}$$

This is the *Gutzwiller trace formula* expressing the quantum mechanical spectrum exclusively in terms of the specific properties of the periodic orbits, namely the stability, the period, and the Maslov index.

Since there are different contributions to the Maslov index, a short recapitulation may be useful. We confine ourselves to discussion of the two-dimensional case:

1. Hard wall reflections increase μ_r by two for Dirichlet boundary conditions. The reflections do not change μ_r for Neumann boundary conditions.
2. Each passage of a conjugate point increases μ_r by one. For a conjugated point the stability prefactor (7.2.119) becomes singular. This implies $S_{AB}^{-1} = 0$, or, expressed in other words, that the upper right corner M_{12} of the monodromy matrix is zero (see Eq. (7.2.86)). Every change of sign of M_{12} on the orbit therefore increases μ_r by one.
3. μ_r is increased by one if the expression $(S_{AA} + 2S_{AB} + S_{BB})$ is negative after a complete orbit. Using the equality $(S_{AA} + 2S_{AB} + S_{BB}) = M_{12}(\mathrm{Tr}(M) - 2)$, following from expression (7.2.86) for the monodromy matrix, this may be alternatively

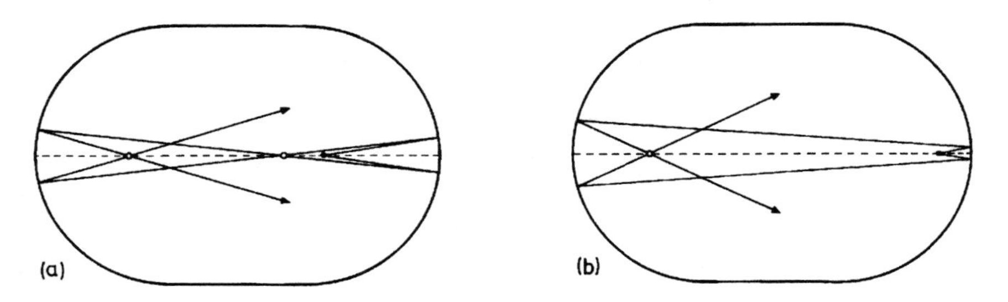

Figure 7.8. The conjugated points for an orbit along the long axis of the stadium billiard with the starting position at a distance between $\frac{R}{2}$ and R from the right end cap (a), and a distance smaller than $\frac{R}{2}$ from the right end cap (b).

expressed as follows: for hyperbolic orbits μ_r increases by one if M_{12} is negative; for elliptic or inverse hyperbolic orbits μ_r increases by one if M_{12} is positive.

It is instructive to calculate the Maslov index for a periodic orbit in a billiard. An inspection of expression (7.2.98) for the monodromy matrix pertaining to the reflection at a circular boundary shows that the upper right corner matrix element is zero for $q^{-1} = 0$, or

$$\frac{\cos\gamma}{q_A} + \frac{\cos\gamma}{q_B} = \frac{2}{R}. \tag{7.2.126}$$

This is exactly the image equation for the concave mirror, where q_A is the object distance, and q_B is the image distance. Remember that the focal length for a concave mirror with radius of curvature R is given by $\frac{R}{2}$. The conjugated points on the orbit are therefore all successive images of the starting point.

As an example we take again the orbit along the long axis in the stadium billiard. The number of reflections is two, yielding for Dirichlet boundary conditions a contribution of 4 to the Maslov index. The number of conjugated points on the orbit depends on the starting position. First, take the starting point at a distance between the focal length $\frac{R}{2}$ and the double focal length R from the right end cap (see Fig. 7.8(a)). After one reflection the first image is found somewhat outside the double focal length. Let us assume that the starting position has been chosen in such a way that this image is also outside the double focal length of the other circular boundary. Under these conditions there is another image yielding a total of two conjugated points after the second reflection. The upper right element of the monodromy matrix is positive after a complete orbit. This is best seen from definition (7.2.82) of the monodromy matrix showing that M_{12} is given by $\delta q_{B\perp}/\delta p_{A\perp}$. If $\delta p_{A\perp}$ and $\delta q_{B\perp}$ have the same sign, which is the case for the situation depicted in Fig. 7.8(a), M_{12} is positive, and there is no additional contribution to the Maslov index. We thus have for the Maslov index $\mu_r = 4 + 2 + 0 = 6$.

Second, our starting point is at a distance smaller than one focal length $\frac{R}{2}$ from the right end cap. From the first reflection there is merely a virtual image, and only the second reflection yields a real image close to the focal point of the second circular boundary. The number of conjugated points is thus only one. But $\delta q_{B\perp}$ and $\delta p_{A\perp}$ have opposite signs after a complete orbit, yielding an additional contribution to the Maslov index. Altogether we again have $\mu_r = 4 + 1 + 1 = 6$ as above. This example shows that the number of conjugated points on an orbit may depend on the starting position, but that the Maslov index μ_r for the complete orbit is an invariant.

7.3 Contributions to the density of states

7.3.1 Smooth part of the density of states

For paths of zero length (see the remark following Eq. (7.2.103)) we need the action $S(q_A, q_B, E)$ in the limit $|q_A - q_B| \to 0$. But in this limit the determinant prefactor entering on the right hand side of Eq. (7.2.102) becomes singular. The stationary phase approximation is thus no longer applicable. We therefore go back to Section 7.2.1 where we obtained for the propagator

$$K_0(q_A, q_B, t) = \left(\frac{m_e}{2\pi\imath\hbar t}\right)^{d/2} \exp\left\{\frac{\imath}{\hbar}\left[\frac{m_e}{2t}(q_B - q_A)^2 - tV\left(\frac{q_A + q_B}{2}\right)\right]\right\}, \tag{7.3.1}$$

valid in the limit of infinitesimally short times (see Eq. (7.2.12)). To denote that only the *direct* path from q_A to q_B will be considered, we have now written $K_0(q_A, q_B, t)$. Equation (7.3.1) has been derived in the limit of infinitely short times, but equivalently we may perform the limit $|q_B - q_A| \to 0$, which is needed in the present context. The contributions of the paths of zero length to the Green function are obtained from the Fourier transform of $K_0(q_A, q_B, t)$ as

$$\begin{aligned} G_0(q_A, q_B, E) &= -\frac{\imath}{\hbar}\int_0^\infty dt \exp\left(\frac{\imath}{\hbar}Et\right) K_0(q_A, q_B, t) \\ &= -\frac{\imath}{\hbar}\left(\frac{m_e}{2\pi\imath\hbar}\right)^{d/2}\int_0^\infty dt\, t^{-d/2} \\ &\quad \times \exp\left(\frac{\imath}{\hbar}\left\{\frac{m_e}{2t}(q_B - q_A)^2 + t\left[E - V\left(\frac{q_A + q_B}{2}\right)\right]\right\}\right). \end{aligned} \tag{7.3.2}$$

The integration can be performed analytically and yields a Hankel function (see Chapter 3 of Ref. [Mag66]). It is, however, more instructive to pursue another course. Generalizing the Fresnel integral (7.1.14) to d dimensions and applying a straightforward shift of variables, we obtain the identity

$$\exp\left[\frac{\imath}{\hbar}\frac{m_e}{2t}(q_B - q_A)^2\right] = \left(\frac{\imath t}{2\pi\hbar m_e}\right)^{d/2} \int \exp\left\{\frac{\imath}{\hbar}\left[p(q_B - q_A) - \frac{p^2}{2m_e}t\right]\right\} dp, \tag{7.3.3}$$

where the integral on the right hand side is over the d-dimensional momentum space. Entering this expression into the integral (7.3.2) we get for the contribution of the short paths to the Green function

$$G_0(q_A, q_B, E) = -\frac{\imath}{\hbar}\left(\frac{1}{2\pi\hbar}\right)^d \int dp \int_0^\infty dt$$
$$\times \exp\left(\frac{\imath}{\hbar}\left\{p(q_B - q_A) + t\left[E - V\left(\frac{q_A + q_B}{2}\right) - \frac{p^2}{2m_e}\right]\right\}\right)$$
$$= \left(\frac{1}{2\pi\hbar}\right)^d \int dp \frac{\exp\left[\frac{\imath}{\hbar} p(q_B - q_A)\right]}{E - \frac{p^2}{2m_e} - V\left(\frac{q_A + q_B}{2}\right)}. \tag{7.3.4}$$

In the t integration we have again used the fact that E is assumed to have an infinitesimally small imaginary part. We can now perform the limit $|q_B - q_A| \to 0$, and calculate the trace

$$g_0(E) = \int G_0(q, q, E)\, dq$$
$$= \int \frac{dp\, dq}{(2\pi\imath\hbar)^d} \frac{1}{E - H(p, q)}, \tag{7.3.5}$$

where

$$H(p, q) = \frac{p^2}{2m_e} + V(q) \tag{7.3.6}$$

is the classical Hamilton function. From the imaginary part of $g_0(E)$ we obtain for the contribution of the paths of length zero to the density of states

$$\rho_0(E) = \int \frac{dp\, dq}{(2\pi\hbar)^d} \delta[E - H(p, q)]. \tag{7.3.7}$$

This equation allows a suggestive interpretation. $\rho_0(E)$ corresponds to the part of the phase space which is accessible to a classical particle with energy E. $\rho_0(E)$ is called the smooth part of the density of states.

For the motion of a particle in a d-dimensional billiard Eq. (7.3.7) yields in particular

$$\rho_0(E) = \frac{V_d}{(2\pi\hbar)^d} \int dp\, \delta\left[E - \frac{p^2}{2m_e}\right], \tag{7.3.8}$$

where V_d is the d-dimensional volume of the billiard. Carrying out the remaining integrations we end with

$$\rho_0(E) = \left(\frac{2m_e}{\hbar^2}\right)^{d/2} \frac{V_d \Omega_d}{2(2\pi)^d} E^{(d-2)/2}, \tag{7.3.9}$$

where Ω_d is the surface of the d-dimensional unit sphere (see Eq. (3.3.18)). This is Weyl's law, which has already been used repeatedly throughout this book. It should be noted that in this approximation the smooth part of the density of states exclusively depends on the volume of the billiard.

In two dimensions $\rho_0(E)$ is independent of the energy,

$$\rho_0(E) = \frac{2m_e}{\hbar^2}\frac{A}{4\pi}, \tag{7.3.10}$$

and the smooth part of the integrated density of states increases linearly with E,

$$n_0(E) = \frac{2m_e}{\hbar^2}\frac{A}{4\pi}E, \tag{7.3.11}$$

where A is the billiard area.

In billiards the wavenumber usually replaces the energy as the variable. The corresponding density of states is then given by

$$\overline{\rho}(k) = \frac{dE}{dk}\rho(E) = \frac{\hbar^2 k}{m_e}\rho(E), \tag{7.3.12}$$

whence follows for the smooth part of the density of states

$$\overline{\rho}_0(k) = \frac{A}{2\pi}k, \tag{7.3.13}$$

and, for the smooth part of the integrated density of states,

$$\overline{n}_0(k) = \frac{A}{4\pi}k^2. \tag{7.3.14}$$

Close to the borders of the classically allowed regions the above calculation of the density of states becomes questionable. In billiards these regions are close to the walls. Here, in addition to the direct path from q_A to q_B, a second one exists, passing from q_A to q_B via an intermediate wall reflection. Qualitatively it can be expected that the regions close to the walls cause a correction term to the density of states proportional to $A_{\text{bound}}k$, where A_{bound} is the area of the boundary region in question. The 'thickness' of this region should be of the order of the wavelength λ, or, alternatively, of the reciprocal wavenumber k^{-1}. The area A_{bound} is thus of the order of Sk^{-1} where S is the circumference of the billiard. We therefore expect a correction to the density of states proportional to S. A quantitative calculation actually yields for two-dimensional billiards with Dirichlet boundary conditions [Bal71]

$$\overline{\rho}_0(k) = \frac{A}{2\pi}k - \frac{S}{4\pi} + \cdots . \tag{7.3.15}$$

Equation (7.3.15) describes the beginning of an asymptotic series. Further details can be found in the book by Baltes and Hilf [Bal76], and in the review article by Eckhardt [Eck88]. We shall come back to this point in the next section.

7.3.2 Oscillatory part of the density of states

In the last section we learnt that there are two contributions to the density of states,

$$\rho(E) = \rho_0(E) + \rho_{\rm osc}(E). \tag{7.3.16}$$

$\rho_0(E)$ results from the paths of length zero and varies only smoothly with the energy, while $\rho_{\rm osc}(E)$ is a sum over the periodic orbits and shows an oscillatory behaviour. It is given by

$$\rho_{\rm osc}(E) = \frac{1}{\pi\hbar}\sum_r{}' \frac{(T_p)_r}{\|M_r - 1\|^{1/2}} \cos\left[\frac{1}{\hbar} S_r(E) - \frac{\mu_r \pi}{2}\right], \tag{7.3.17}$$

where the prime denotes that the orbits of length zero have been omitted. As to the periodic orbits the separation into two parts seems quite natural, but it is not free from ambiguities. Berry and Howls [Ber94] derived an expression for the density of states in a two-dimensional billiard in powers of $E^{-1/2}$,

$$\rho(E) = \sum_{n=0}^{\infty} \frac{a_n}{E^{n/2}}, \tag{7.3.18}$$

from which the periodic orbits have completely disappeared. Equation (7.3.18) may be considered as a generalization of Weyl's law. The expression is, however, divergent, and should therefore be replaced by

$$\rho(E) = \sum_{n=0}^{N} \frac{a_n}{E^{n/2}} + R_N, \tag{7.3.19}$$

where R_N is a remainder which has to be determined. A comparison of Eqs. (7.3.19) and (7.3.16) shows that the periodic orbits must be present somewhere in the remainder. For further details see the original work.

The Gutzwiller trace formula is affected by serious divergence problems, too. They are caused by the exponential proliferation of the number of periodic orbits with the length. As an illustrative example let us consider a Sinai billiard. In the repeated zone scheme (see Fig. 7.9) a periodic orbit corresponds to an infinite sequence of segments connecting the disks, which repeats itself after say k segments. For large k values the length of the periodic orbit is given by $l = k\langle l_s \rangle$ where $\langle l_s \rangle$ is the average segment length. The number of possible arrangements of k segments is about a^k, if a is the number of neighbouring disks visible from a given disk. Not all of these trajectories will be geometrically allowed, however. This will slightly reduce the average value of a. Quite pictorially the removal of the geometrically forbidden orbits is called 'pruning' by the specialists. Since in a periodic arrangement of k segments

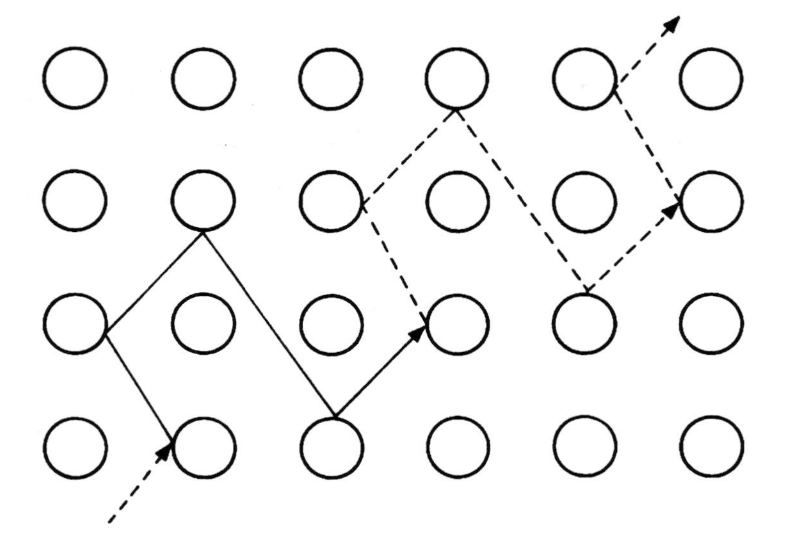

Figure 7.9. A periodic orbit of a Sinai billiard in the repeated zone scheme.

every segment can be chosen as the starting element, every arrangement is counted k-fold in this way. The number of different arrangements is therefore given by a^k/k. With $k = l/\langle l_s \rangle$ we obtain the density of $n(l)$ of periodic orbits

$$n(l) \sim \frac{e^{hl}}{l}, \tag{7.3.20}$$

where $h = \ln a/\langle l_s \rangle$ is the *topological entropy*. Alternatively the number $N(l) = \int_0^l n(l)\,dl$ of periodic orbits with a length smaller than l may be considered, which, in the leading term, shows the same l dependence as $n(l)$:

$$N(l) \sim \frac{e^{hl}}{hl} + \cdots . \tag{7.3.21}$$

The unwritten terms are of $\mathcal{O}(e^{hl})/l^2$. This is easily verified using an integration by parts.

The calculation for the Sinai billiard, which has only been sketched in this section, can be found in Appendix H of Ref. [Ber81]. More refined techniques to calculate topological entropies have been developed by Dahlqvist [Dah95].

We might hope that the exponential proliferation of orbits with length is compensated by a corresponding decrease of weight factors in the periodic orbit sum (7.3.17), but this is not the case. Let us consider once more a two-dimensional billiard. Taking again the wavenumber as the variable, and using Eq. (7.3.12), we obtain for the density of states

$$\overline{\rho}(k) = \overline{\rho}_0(k) + \overline{\rho}_{\text{osc}}(k), \tag{7.3.22}$$

where the smooth contribution is given by

$$\overline{\rho}_0(k) = \frac{A}{2\pi}k - \frac{S}{4\pi} + \cdots \tag{7.3.23}$$

(see Eq. (7.3.15)), and where the oscillatory contribution reads

$$\overline{\rho}_{\text{osc}}(k) = \frac{\hbar k}{\pi m_e} \sum_r{}' \frac{(T_p)_r}{\|M_r - 1\|^{1/2}} \cos\left(kl_r - \frac{\mu_r \pi}{2}\right). \tag{7.3.24}$$

The sum is over all orbits including repetitions. The period $(T_p)_r$ of a primitive orbit can be expressed in terms of the length l_p of the primitive orbit as

$$(T_p)_r = \frac{l_p}{v} = \frac{m_e l_p}{\hbar k}. \tag{7.3.25}$$

We then obtain for the oscillatory part of the density of states

$$\overline{\rho}_{\text{osc}}(k) = \frac{1}{\pi} \sum_{p,n}{}' \frac{l_p}{\|M_p^n - 1\|^{1/2}} \cos\left[n\left(kl_p - \frac{\mu_p \pi}{2}\right)\right], \tag{7.3.26}$$

where we have replaced the sum over the orbits by a double sum over the primitive orbits and the number of repetitions. M_p and μ_p are monodromy matrix and Maslov index, respectively, of the primitive orbit. The monodromy matrix for an n-fold repeated orbit is the nth power of the monodromy matrix of the primitive orbit, and the actions and the Maslov indices are additive.

For two-dimensional billiards the determinant prefactor can be calculated explicitly. For a hyperbolic orbit, the eigenvalues of the monodromy matrix are given by e^{α} and $e^{-\alpha}$. The determinant entering into Eq. (7.3.26) consequently reads

$$\begin{aligned}\|M^n - 1\|^{1/2} &= \|(e^{n\alpha} - 1)(e^{-n\alpha} - 1)\|^{1/2} \\ &= 2\sinh\frac{n\alpha}{2}.\end{aligned} \tag{7.3.27}$$

The elliptic and the inverse-hyperbolic case are treated in the same way. We thus obtain for the three cases

$$\|M^n - 1\|^{1/2} = \begin{cases} 2\left|\sin\dfrac{n\alpha}{2}\right| & \text{elliptic case} \\ 2\sinh\dfrac{n\alpha}{2} & \text{hyperbolic case} \\ 2\cosh\dfrac{n\alpha}{2} & \text{inverse-hyperbolic case} \end{cases} \tag{7.3.28}$$

Although in the hyperbolic systems the exponentially decreasing weight factors help to reduce the divergences, this is not sufficient to compensate the exponential increase in the number of periodic orbits with length, as has been shown by Eckhardt and Aurell [Eck89]. We have to accept that the Gutzwiller trace formula cannot be used in its present form. These divergences were a hindrance to the application of the trace formula for nearly two decades. Finally in the eighties, different techniques were developed to circumvent these problems. We shall return to these questions and a number of illustrative examples in Chapter 8.

7.3.3 Bouncing-ball contributions

The stationary phase approximation, which was applied in Section 7.2.6 to calculate the trace of the Green function, fails for families of nonisolated orbits such as the bouncing ball in the stadium billiard. Since the bouncing balls play a dominant role in the spectra of billiards, they have to be dealt with separately. We shall follow the paper of Sieber *et al.* [Sie93]. As an example we consider the bouncing ball in the quarter stadium.

We start with the expression

$$K(q_A, q_B, t) = \frac{1}{2\pi\imath\hbar} \sum_r \left| -\frac{\partial^2 W_r}{\partial q_A \partial q_B} \right|^{1/2} \exp\left[\frac{\imath}{\hbar} W_{BA,r}(t) - \imath \frac{\nu_r \pi}{2} \right] \tag{7.3.29}$$

for the quantum mechanical propagator in two dimensions (see Section 7.2.3). Dividing the sum over the trajectories into a contribution $K_{\text{bb}}(q_A, q_B, t)$ from the bouncing ball and a contribution $K_r(q_A, q_B, t)$ from all other orbits, we obtain

$$K(q_A, q_B, t) = K_r(q_A, q_B, t) + K_{\text{bb}}(q_A, q_B, t). \tag{7.3.30}$$

$K_r(q_A, q_B, t)$ is treated by applying the techniques described in the preceding sections. Our present concern is the bouncing-ball contribution. The paths contributing to $K_{\text{bb}}(q_A, q_B, t)$ can best be found by the method of images. There are two different types of paths with the lengths

$$l_e^n(q_A, q_B) = \sqrt{(x_B - x_A)^2 + (y_A - y_B + 2an)^2}, \tag{7.3.31}$$

$$l_o^n(q_A, q_B) = \sqrt{(x_B - x_A)^2 + (y_A + y_B + 2an)^2}, \tag{7.3.32}$$

where n is an arbitrary integer. a is the width of the quarter stadium. The indices 'e' and 'o' express the fact that there is an even and an odd number of reflections for the respective paths.

The principal function associated with a path of the length l is given by

$$W_{BA}(t) = \frac{m_e}{2t} l^2 \tag{7.3.33}$$

(see Eq. (7.2.13)) Thus the stability factor entering on the right hand side of Eq. (7.3.29) is identical for all orbits and is given by

$$\left| -\frac{\partial^2 W_{BA}}{\partial q_A \partial q_B} \right| = \left(\frac{m_e}{t} \right)^2. \tag{7.3.34}$$

The bouncing-ball contribution to the propagator thus reads

$$K_{\rm bb}(q_A, q_B, t) = \frac{m_e}{2\pi\imath\hbar t}\sum_{n=-\infty}^{\infty}\left(\exp\left\{\frac{\imath m_e}{2\hbar t}[l_e^n(q_A, q_B)]^2\right\} - \exp\left\{\frac{\imath m_e}{2\hbar t}[l_o^n(q_A, q_B)]^2\right\}\right), \tag{7.3.35}$$

where the minus sign with the second term accounts for the fact that for an odd number of reflections there is an additional phase shift of π. The Fourier transform of the propagator can be performed analytically yielding

$$G_{\rm bb}(q_A, q_B, E) = \frac{m_e}{2\imath\hbar^2}\sum_{n=-\infty}^{\infty}\{H_0^{(1)}[kl_e^n(q_A, q_B)] - H_0^{(1)}[kl_o^n(q_A, q_B)]\} \tag{7.3.36}$$

for the Green function, where $H_0^{(1)}(x)$ is a Hankel function. For the calculation of the density of states from the trace of the Green function we again take the wavenumber as the variable instead of the energy. Using Eq. (7.3.12) we obtain for the bouncing-ball contribution $\overline{\rho}_{\rm bb}(k)$ to the density of states

$$\overline{\rho}_{\rm bb}(k) = -\frac{k\hbar^2}{\pi m_e}\,{\rm Im}\left[\int_0^b dx\int_0^a dy\, G_{\rm bb}(q, q, E)\right], \tag{7.3.37}$$

where b is the length of the straight part of the quarter stadium. The integrations can be easily performed [Sie93] and yield

$$\overline{\rho}_{\rm bb}(k) = \frac{abk}{2\pi}\sum_{n=-\infty}^{\infty} J_0(2ka|n|) - \frac{b}{2\pi}. \tag{7.3.38}$$

Using the asymptotic approximation for the Bessel functions we obtain

$$\overline{\rho}_{\rm bb}(k) = \frac{abk}{2\pi} - \frac{b}{2\pi} + \frac{b}{\pi}\sqrt{\frac{ak}{\pi}}\sum_{n=1}^{\infty}\frac{1}{\sqrt{n}}\cos\left(2ank - \frac{\pi}{4}\right). \tag{7.3.39}$$

In the first two terms on the right hand side we recognize Weyl's formula, including the surface term (see Eq. (7.3.15)). The third term is a periodic orbit expansion of the fluctuating part of $\overline{\rho}_{\rm bb}(k)$. Note that $2an$ is the length of the n-fold repeated bouncing-ball orbit!

The bouncing-ball contribution to the density of states has been demonstrated in the spectrum of a superconducting microwave resonator in the shape of a quarter stadium [Grä92]. Figure 7.10 shows the experimentally obtained density of states, where the smooth part of the spectrum has been subtracted. The remaining oscillations directly reflect the bouncing-ball contributions to the integrated density of states. The solid line has been obtained by calculating

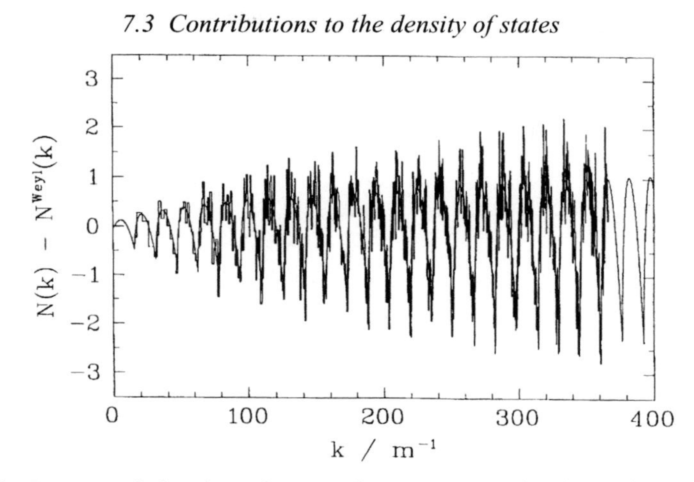

Figure 7.10. Integrated density of states for a superconducting microwave cavity in the shape of a quarter stadium billiard. The smooth part, calculated from the Weyl formula, has been subtracted. The oscillations reflect the presence of the bouncing-ball orbit [Grä92] (Copyright 1992 by the American Physical Society).

the bouncing-ball contribution to the integrated density of states from Eq. (7.3.39). A quantitative agreement between experiment and theory is found.

Meanwhile there are an increasing number of attempts to go 'beyond Gutzwiller' in the description of the spectra [Gas93, Gas95]. Often *diffraction* orbits have to be considered. In Sinai billiards waves may creep along the central circle and are thus able to penetrate into the region of the geometrical shadow. Corrections due to diffractive orbits have been calculated by several authors [Vat94, Pav95, Pri96]. For more than a century these have been well-known effects in optics. Their influence on the spectra, however, has only now been recognized. (The author has a suspicion that some of Fraunhofer's and Kirchhoff's papers, which are long forgotten, address similar questions!).

The so-called *ghost* orbits [Kuś93] are another phenomenon which sometimes has to be considered. These orbits are observed in the spectra when a classical orbit eventually disappears depending on some control parameter. In such a situation the corresponding contribution to the trace formula does not vanish immediately, but the action acquires an imaginary part causing an exponential damping of the contribution. It may be observable nevertheless, if one is still close to the critical point. A more detailed discussion of all these refinements would go beyond the scope of an introductory presentation.

8
Applications of periodic orbit theory

In the preceding chapter we derived semiclassical expressions for the quantum mechanical propagator, the Green function, and the density of states. Now we are going to discuss a number of consequences and applications, in particular of the Gutzwiller trace formula. Anybody working in the field should be familiar with the basic ideas applied in the derivation, but an understanding of the implications of periodic orbit theory is also possible for readers not willing to enter too deeply into the details. For this reason knowledge of the preceding chapter should be dispensable for the major part of the present one.

We start with a description of the techniques applied to extract the contributions of the different periodic orbits from the spectra and wave functions. The spectra of billiards and hydrogen atoms in strong magnetic fields suggest a semiclassical interpretation. The most spectacular manifestation of periodic orbits is the scarring phenomenon, found in many wave functions of chaotic billiards.

As the quantum mechanical spectrum is uniquely determined by the periodic orbits, i.e. by individual system properties, the success of random matrix theory in description of numerous spectral correlations requires an explanation. This is given by Berry in his semiclassical theory of spectral rigidity [Ber85].

It is comparatively easy to analyse a given spectrum in terms of periodic orbit theory. This is not true for the reverse procedure, namely the calculation of a spectrum from the periodic orbits of the system. Powerful resummation techniques had to be developed to overcome the exponential increase in the number of periodic orbits with length.

The chapter ends with a discussion of non-Euclidean billiards on a metric of constant negative curvature. The spectra of billiards in such metrics allow a periodic orbit expansion by means of the Selberg trace formula, which is exact, in contrast to its Euclidean equivalent, the Gutzwiller trace formula. In this respect billiards on a metric with constant negative curvature are simpler than Euclidean ones.

8.1 Periodic orbit analysis of spectra and wave functions

8.1.1 Periodic orbits in the spectra

The main result of the last chapter is the Gutzwiller trace formula expressing the density of states in terms of sums over classical periodic orbits,

$$\rho(E) = \rho_0(E) + \rho_{\text{osc}}(E). \tag{8.1.1}$$

$\rho_0(E)$ is the smooth part of the density of states and is obtained from the Weyl formula (7.3.8), whereas the oscillatory part $\rho_{\text{osc}}(E)$ is given by

$$\rho_{\text{osc}}(E) = \frac{1}{\pi\hbar} \sum_{p,n}{}' \frac{T_p}{\|M_p^n - 1\|^{1/2}} \cos\left\{ n \left[\frac{1}{\hbar} S_p(E) - \frac{\mu_p \pi}{2} \right] \right\}. \tag{8.1.2}$$

The sum is over all primitive periodic orbits p and their repetitions n. T_p is the period for a primitive orbit, $S_p(E)$ its action, and μ_p its Maslov index counting the number of classical turning points, conjugated points, etc. M_p is the monodromy matrix describing the stability of the orbit.

In billiard systems the wavenumber is usually taken as the variable. In terms of k the density of states reads

$$\overline{\rho}(k) = \overline{\rho}_0(k) + \overline{\rho}_{\text{osc}}(k). \tag{8.1.3}$$

The two leading terms of the smooth part are given by

$$\overline{\rho}_0(k) = \frac{A}{2\pi} k - \frac{S}{4\pi}, \tag{8.1.4}$$

where A and S are area and circumference of the billiard, respectively. The significance of an orbit for the oscillatory part of the spectral density depends on its stability. In two-dimensional billiards the orbit may be either elliptic, hyperbolic, or inverse-hyperbolic. An elliptic orbit is stable, whereas hyperbolic and inverse-hyperbolic orbits are exponentially unstable. If all orbits are hyperbolically unstable, as is the case for the stadium billiard, the oscillatory part of the density of states reads

$$\overline{\rho}_{\text{osc}}(k) = \frac{1}{2\pi} \sum_{p,n}{}' \frac{l_p}{\sinh \dfrac{n\alpha_p}{2}} \cos\left[n \left(k l_p - \frac{\mu_p \pi}{2} \right) \right], \tag{8.1.5}$$

where l_p is the length of the primitive orbit, and α_p its instability exponent.

The Gutzwiller trace formula only holds for isolated orbits. Families of orbits, such as the bouncing ball, have to be treated separately and give rise to an additional contribution $\overline{\rho}_{bb}(k)$ to the density of states (see Section 7.3.3).

The bouncing-ball contribution can also be written as a sum over periodic orbits. However, the weight factors differ from those entering into Eq. (8.1.5)

For the following discussion we rewrite expression (8.1.3) for the density of states as

$$\overline{\rho}(k) = \overline{\rho}_0(k) + \sum_r A_r e^{\imath k l_r}, \tag{8.1.6}$$

where the sum now runs over all orbits, including repetitions, as well as over positive and negative directions. Stability factors and Maslov phase factors are incorporated into the prefactors A_r.

The periodic orbit sum (8.1.6) is divergent and therefore not suited to calculating the spectrum from the periodic orbits. But the inverse procedure, namely to extract the contributions of the different periodic orbits from the spectra, is straightforward. For billiards the Fourier transform of the fluctuating part of the density of states,

$$\begin{aligned} \hat{\rho}_{\mathrm{osc}}(l) &= \int \overline{\rho}_{\mathrm{osc}}(k) e^{\imath k l}\, dk \\ &= \sum_r A_r \delta(l - l_r), \end{aligned} \tag{8.1.7}$$

directly yields the contributions of the orbits to the spectrum [Stö90]. Each orbit gives rise to a delta peak at an l value corresponding to its length, and a weight corresponding to the stability factor of the orbit.

As an example Fig. 8.1 shows the squared modulus of the Fourier transform of the spectrum of a microwave resonator shaped as a quarter Sinai billiard [Stö95]. Each peak corresponds to a periodic orbit of the billiard. For the bouncing-ball orbit, labelled ‘1’, at least two peaks associated with repeated orbits are clearly visible. The smooth part of the spectrum is responsible for the increase of $|\hat{\rho}(k)|^2$ at small lengths.

The technique can be extended to all cases in which the action scales with the energy,

$$S_r(E) = S_r(E_0)\left(\frac{E}{E_0}\right)^{\gamma}. \tag{8.1.8}$$

With $k = E^{\gamma}$ as the new variable we once more obtain an equation of the type (8.1.6) allowing the contributions of the periodic orbits to be extracted by means of a Fourier transformation.

Frequently it is not the energy or frequency which is varied in the experiment, but some other parameter, such as an applied magnetic or electric field. An example of the latter type are the resonant tunnelling experiments of Fromhold and coworkers [Fro94] (see Section 2.3.3). An electron gas is

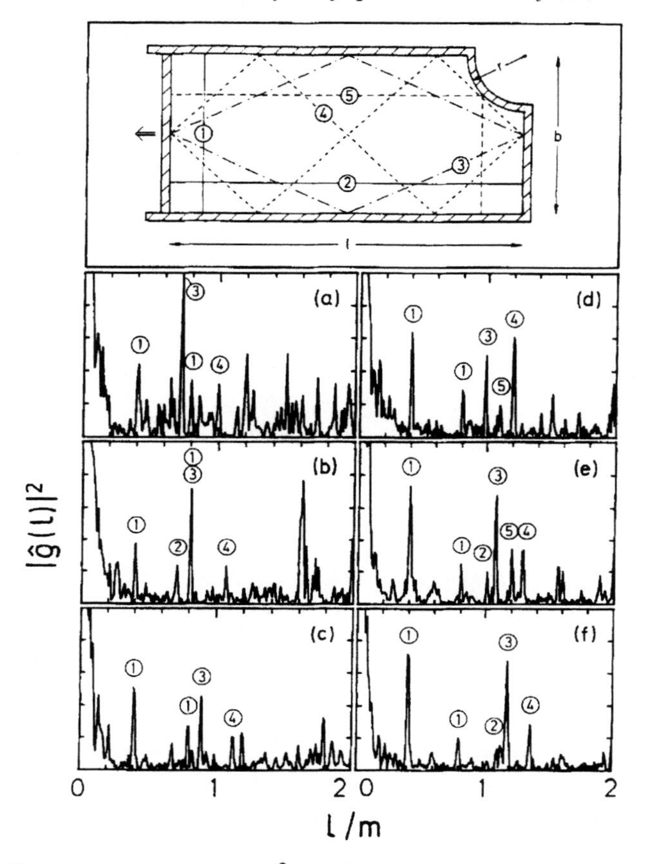

Figure 8.1. Squared modulus $|\hat{p}(l)|^2$ of the Fourier transform of the spectrum of a quarter Sinai billiard ($b = 20$ cm, $r = 7$ cm) for different billiard lengths $l/\mathrm{cm} = 30$ (a), 35 (b), 40 (c), 45 (d), 50 (e), 55 (f). Each resonance can be associated with a periodic orbit [Stö95] (Copyright 1995 by the American Physical Society).

confined to a two-dimensional layer within two tunnelling barriers. The motion of the electrons between the barriers is governed by the Hamiltonian

$$\mathscr{H} = \frac{1}{2m_e}\left(p_x^2 + p_z^2\right) + V(x, z), \tag{8.1.9}$$

where

$$V(x, z) = \frac{1}{2m_e}\left[\hbar k_y - \frac{eB}{c}(x \sin\theta - z\cos\theta)\right]^2 + e(U_o - xF) \tag{8.1.10}$$

is an effective potential [Fro95b]. The first term on the right hand side describes the interaction of the electrons with a magnetic field, where θ is the angle between the magnetic field and the normal to the layer planes. The second term describes the interaction with the electric field produced by the voltage applied. For the special cases $\theta = 0°$ and $\theta = 90°$ the x and the z dependences can be separated, and the Hamiltonian is integrable. It is chaotic for any other angle.

The experimental results have already been presented in Section 2.3.3. Figure 2.23 shows the second derivative of the tunnelling current through the double barrier as a function of the voltage applied for different orientations of the magnetic field [Fro94]. To understand the patterns observed, we have to remember that transmission maxima are observed whenever an eigenstate of the region between the barriers is in resonance with the energy E_L of the lowest Landau level of the emitter region (see Section 2.3.3). But this is only a necessary condition. In addition there must be an overlap of the respective wave functions. Thus the experiment yields a weighted density of states $\rho_w(E_L, F)$ for fixed E_L as a function of the field strength F, where the weight factors are given by the overlap matrix elements. Therefore not all periodic orbits are accessible by the electrons injected into the tunnelling structure. The experiment demonstrates that in a given voltage range one periodic orbit is always dominating. The authors have been able to associate each of the oscillations found in the measurements with individual orbits depicted in the inserts of Fig. 2.23.

For a quantitative comparison with the semiclassical theory Fromhold *et al.* [Fro95b] calculated the spectrum of the tunnelling structure for a typical parameter set as realized in the experiment. In contrast to the situation found in billiard systems the Hamiltonian (8.1.9) does *not* scale with the energy. But an extraction of the periodic orbits by means of the Fourier transformation is nevertheless possible, if the transformation is limited to an energy window of a sufficiently small width. In this case we may expand the action at the centre E_0 of the window up to the linear term,

$$\begin{aligned} S_r(E) &= S_r(E_0) + \left.\frac{\partial S_r}{\partial E}\right|_{E=E_0} (E - E_0) \\ &= S_r(E_0) + T_r(E - E_0), \end{aligned} \tag{8.1.11}$$

where we have used the fact that the derivative of the action with respect to the energy gives the period T_r of the orbit (see Eq. (7.2.34)). Figure 8.2 shows the Fourier transform of the calculated spectrum. All peaks in the Fourier transform can be associated with periodic orbits. Most of them are already known from the experiment (see Fig. 2.23).

In mesoscopic systems and quantum dots the conductance is usually measured as a function of the magnetic field. Applying a similar Fourier transform

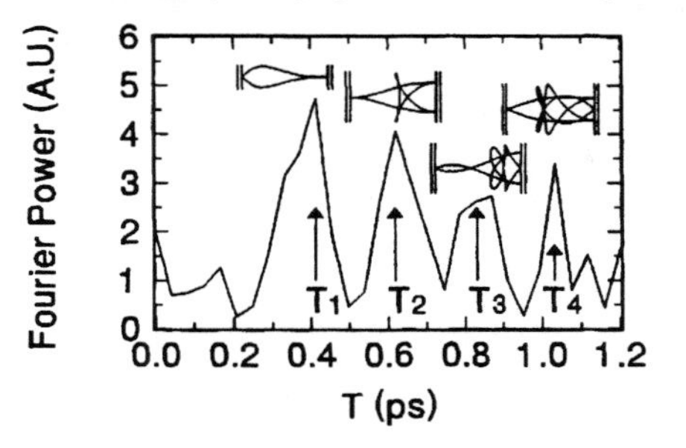

Figure 8.2. Fourier transform of one calculated spectrum for the tunnelling structure used in the experiments shown in Fig. 2.22, with $B = 11.4$ T and $\theta = 20°$ [Fro95b] (Copyright 1995 by the American Physical Society).

technique as described above, Marcus *et al.* [Mar92] extracted the contributions of the periodic orbits to the magnetoresistance fluctuations in mesoscopic billiards (see Section 2.3.2). The Hall resistance measurements in antidot structures by Weiss *et al.* [Wei93] also showed the influence of the periodic orbits, as has already been discussed in Section 2.3.1.

8.1.2 Hydrogen atom in a strong magnetic field

The hydrogen atom in a strong magnetic field is one of the paradigms of quantum chaos. The system becomes chaotic as soon as the interaction of the electron with the magnetic field is of the same order of magnitude as the Coulomb interaction with the nucleus. As the maximum attainable field strengths are of the order of some tesla in the laboratory, the hydrogen atom has to be excited to principal quantum numbers of the order of 100 to attain the chaotic region. We have already met this situation in Section 4.2.2 when discussing the behaviour of the hydrogen atom in strong microwave fields. Atoms in these highly excited states are called *Rydberg atoms* [Gal94]. All of them behave essentially hydrogen-like, in particular the alkaline ones.

Hydrogen atoms in strong magnetic fields are not of merely academic interest. They are found in astrophysics in exotic stellar objects. The magnetic fields in white dwarfs are of the order of 10^2 to 10^5 T, in neutron stars even of the order of 10^7 to 10^9 T. Under these extreme conditions the hydrogen atoms are chaotic in the ground state. A recent compilation of the state of the art of atoms in strong magnetic fields, with special emphasis on the astrophysical implications, can be found in Ref. [Rud94].

Laboratory experiments with hydrogen atoms in strong magnetic fields have been performed in particular by Welge and his group [Hol86, Mai86]. Since the experimental technique is described in Ref. [Rot86], it will be only sketched here. Hydrogen atoms in an atomic beam are excited by one laser to the $n = 2$ state, and by a second to the Rydberg state desired. Subsequently the excited atoms are field-ionized in an electric field. As a function of the wavelength of the second laser the photoelectron yield gives the spectrum of the Rydberg states of the hydrogen atoms. Essentially the same technique has been used by Kleppner and coworkers [Cou95] in their studies of the Rydberg states of lithium atoms.

Parallel to the experimental progress, calculating techniques have improved more and more, also motivated by astrophysical interest. Figure 8.3 shows the comparison of an experimental and a theoretical spectrum of a lithium atom in a strong magnetic field closely below and above the ionization [Iu91]. The agreement between the two spectra is convincing. This is the more astonishing as the experiments were performed with lithium atoms, whereas the calculations were done for hydrogen atoms. Only the nuclear mass has been replaced by that of ^{7}Li.

The level of agreement is such that we may safely state that the system is understood both from the experimental and the theoretical point of view. We may even come to the conclusion that there is nothing left to do and turn to other topics. But can we really be content with the fact that the Schrödinger equation works well for a chaotic system without really knowing what is happening? Here periodic orbit theory comes in, and we shall see again that spectra looking completely chaotic, as shown in Fig. 8.3, allow a simple semiclassical interpretation.

Since the essential ideas have been developed by Wintgen [Win87], we shall follow his presentation. We shall not go again into the random matrix aspects of the spectra [Hön89], which have already been discussed in Sections 3.2.1 and 3.2.4. The state of knowledge up to 1989 has been compiled in two reviews by Friedrich and Wintgen [Fri89] and by Hasegawa *et al.* [Has89].

The Schrödinger equation for the hydrogen atom in a strong magnetic field is given by

$$\left[-\frac{\hbar^2}{2m_e}\Delta - \frac{e^2}{r} - \hbar\omega L_z + \frac{1}{2}m_e\omega^2\left(x^2+y^2\right)\right]\psi = E\psi, \tag{8.1.12}$$

where m_e is the reduced electron mass, and where

$$\omega = \frac{1}{2}\omega_0 = \frac{eB}{2m_e c} \tag{8.1.13}$$

is half the cyclotron frequency. The Zeeman interaction with the electron spin has not been taken into account. Introducing dimensionless variables $r \to ar$, $E \to (e^2/a)E$, where $a = \hbar^2/m_e e^2$ is the Bohr radius, Eq. (8.1.12) reduces to

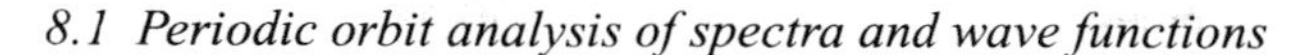

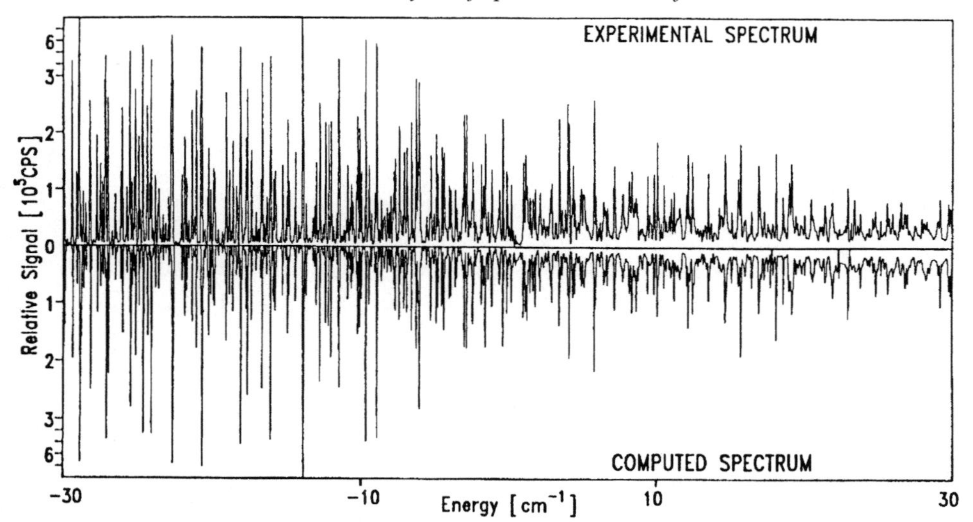

Figure 8.3. Experimental 0^--spectrum of a lithium atom in a magnetic field of 6.113 T (top), and calculated spectrum of a hydrogen atom, where only the nuclear mass has been replaced by that of ^{7}Li (bottom). To facilitate the comparison, the spectra have been convoluted with a Gaussian and displayed as mirror images of each other [Iu91] (Copyright 1991 by the American Physical Society).

$$\left[-\frac{1}{2}\Delta - \frac{1}{r} - \omega' L_z + \frac{1}{8}\gamma^2 \left(x^2 + y^2\right)\right]\psi = E\psi, \tag{8.1.14}$$

where $\omega' = \hbar\omega a/e^2$ is the rescaled half cyclotron frequency, and where

$$\gamma = \frac{\hbar^3 B}{m_e^2 c e^3} \tag{8.1.15}$$

is the magnetic induction in reduced units. $\gamma = 1$ corresponds to a magnetic field $B_0 = 2.35 \times 10^5$ T. The Hamilton operator is rotationally symmetric about the z axis and has a reflection symmetry with respect to the xy plane. Each state can therefore be labelled by m^π, where m is the magnetic quantum number, and π is the parity with respect to the mirror symmetry. The term $-\omega' L_z$ only produces a constant energy offset for each subspectrum, and will therefore be dropped in the following. Introducing cylindrical polar coordinates

$$x = \rho\cos\phi, \quad y = \rho\sin\phi, \quad z = z, \tag{8.1.16}$$

the ϕ dependence can be separated, and Eq. (8.1.14) reduces to a two-dimensional Schrödinger equation for ρ, z which cannot be further reduced, and must be solved numerically.

For a comparison with periodic orbit theory we need the action on a periodic orbit. From the Hamiltonian (8.1.14) we get

$$S(E, \gamma) = \hbar \oint \sqrt{2\left[E + \frac{1}{r} - \frac{\gamma^2}{8}(x^2 + y^2)\right]}\, ds, \qquad (8.1.17)$$

where s is the arc length on the trajectory. Substituting the variables $x \to \gamma^{-2/3}x$, $y \to \gamma^{-2/3}y$, $z \to \gamma^{-2/3}z$, we find that $S(E, \gamma)$ scales according to

$$S(E, \gamma) = \gamma^{-1/3} S(\gamma^{-2/3}E, 1). \qquad (8.1.18)$$

We have again found a scaling behaviour for the action similar to that in the billiard systems, but now the action scales with the magnetic field. A Fourier transform with $k = \gamma^{-1/3}$ as the variable should therefore again extract the contributions of the different periodic orbits from the spectrum. Some care has to be taken, however, since γ enters in addition via the scaled energy $\overline{E} = \gamma^{-2/3}E$. The method only works properly if $\overline{E}$ is held constant.

Wintgen thus succeeded in extracting the periodic orbits from his calculated spectra [Win87]. Corresponding experimental results have been obtained by the Welge group [Hol88], which will be presented instead. In the experiment the magnetic induction was scanned linearly on a $\gamma^{-1/3}$ scale. Simultaneously E was adjusted by a variation of the laser wavelength to hold $\overline{E}$ constant. Figure 8.4 shows the result. At the top the spectrum is shown, in the lower part

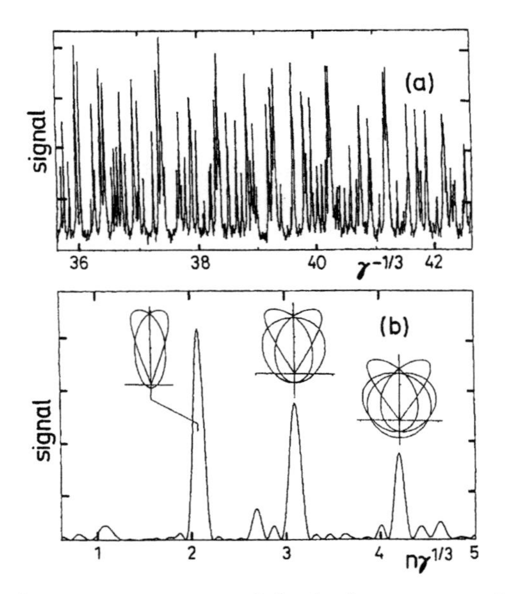

Figure 8.4. (a) Scaled energy spectrum of the hydrogen atom in a strong magnetic field for $\overline{E} = -0.45$ as a function of $\gamma^{-1/3}$. (b) Fourier transform of the same spectrum. The resonances correspond to the periodic orbits shown as insets in a (ρ, z) projection. The z coordinate is vertical [Hol88] (Copyright 1988 by the American Physical Society).

the corresponding Fourier transform is exhibited. Each of the three peaks observed can be associated with one periodic orbit. They are all of the same type: the electron performs a spiral motion under the influence of the Lorentz force. Simultaneously it departs periodically from the hydrogen nucleus and reapproaches under the influence of the Coulomb interaction. This is repeated several times until the orbit closes.

8.1.3 Scars

The discovery of the scarring phenomenon by E. Heller [Hel84] in 1984 was a surprise. Intuitively we would have expected that in the semiclassical limit the quantum mechanical probability of finding a particle somewhere in a billiard gradually turns into the corresponding classical probability. Exactly this behaviour has been found for the harmonic oscillator in Section 3.2.3 (see Fig. 3.13). According to mathematical theorems developed by Shnirelman [Shn74], Zelditch [Zel87], and Colin de Verdière [Col85] the classical and the quantum mechanical probability should indeed become equal in the semiclassical limit. The scars seem to contradict this picture: there are wave functions showing high amplitudes close to unstable periodic orbits and only low or no amplitudes elsewhere. As an example Fig. 8.5(a) shows scarred wave functions of the stadium billiard taken from Heller's paper [Hel84]. A classical particle, on the other hand, starting with only a minute deviation from an unstable periodic orbit, will fill the complete accessible phase space after a short time (see Fig. 7.7).

In the following years it became evident that scarring is a common phenomenon in all chaotic systems. Scarred billiard wave functions have already been displayed repeatedly in previous chapters. Three examples taken from other systems are shown in Figs. 8.5(b) to (d). Scarred wave functions have been observed up to quantum numbers of about one million [Li97]. This is not in contradiction with the above mentioned mathematical theorems, which only state that the fraction of scarred wave functions must vanish in the limit of large quantum numbers.

The first attempt to explain the scars within the framework of periodic orbit theory was made by Bogomolny [Bog88]. Further improvement was attained by Berry [Ber89], and Agam and Fishman [Aga93,Aga94]. We shall discuss Bogomolny's approach in detail, but will only sketch the ideas underlying the other papers.

Bogomolny considers the average of the squared modulus of the wave function over some energy range

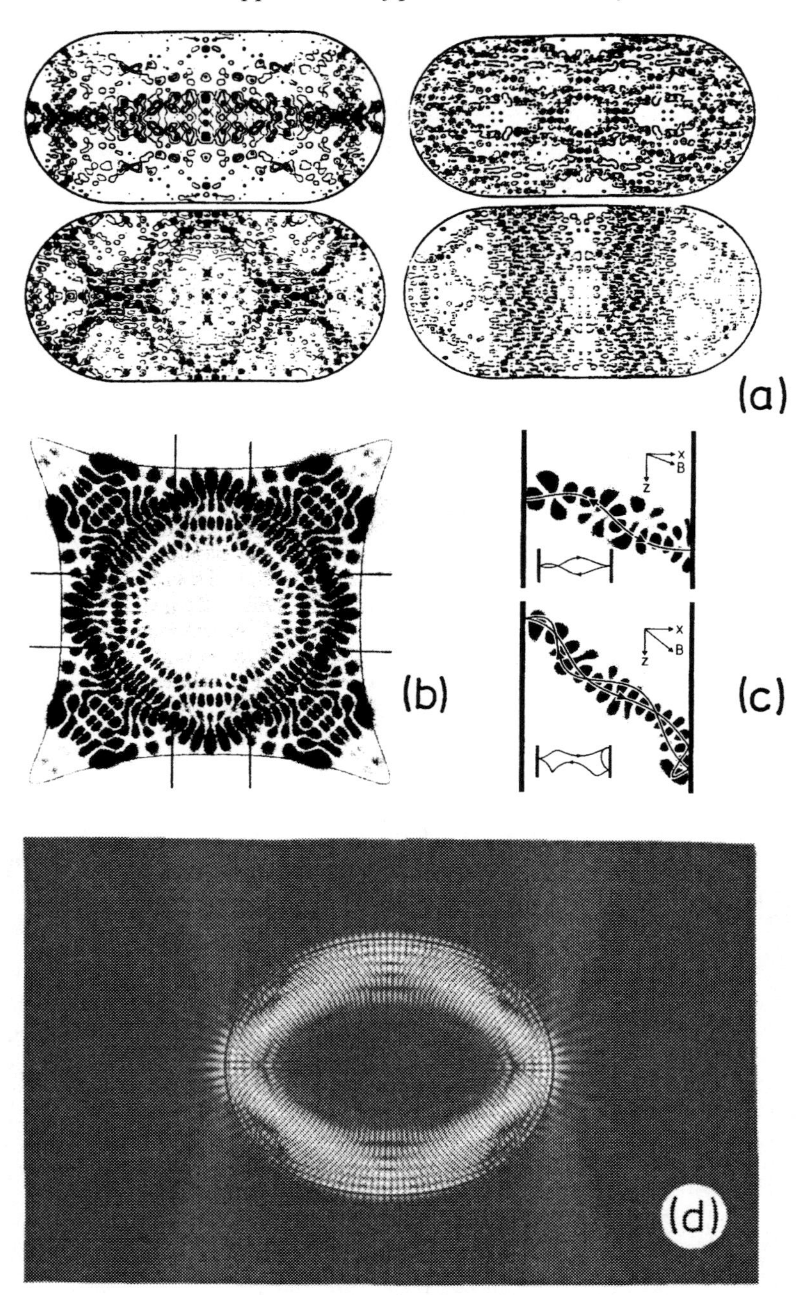

Figure 8.5. (a) Scarred wave functions of a stadium billiard [Hel84], (b) a hydrogen atom in a strong magnetic field [Win89], and (c) an electron confined between two barriers with an applied magnetic field [Fro95a] (Copyright 1984–89 by the American Physical Society). (d) Scarred light distribution in a distorted glass fibre cavity [Nöc97] (Copyright 1997 by MacMillan Magazines).

$$\langle|\psi(q)|^2\rangle = \int w(E)\sum_n |\psi(q_n)|^2\delta(E-E_n)\,dE, \tag{8.1.19}$$

where $w(E)$ is a window function of the type used earlier. The approach only yields information on the energy averaged contributions of the periodic orbits to the wave functions. The scarring of individual wave functions will not be explained.

The expression entering the integral on the right hand side of Eq. (8.1.19) can be expressed in terms of the Green function. From the expansion

$$G(q_A, q_B, E) = \sum_n \frac{\psi_n^*(q_A)\psi_n(q_B)}{E-E_n} \tag{8.1.20}$$

of the Green function in terms of the eigenfunctions we obtain

$$\sum_n \delta(E-E_n)\psi_n^*(q_A)\psi_n(q_B) = -\frac{1}{\pi}\,\mathrm{Im}[G(q_A, q_B, E)] \tag{8.1.21}$$

(see Section 3.1.5). Equation (8.1.19) may therefore be written as

$$\langle|\psi(q)|^2\rangle = -\frac{1}{\pi}\,\mathrm{Im}\left(\int w(E)G(q, q, E)\,dE\right). \tag{8.1.22}$$

Just as the density of states, the semiclassical Green function can be decomposed into a smooth part from the trajectories of length zero, and an oscillatory one from all other trajectories,

$$G(q, q, E) = G_0(q, q, E) + G_{\mathrm{osc}}(q, q, E). \tag{8.1.23}$$

In Section 7.3.1 we have derived for the smooth part the expression

$$G_0(q, q, E) = \left(\frac{1}{2\pi\imath\hbar}\right)^d \int dp \frac{1}{E-H(p, q)}, \tag{8.1.24}$$

where $H(p, q)$ is the classical Hamilton function, and d is the dimension of space. Inserting expression (8.1.23) into Eq. (8.1.22), we obtain for the smooth contribution to the quadratically averaged wave functions

$$\langle|\psi(q)|^2\rangle_0 = \langle\rho_0(q, E)\rangle, \tag{8.1.25}$$

where

$$\rho_0(q, E) = \left(\frac{1}{2\pi\imath\hbar}\right)^d \int dp\,\delta[E-H(p, q)] \tag{8.1.26}$$

is the classical probability density of finding a particle with energy E at point q. This is exactly the behaviour expected in the semiclassical limit. But there is an additional contribution from the oscillatory part of the Green function, and we shall see that this part is responsible for the occurrence of scars.

In Section 7.2.4 we have obtained for $G_{\mathrm{osc}}(q, q, E)$ the expression

$$G_{\rm osc}(q,q,E) = -\frac{\imath}{\hbar}\left(\frac{1}{2\pi\imath\hbar}\right)^{(d-1)/2}\sum_r{}' |\Delta_r|^{1/2}\exp\left[\frac{\imath}{\hbar}S_r(q,q,E) - \imath\frac{\nu_r\pi}{2}\right], \tag{8.1.27}$$

where the sum is over all closed trajectories, including the nonperiodic ones, which start and end at q. To perform the energy average we expand the action at $E = E_0$ up to the linear term,

$$S_r(q,q,E) = S_r(q,q,E_0) + T_r(q,E_0)(E-E_0) + \cdots \tag{8.1.28}$$

where T_r is the passage time of a particle on the trajectory. Entering the expansion into Eq. (8.1.27) we get

$$\langle G_{\rm osc}(q,q,E)\rangle = -\frac{\imath}{\hbar}\left(\frac{1}{2\pi\imath\hbar}\right)^{(d-1)/2}\sum_r{}' |\Delta_r|^{1/2}\exp\left[\frac{\imath}{\hbar}S_r(q,q,E_0) - \imath\frac{\nu_r\pi}{2}\right]\hat{w}[T_r(q,E_0)], \tag{8.1.29}$$

where

$$\hat{w}(T) = \int w(E)\exp\left[\frac{\imath}{\hbar}T(E-E_0)\right]dE \tag{8.1.30}$$

is the Fourier transform of the window function. For a width ΔE of the window function the width of $\hat{w}(T)$ is of the order of $\hbar/\Delta E$. The factor $\hat{w}(T)$ thus causes a cut-off of the sum for $T_r \geqslant \hbar/\Delta E$.

The sum in Eq. (8.1.29) is still over all closed trajectories from q to q, but again the periodic orbits are the dominant contribution. This can be realized as follows. Take an arbitrary closed trajectory with the action $S_r(q,q,E_0)$. Now take another closed trajectory starting and ending at a slightly shifted position $q+\Delta q$. The action for the modified trajectory is given in linear approximation by

$$\begin{aligned} S_r(q+\Delta q, q+\Delta q, E_0) &= S_r(q,q,E_0) + \Delta q\left(\frac{\partial S_r}{\partial q_B} + \frac{\partial S_r}{\partial q_A}\right)\Bigg|_{q_A=q_B=q} \\ &= S_r(q,q,E_0) + \Delta q\left(p_B|_q - p_A|_q\right), \end{aligned} \tag{8.1.31}$$

where $p_A|_q$ and $p_B|_q$ are the momenta of the trajectory at the beginning and the end, respectively (see Eq. (7.2.34)). A shift of the position by Δq thus leads to an additional phase $\Delta q(p_B - p_A)/\hbar$ in the exponential in Eq. (8.1.29). As a function of Δq this results in oscillatory contributions to $\langle|\psi(q)|^2\rangle$. Most of these contributions will average to zero. Exceptions are the contributions from the trajectories obeying $p_B|_q = p_A|_q$. But these are the periodic orbits.

For a given periodic orbit we now introduce a coordinate system, where one coordinate $q_\parallel$ is parallel to the orbit, and $(d-1)$ components $q_\perp$ are perpendicular to it. We expand the action $S_r(q, q, E_0)$ up to the quadratic term in the perpendicular components, as we did in Section 7.2.6,

$$S_r(q, q, E_0) = S_r(E_0) + \frac{1}{2}(q_\perp)^T(S_{AA,r} + 2S_{AB,r} + S_{BB,r})q_\perp. \tag{8.1.32}$$

S_{AA}, S_{AB}, S_{BB} are the matrices of the second derivatives of the action with respect to the start and end coordinates. We then obtain for the energy average of the squared wave function

$$\begin{aligned}\langle|\psi(q)|^2\rangle = \langle\rho_0(q, E_0)\rangle - \mathrm{Im}\Bigg\{&\frac{\imath}{\hbar}\left(\frac{1}{2\pi\imath\hbar}\right)^{(d-1)/2}\sum_r{}' |\Delta_r|^{1/2} \\ &\times \exp\left(\frac{\imath}{\hbar}\left[S_r(E_0) + \frac{1}{2}(q_\perp)^T(S_{AA,r} + 2S_{AB,r} + S_{BB,r})q_\perp\right] - \imath\frac{\nu_r\pi}{2}\right) \\ &\times \hat{w}[T_r(q, E_0)]\Bigg\}.\end{aligned} \tag{8.1.33}$$

This is the main result of Bogomolny's paper. The energy average of $|\psi(q)|^2$ is expressed as a sum over a smooth part reflecting the classical probability and additional contributions from the periodic orbits. The phases of the exponential functions are stationary along the orbits, and show oscillatory behaviour perpendicular to the orbit which depends quadratically on the perpendicular coordinates $q_\perp$. Semiclassical quantum mechanics 'sews the wave flesh on the classical bones', to use a metaphor from the review article by Berry and Mount [Ber72]. Usually the number of orbits entering into the sum is large. Only if the width is increased up to about $\hbar/T_{\min}$, where $T_{\min}$ is the period of the shortest periodic orbit, do the number of terms reduce to perhaps one or two. But in this limit the information on individual wave functions is completely lost due to energy smoothing. Thus Eq. (8.1.33) gives as yet no explanation for the scarring of individual wave functions.

The complementary work of Berry [Ber89] will be discussed only briefly. He starts with the Wigner function $W_n(p, q)$, which is constructed from an eigenfunction $\psi_n(q)$ as

$$W_n(p, q) = \left(\frac{1}{2\pi\imath\hbar}\right)^d \int d\overline{q}\, \psi_n^*(q + \overline{q})\psi_n(q - \overline{q}) \exp\left(-\frac{2\imath p\overline{q}}{\hbar}\right). \tag{8.1.34}$$

The Wigner function is often used for a semiclassical analysis of wave functions, since in the limit of large quantum numbers $W_n(p, q)$ approaches the classical phase space density of the classical orbits contributing to $\psi_n(q)$. For a more detailed discussion of the Wigner function see Chapter 15.4 of Ref. [Gut90]. A good overview can also be found in the review article by Takahashi [Tak89].

From the Wigner function (8.1.34) we obtain the spectral Wigner function as

$$W(p, q, E) = (2\pi\imath\hbar)^d \sum_n \delta(E - E_n) W_n(p, q). \qquad (8.1.35)$$

Using once more relation (8.1.21) the spectral Wigner function can be expressed in terms of the Green function as

$$W(p, q, E) = -\frac{1}{\pi} \operatorname{Im}\left[\int d\bar{q}\, G(q+\bar{q}, q-\bar{q}, E) \exp\left(-\frac{2\imath p\bar{q}}{\hbar}\right)\right]. \qquad (8.1.36)$$

Inserting the semiclassical expression (8.1.23) for the Green function and evaluating the integral in the stationary phase approximation, Berry obtains the following expression for the spectral Wigner function

$$W(p, q, E) = \delta[E - H(p, q)] + \sum_j W_j(p, q, E). \qquad (8.1.37)$$

The first term on the right hand side gives the classical probability density of finding a particle with energy E at a given phase space position. The second term is a sum over the classical periodic orbits. Explicit expressions for the $W_j(p, q, E)$ are given in the original work. Here it may suffice to note that in the semiclassical limit the W_j concentrate along the classical orbits. If Eq. (8.1.37) is integrated over p, Bogomolny's scar formula (8.1.33) is recovered.

Bogomolny's and Berry's calculations, in common, do not allow statements on individual wave functions. In this respect Agam and Fishman [Aga93] proceeded further. They showed that individual Wigner functions $W_n(q)$ can be calculated from an essentially finite number of periodic orbits, the term 'essentially' meaning that the contribution from orbits with lengths exceeding a given cut-off becomes negligibly small. For a hyperbola billiard the authors [Aga94] could demonstrate that in all wave functions, where the scar formula predicts a specific orbit, the corresponding scar is actually found. Unfortunately the derivation is too technically complicated to be reproduced here.

8.2 Semiclassical theory of spectral rigidity

8.2.1 Rigidity for integrable systems

In Chapter 3 we found that random matrix theory can explain numerous spectral correlations in chaotic systems, such as number variance and spectral rigidity. In Chapter 7 we derived the Gutzwiller trace formula, expressing the spectrum of a quantum mechanical system in terms of its periodic orbits. Since the periodic orbits are individual system properties, the great success of a universal approach, such as random matrix theory, is somewhat mysterious.

This apparent contradiction is resolved by the fact that in the limit of long orbits the individual system properties are no longer of relevance, and statistical methods have to be applied to the periodic orbits. As every orbit of length l adds an oscillatory contribution with a period $\Delta k = 2\pi/l$ to the density of states (see Eq. (8.1.6)), the long orbits determine the short distance behaviour of the eigenvalues. This is the range where random matrix theory works well. The short orbits, on the other hand, are responsible for the long distance behaviour. We can thus expect that at wave number differences of the order of $\Delta k \geqslant 2\pi/l_{\text{min}}$, where l_{min} is the length of the shortest periodic orbit, random matrix theory will no longer be applicable. Deviations from random matrix behaviour in the spectra due to periodic orbits have already been presented. The oscillations in the density of states of the stadium billiard presented in Section 7.3.3 (see Fig. 7.10) are one relevant example.

Another example, where periodic orbits disturb the universal correlations, has been found by Casati *et al.* [Cas85] in the spectra of rectangular billiards. Figure 8.6 shows the spectral rigidity taken in two different energy windows of the spectra. For small L values the curves closely follow the straight line expected for uncorrelated eigenvalues (see Section 3.2.4), but for higher L values saturation behaviour is found, with a saturation value proportional to the square root of the energy. This clear indication of nonuniversal behaviour triggered Berry's work on the semiclassical theory of spectral rigidity [Ber85], which is one of the landmarks of quantum chaos research.

To understand the saturation behaviour displayed in Fig. 8.6 we go back to expression (7.1.63) for the spectral form factor,

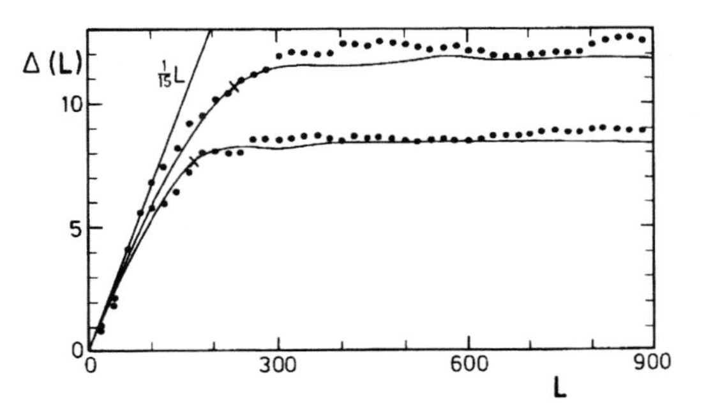

Figure 8.6. Spectral rigidity for a rectangular billiard in the energy range $7850 \leqslant E \leqslant 8640$ (lower curve) and $15\,700 \leqslant E \leqslant 16\,490$ (upper curve). The curves have been obtained by superimposing the results from different billiards with side ratios a/b in the range of 0.95 to 1.1. The energies are given in units of the mean level spacings. The points are numerical results by Casati *et al.* [Cas85], the solid lines are semiclassical calculations by Berry [Ber85].

$$K(t) = \frac{ab}{2\pi} \sum_{n,m}{}' \frac{\epsilon_{nm}}{l_{nm}} \delta(l_{nm} - vt), \tag{8.2.1}$$

derived in Section 7.1.2 for a rectangular billiard. v is the velocity of the particle, and the l_{nm} are lengths of the periodic orbits in the rectangle, given by

$$l_{nm} = 2\sqrt{(an)^2 + (bm)^2}. \tag{8.2.2}$$

ϵ_{nm} takes a value of four, if both n and m are different from zero, and two, if either n or m is zero. The sum is over positive and negative values of n and m with the exception of the term $n = m = 0$.

In Section 7.1.2 we replaced the sum by an integral and obtained $K(t) = 1$, corresponding to completely uncorrelated eigenvalues. However, the saturation behaviour found numerically shows that some long range correlations must be present in the spectrum. This discrepancy is resolved if we bear in mind that the replacement of the integral by a sum becomes wrong in the range of short orbits. This can be taken into account qualitatively by excluding a small circle of radius $l_{\min}$ from the integration, where $l_{\min}$ is the length of the shortest orbit. We then obtain

$$\begin{aligned} K(t) &= \frac{2ab}{\pi} \int_{l > l_{\min}} dn\, dm\, \frac{1}{l_{nm}} \delta(l_{nm} - vt) \\ &= \int_{l_{\min}}^{\infty} dl\, \delta(l - vt). \end{aligned} \tag{8.2.3}$$

In the second step polar coordinates $n = (l/2a)\cos\phi$, $m = (l/2b)\sin\phi$ have been introduced, and the trivial ϕ integration has already been performed. The remaining integration yields

$$K(t) = \begin{cases} 0 & t < \dfrac{l_{\min}}{v} \\ 1 & t > \dfrac{l_{\min}}{v} \end{cases}. \tag{8.2.4}$$

In contrast to our earlier result we have now obtained a correlation hole for short times (see Section 3.2.5).

The spectral rigidity is obtained from the relation

$$\Delta_3(L) = \frac{\hbar}{\pi} \int_0^{\infty} \frac{dt}{t^2} K(t) G\left(\frac{Lt}{2\hbar}\right), \tag{8.2.5}$$

derived in Section 3.2.5 (see Eq. (3.2.91)). Here $G(x)$ is given by

$$G(x) = 1 - [f(x)]^2 - 3[f'(x)]^2, \tag{8.2.6}$$

where

$$f(x) = \frac{\sin x}{x}. \tag{8.2.7}$$

$G(x)$ is plotted in Fig. 8.7. It takes small values for $x \leqslant \pi$ and becomes constant for large x. $G(x)$ was denoted the *orbit selection function* by Berry, since it selects the orbits with periods larger than $\hbar/L$ from the integral (8.2.5). Inserting expression (8.2.4) for $K(t)$ into Eq. (8.2.5) we get

$$\Delta_3(L) = \frac{L}{2\pi}\int_{x_{\min}}^{\infty} \frac{dx}{x^2}\, G(x), \tag{8.2.8}$$

where $x_{\min} = Lt_{\min}/2\hbar$. Here $t_{\min} = l_{\min}/v$ is the period of the shortest periodic orbit. As we are only interested in the qualitative behaviour of $\Delta_3(L)$, we approximate $G(x)$ somewhat crudely by the step function

$$G(x) \approx \begin{cases} 0 & x < \pi \\ 1 & x > \pi \end{cases}. \tag{8.2.9}$$

We can now perform the integration and get

$$\Delta_3(L) \approx \begin{cases} \dfrac{L}{2\pi^2} & L < \dfrac{2\pi\hbar}{t_{\min}} \\[2ex] \dfrac{\hbar}{\pi t_{\min}} & L > \dfrac{2\pi\hbar}{t_{\min}} \end{cases}. \tag{8.2.10}$$

For small L values we have again obtained the linear increase of $\Delta_3(L)$ with L, known from random matrix theory for integrable systems. The correct slope $1/15$ is not reproduced, but this is only due to the poor approximation (8.2.9) for $G(x)$. Beyond the critical L value of $2\pi\hbar/t_{\min}$ the rigidity saturates at a value proportional to $t_{\min}^{-1}$. As in billiards the period is proportional to $E^{-1/2}$, this is exactly the energy dependence of the saturation values found in the data shown in Fig. 8.6.

For a quantitative calculation we have to enter expression (8.2.1) for $K(t)$ directly into Eq. (8.2.5) without approximating the sums by integrals. We then

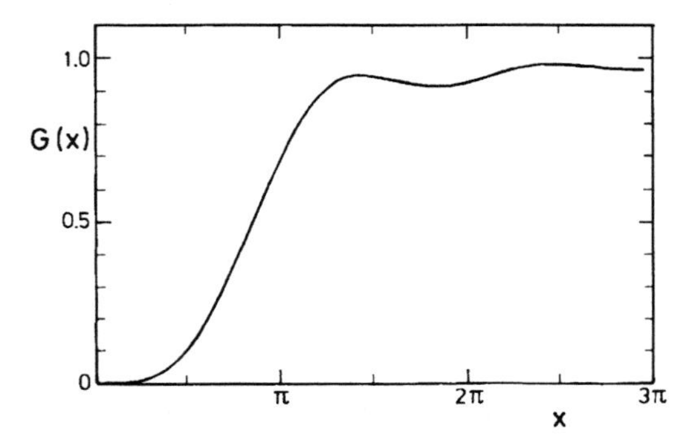

Figure 8.7. The orbit selection function $G(x)$ defined by Eq. (8.2.6) [Ber85].

get a rapidly converging sum for $\Delta_3(L)$ which can be easily summed up. In this way Berry obtained the solid lines shown in Fig. 8.6 which reproduce the numerical data quantitatively.

8.2.2 Semiclassical sum rule

To determine the spectral form factor for the rectangular billiard, the diagonal approximation has been applied at one crucial point (see Section 7.1.2). This is not permissible for nonintegrable systems. We have therefore to go back to the spectral autocorrelation function

$$C(E) = \left\langle \left[\rho\left(\overline{E} + \frac{1}{2}E\right) - 1\right]\left[\rho\left(\overline{E} - \frac{1}{2}E\right) - 1\right]\right\rangle, \tag{8.2.11}$$

to calculate the semiclassical spectral form factor for nonintegrable systems. The brackets denote an average over $\overline{E}$ within a window of width $\Delta\overline{E}$ (see Section 3.2.5). Remember that the mean level spacing $\rho_0(E)$ has been normalized to one. Using the periodic orbit expansion (8.1.2)

$$\rho(E) = 1 + \sum_n{}' A_n \exp\left[\frac{\imath}{\hbar} S_n(E)\right], \tag{8.2.12}$$

for the density of states, where the prefactor is given by

$$A_n = \frac{1}{2\pi\hbar}\frac{(T_p)_n}{\|M_n - 1\|^{1/2}} \exp\left(-\imath\frac{\mu_n\pi}{2}\right), \tag{8.2.13}$$

we get for the spectral autocorrelation function

$$C(E) = \sum_{n,m}{}' A_n A_m^* \left\langle \exp\left\{\frac{\imath}{\hbar}\left[S_n\left(\overline{E} + \frac{E}{2}\right) - S_m\left(\overline{E} - \frac{E}{2}\right)\right]\right\}\right\rangle. \tag{8.2.14}$$

Expanding the action with respect to the energy up to the linear term,

$$S_n\left(\overline{E} \pm \frac{E}{2}\right) = S_n \pm \frac{E}{2} T_n, \tag{8.2.15}$$

where S_n is an abbreviation for $S_n(\overline{E})$, and where $T_n = dS_n/dE$ is the period of the nth orbit, we obtain for the spectral autocorrelation function

$$C(E) = \sum_{n,m}{}' A_n A_m^* \left\langle \exp\left[\frac{\imath}{\hbar}\left(S_n - S_m + \frac{T_n + T_m}{2} E\right)\right]\right\rangle. \tag{8.2.16}$$

The spectral form factor $K(t)$ is obtained as usual from the Fourier transform of the spectral autocorrelation function. As is customary in this context, all times will be expressed in units of $T_H = 2\pi\hbar$. T_H is related to the mean level spacing via the uncertainty relation, and is therefore called the *Heisenberg time*. Note that due to the normalization of the mean level spacing to one, $\hbar$ has acquired the dimension of a time. For the redefined spectral form factor $\hat{K}(\tau) = K(\tau T_H)$ we obtain from Eq. (3.2.85)

$$\hat{K}(\tau) = \int_{-\infty}^{\infty} C(E) e^{-2\pi \imath E\tau} \, dE. \tag{8.2.17}$$

Inserting expression (8.2.16) for $C(E)$ we get

$$\hat{K}(\tau) = \sum_{n,m}{}' A_n A_m^* \left\langle \exp\left[\frac{\imath}{\hbar}(S_n - S_m)\right] \right\rangle \delta\left(\tau - \frac{\tau_n + \tau_m}{2}\right), \tag{8.2.18}$$

where $\tau_n = T_n/T_H$ is the period of the nth orbit in units of the Heisenberg time.

For integrable systems it is allowable to take only the diagonal terms of the sum, by arguing that due to the averaging over an energy window all nondiagonal contributions will vanish. But this argument cannot be generalized to nonintegrable systems because of the exponential proliferation of periodic orbits with length, as already discussed in Section 7.1.2.

Nevertheless, let us see what will happen if only the diagonal terms are taken. We have to distinguish between systems with and without time-reversal symmetry. In the first case all orbits occur in pairs with stability factors which are complex conjugates of each other. Consequently each diagonal term occurs twice in the sum (8.2.18). In systems without time-reversal symmetries the amplitude factors are uncorrelated. The diagonal part of the sum is thus given by

$$\hat{K}_D(\tau) = g \sum_n{}' |A_n|^2 \delta(\tau - \tau_n), \tag{8.2.19}$$

where $g = 1$ for systems without, and $g = 2$ for systems with time-reversal symmetry.

For short orbits Eq. (8.2.19) gives the length spectrum of the periodic orbits. This is the range discussed in Section 8.1.1. In the present context the range of long orbits is of interest, where the individual contributions to the sum (8.2.19) can no longer be resolved.

Hannay and Ozorio de Almeida [Han84] found that the sum (8.2.19) allows a simple semiclassical interpretation. They considered the classical probability $P_n(E, t)$ for a particle with energy E to return on a given periodic orbit n to its starting position after the time t. This probability can be calculated classically, following the lines sketched in Section 7.2.3, and yields

$$P_n(E, t) = \frac{(T_p)_n}{\|M_n - 1\|} \delta(t - T_n), \tag{8.2.20}$$

where T_n is the period of the orbit n, and $(T_p)_n$ is the primitive period of the orbit. A derivation of this result can be found in the appendix of Ref. [Arg93]. Using this result, the sum in Eq. (8.2.19) may be written as

$$\begin{aligned}
\sum_n{}' |A_n|^2 \delta(\tau - \tau_n) &= 2\pi\hbar \sum_n{}' |A_n|^2 \delta(t - T_n) \\
&= \frac{1}{2\pi\hbar} \sum_n{}' \frac{(T_p)_n^2}{\|M_n - 1\|} \delta(t - T_n) \\
&= \frac{1}{2\pi\hbar} \sum_n{}' (T_p)_n P_n(E, t) \\
&\approx \frac{1}{2\pi\hbar} \sum_n{}' T_n P_n(E, t) \\
&= \frac{t}{2\pi\hbar} \sum_n{}' P_n(E, t) \\
&= \tau P(E, t).
\end{aligned} \tag{8.2.21}$$

In the second step we used Eq. (8.2.13), and in the fourth step we have replaced $(T_p)_n$ by T_n. The error produced becomes negligibly small for large lengths, since the number of periodic orbits increases exponentially with length, whereas the number of repeated orbits increases only algebraically with length.

$P(E, t)$ is the probability for a particle to return to its starting position on *any* periodic orbit at the time t. It may be written as

$$P(E, t) = \int dp\, dq\, \delta[H(p, q) - E] \delta^{2d-1}[q - q(t)]. \tag{8.2.22}$$

$q(t)$ is the position of a particle on the energy surface at time t, which started at $t = 0$ at position q. The first delta function restricts the integral to the energy surface. The second $(2d - 1)$-dimensional delta function ensures that there will be only contributions from those points q on the energy surface, for which a periodic orbit with period t exists.

We now come to the key point of the argument: in chaotic systems the phase space distribution of the very long orbits is uniform on the energy surface [Han84]. The trajectory for the stadium billiard shown in Fig. 7.7 demonstrates this *principle of uniformity*. This allows us to approximate the $(2d - 1)$-dimensional delta function in Eq. (8.2.22) by its average, which is just the reciprocal of the area Ω_E of the energy surface. Consequently we obtain

$$\begin{aligned}
P(E, t) &\approx \int dp\, dq\, \delta(H(p, q) - E) \langle \delta^{2d-1}[q - q(t)] \rangle \\
&= \frac{1}{\Omega_E} \int dp\, dq\, \delta[H(p, q) - E] = 1.
\end{aligned} \tag{8.2.23}$$

Collecting the results, we get the remarkably simple expression

$$\sum_n{}' |A_n|^2 \delta(\tau - \tau_n) = \tau. \tag{8.2.24}$$

This is the sum rule of Hannay and Ozorio de Almeida [Han84]. Generalizations of this result can be found in papers by Argaman *et al.* [Arg93] and Eckhardt *et al.* [Eck95].

We have thus obtained

$$\hat{K}_D(\tau) = g\tau \tag{8.2.25}$$

for the diagonal part of the spectral form factor. Since the spectral form factor is given by the Fourier transform of the spectral autocorrelation function, it is obvious that $\hat{K}(\tau)$ cannot increase in an unlimited way. This shows that for chaotic systems the diagonal part $\hat{K}_D(t)$ cannot describe $\hat{K}(\tau)$ correctly for large τ. Calculations beyond the diagonal approximation can be found in the recent publication by Bogomolny and Keating [Bog96a].

8.2.3 Rigidity for nonintegrable systems

The sum rule of Hannay and Ozorio de Almeida enabled us to calculate the spectral form factor for small τ values. The long time limit becomes accessible from another angle [Ber85]. We start with a Lorentzian smoothed density of states

$$\rho_\epsilon(E) = \sum_n \delta_\epsilon(E - E_n) = \frac{\epsilon}{\pi} \sum_n \frac{1}{(E - E_n)^2 + \epsilon^2}. \tag{8.2.26}$$

Squaring both sides of Eq. (8.2.26) we get

$$\begin{aligned} [\rho_\epsilon(E)]^2 &= \frac{\epsilon^2}{\pi^2} \sum_{n,m} \frac{1}{[(E - E_n)^2 + \epsilon^2][(E - E_m)^2 + \epsilon^2]} \\ &= \frac{\epsilon^2}{\pi^2} \sum_n \frac{1}{[(E - E_n)^2 + \epsilon^2]^2}. \end{aligned} \tag{8.2.27}$$

In the second step the double sum has been restricted to the diagonal terms. This is allowed as long as the width of the Lorentzians is small as compared to the mean level spacing, i.e. for $\epsilon \leq 1$.

Equation (8.2.27) may be written as

$$2\pi\epsilon[\rho_\epsilon(E)]^2 = \sum_n \overline{\delta}_\epsilon(E - E_n), \tag{8.2.28}$$

where we have introduced the function

$$\overline{\delta}_\epsilon(x) = \frac{2\epsilon^3}{\pi} \frac{1}{(x^2 + \epsilon^2)^2}. \tag{8.2.29}$$

$\overline{\delta}_\epsilon(x)$ is normalized to one,

$$\int_{-\infty}^{\infty} \overline{\delta}_\epsilon(x)\, dx = 1, \tag{8.2.30}$$

and may therefore be interpreted as another type of a smoothed delta function. If both sides of Eq. (8.2.28) are averaged over an energy window sufficiently large to suppress any oscillatory contribution to the density of states, the right hand side reduces to the smooth part of the density of states, $\rho_0(E) = 1$. On the left hand side only the quadratic average of the oscillatory part survives, as is easily seen. We consequently end with

$$2\pi\epsilon\langle[\rho_{\text{osc},\epsilon}(E)]^2\rangle = \rho_0(E) = 1. \tag{8.2.31}$$

The equation expresses the smooth part of the density of states in terms of its oscillatory part. Both parts are thus closely linked, a feature which has been termed 'analytic bootstrap' by Berry.

A periodic orbit expansion of $\rho_{\text{osc},\epsilon}(E)$ is obtained from Eqs (8.2.12) and (8.2.15) by means of a Lorentzian folding

$$\rho_{\text{osc},\epsilon}(E) = \sum_n{}' A_n \exp\left[\frac{\imath}{\hbar} S_n(E) - \frac{\epsilon}{\hbar}|T_n|\right]. \tag{8.2.32}$$

Entering this expression into Eq. (8.2.31) we get

$$2\pi\epsilon\left\langle \sum_{n,m}{}' A_n A_m^* \exp\left\{\frac{\imath}{\hbar}[S_n(E) - S_m(E)] - \frac{\epsilon}{\hbar}(|T_n| + |T_m|)\right\}\right\rangle = 1. \tag{8.2.33}$$

A comparison with expression (8.2.18) shows that the left hand side may be interpreted as the Laplace transform of the spectral form factor:

$$2\pi\epsilon\int_0^{\infty} \hat{K}(\tau)e^{-2\pi\epsilon\tau}\, d\tau = 1. \tag{8.2.34}$$

This relation must hold for all ϵ values $\leqslant 1$. This is only possible if

$$\hat{K}(\tau) \to 1 \quad \text{for } \tau \geqslant 1. \tag{8.2.35}$$

In combination with Eq. (8.2.25) we have arrived at the following scenario. For $\tau < 1$, i.e. for times smaller than the Heisenberg time it is possible to apply the diagonal approximation. This leads to a linear increase of $\hat{K}(\tau)$ with slopes 1 and 2 for systems without and with time-reversal symmetry, respectively. For $\tau \gg 1$ $\hat{K}(\tau)$ approaches the constant value 1. The exact dependence of $\hat{K}(\tau)$ in the transition region cannot be obtained in this way.

For the calculation of the asymptotic behaviour of spectral rigidity a knowledge of $\hat{K}(\tau)$ in the transition region is not necessary [Ber85]. The spectral rigidity in terms of $\hat{K}(\tau)$ reads

$$\Delta_3(L) = \frac{1}{2\pi^2} \int_0^\infty \frac{d\tau}{\tau^2} \hat{K}(\tau) G(\pi L \tau), \tag{8.2.36}$$

where expression (8.2.5) has been used once more. We know that $G(x) \to 1$ for $x \geqslant \pi$ (see Eq. (8.2.9)) and that $\hat{K}(\tau)/\tau = g$ for $\tau \leqslant 1$. This suggests splitting the integral into two parts and approximating the spectral rigidity by

$$\Delta_3(L) \approx \frac{1}{2\pi^2} \left[g \int_0^1 \frac{d\tau}{\tau} G(\pi L \tau) + \int_1^\infty d\tau \, \frac{\hat{K}(\tau)}{\tau^2} \right] \tag{8.2.37}$$

which should hold for large L. In this limit the first integral can be carried out and yields [Ber85]

$$\int_0^1 \frac{d\tau}{\tau} G(\pi L \tau) \approx \ln(\pi L) + \gamma + \ln 2 - \frac{9}{4}, \tag{8.2.38}$$

where $\gamma = 0.577\ldots$ is the Euler constant. In the second integral we replace $\hat{K}(\tau)$ by its asymptotic value one and obtain

$$\int_1^\infty d\tau \, \frac{\hat{K}(\tau)}{\tau^2} \approx 1. \tag{8.2.39}$$

We thus get for the spectral rigidity

$$\Delta_3(L) \approx \frac{g}{2\pi^2} \ln L + \text{const.} \tag{8.2.40}$$

This is exactly the asymptotic behaviour we found in Section 3.2.4 for the spectral rigidity for the GOE ($g = 2$) and the GUE ($g = 1$). If we had proceeded a bit more carefully we would even have obtained the correct value of the constant for the GUE.

In this way Berry succeeded in linking random matrix theory to periodic orbit theory on the level of the two-point correlation function. The two essential ingredients for this result are (i) the *principle of uniformity* from which the short time behaviour of $\hat{K}(\tau)$ is determined, and (ii) the analytic bootstrap yielding the asymptotic behaviour.

8.3 Periodic orbit calculation of spectra

8.3.1 Dynamical zeta function

In the preceding sections we have discussed techniques for extracting the contributions of periodic orbits from spectra and wave functions. The divergence of the trace formula has not been a problem in this context. The calculation of the spectra from the periodic orbits, however, can only work if we somehow succeed in enforcing the convergence of the sum over the periodic orbits. One realization is the introduction of convergence-generating factors.

As the number of orbits increases exponentially with length, a cut-off over-compensating this increase has to be introduced. This can be done with a Gaussian smoothed density of states,

$$\overline{\rho}_\epsilon(k) = \sum_n \delta_\epsilon(k - k_n), \tag{8.3.1}$$

where

$$\delta_\epsilon(k) = \frac{1}{\sqrt{2\pi}\epsilon} \exp\left[-\frac{1}{2}\left(\frac{k}{\epsilon}\right)^2\right]. \tag{8.3.2}$$

We then obtain from Eq. (8.1.6) the periodic orbit expansion

$$\overline{\rho}_\epsilon(k) = \overline{\rho}_0(k) + \sum_r{}' A_r e^{ikl_r} \exp\left[-\frac{1}{2}(l_r\epsilon)^2\right] \tag{8.3.3}$$

for the smoothed density of states. The additional Gaussian factor on the right hand side ensures the convergence of the sum. The Lorentzian smoothing of the delta functions applied in the last section would not have been sufficient for this purpose.

If the individual eigenvalues are to be resolved, ϵ must be of the order of the mean distance Δk between neighbouring eigenvalues. For the right hand side of the equation orbits up to a length of $(\Delta k)^{-1}$ have thus to be considered. Since the number of orbits increases exponentially with l, whereas Δk decreases with k^{-1}, it is evident that we shall not proceed very far in this direction.

The situation is simpler if only the Gaussian smoothed spectrum $\overline{\rho}_\epsilon(k)$ is calculated. If we confine ourselves to moderate values of ϵ, a comparatively small number of orbits may suffice for the calculation. In this way Wintgen succeeded in reproducing the Gaussian smoothed spectrum of the hydrogen atom in a strong magnetic field [Win88]. Figure 8.8 shows the result of his calculation. In the upper part only 2 periodic orbits have been considered. For this case the smoothing is still considerably larger than the mean level distance. In the lower part 13 orbits have been considered. Here the lowest eigenvalues are reproduced quite well.

Another example is given in Fig. 8.9, namely the Gaussian smoothed spectrum of an octagon billiard on a metric with a constant negative curvature [Aur88]. 10 000 (!) periodic orbits have been considered in the calculation of the semiclassical smoothed density of states, but in spite of this effort we are still far from resolution of the individual eigenvalues. Only the smoothed density of states has been reproduced correctly. Non-Euclidean billiards will be discussed in Sections 8.4.

The last example shows in particular that it is extremely ineffective to calculate the quantum mechanical spectrum by a direct summation of the

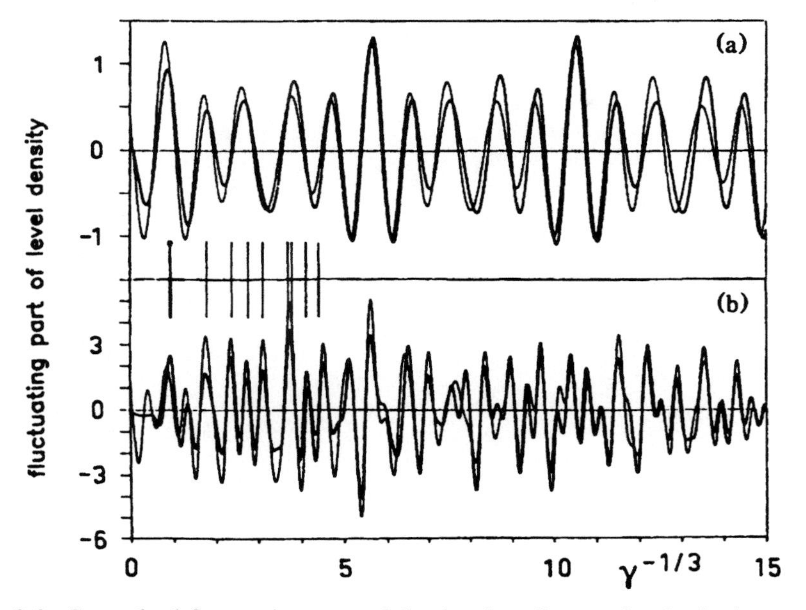

Figure 8.8. Smoothed fluctuating part of the density of states for the hydrogen atom in a strong magnetic field as a function of $\gamma^{-1/3}$ (see Section 8.1.2). The thick lines correspond to the exact quantum mechanical results, the thin lines have been obtained from semiclassical calculations including 2 (a) and 13 (b) orbits, respectively. The lowest exact eigenvalues are marked by vertical lines [Win88] (Copyright 1988 by the American Physical Society).

periodic orbits. There is no other way left but to improve the convergence properties of the trace formula by a suitable rearrangement of terms. Again two-dimensional billiards will be considered to illustrate the technique. Moreover, we assume that all orbits are hyperbolic. According to Eq. (8.1.5) the oscillatory part of the density of states is then given by

$$\overline{\rho}_{\mathrm{osc}}(k) = \frac{1}{2\pi}\,\mathrm{Re}\left\{ {\sum_{p,n}}' \frac{l_p}{\sinh \dfrac{n\alpha_p}{2}} \exp\left[\imath n\left(k l_p - \frac{\mu_p \pi}{2}\right)\right]\right\}. \tag{8.3.4}$$

As the number of orbits exponentially increases with e^{hl}, where h is the topological entropy (see Eq. (7.3.20)), the sum diverges for real values of k. But if we attribute to k a positive imaginary part $\imath h$, the exponential increase of the number of orbits is compensated by an additional factor e^{-nhl_p} on the right hand side, and the sum becomes convergent. The line $\mathrm{Im}(k) = h$ is therefore called the *entropy barrier*. Only beyond this barrier is the trace formula mathematically well-defined. The real barrier is somewhat closer to the real axis since the prefactor $(\sinh n\alpha_p/2)^{-1}$ slightly improves the convergence [Eck89]. The essential point is the following: there *are* regions in the complex

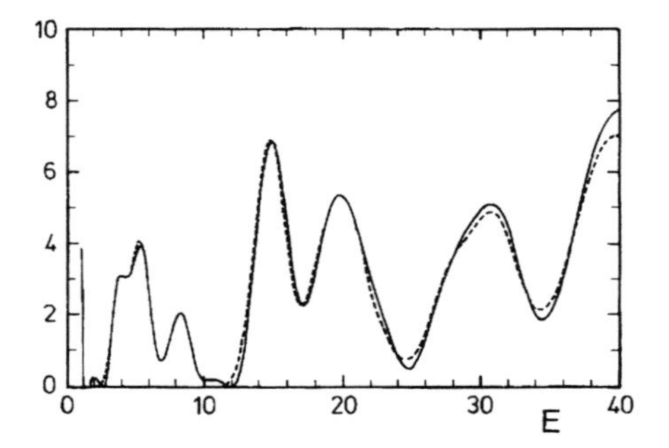

Figure 8.9. Gaussian smoothed density of states for an octagon billiard on a metric with constant negative curvature in the range of the first 40 eigenvalues. The solid line corresponds to the exact quantum mechanical calculation, the dashed line has been obtained by means of the Selberg trace formula, which replaces the Gutzwiller trace formula in metrics with a constant negative curvature, see Section 8.4.1. 10 000 primitive periodic orbits have been taken into account [Aur88] (with kind permission from Elsevier Science).

plane, where the sum is convergent. If we succeed in rearranging the sum in such a way that it converges on the real axis too, then the principle of analytic continuation guarantees the correct value for the sum in the resulting expression. In another context we have argued similarly in Section 3.1.5.

Let us start by expanding the stability prefactor into a geometrical series,

$$\left(\sinh\frac{n\alpha_p}{2}\right)^{-1} = 2\sum_{m=0}^{\infty}\exp\left[-\left(m+\frac{1}{2}\right)n\alpha_p\right]. \tag{8.3.5}$$

If this expression is inserted into Eq. (8.3.4), the sum over n is once more a geometric series which can be summed up:

$$\overline{\rho}_{\rm osc}(k) = \frac{1}{\pi}\,\mathrm{Re}\left\{\sum_{p,m}{}' \frac{l_p}{1-\exp\left[\imath\left(kl_p - \frac{\mu_p\pi}{2}\right) - \left(m+\frac{1}{2}\right)\alpha_p\right]}\right\}. \tag{8.3.6}$$

This may be written in the form

$$\overline{\rho}_{\rm osc}(k) = -\frac{1}{\pi}\,\mathrm{Im}\left[\frac{d}{dk}\ln Z(k)\right], \tag{8.3.7}$$

where

$$Z(k) = \prod_{p,m}\left\{1-\exp\left[\imath\left(kl_p - \frac{\mu_p\pi}{2}\right) - \left(m+\frac{1}{2}\right)\alpha_p\right]\right\} \tag{8.3.8}$$

is the so-called *dynamical zeta function* [Vor88]. Reminiscence of the Riemann zeta function is not accidental, as we shall see in the next section. The last two equations show that the zeros of $Z(k)$ yield the spectrum $\rho(k)$. All zeros must therefore be real, although this is not evident from expression (8.3.8).

As yet we have not come closer to our final goal, namely to improve the convergence of the trace formula. We have only transformed the divergent sum (8.3.4) into the equally divergent infinite product (8.3.8). Cvitanovich [Cvi88] noticed that the infinite product can be made convergent if it is expanded,

$$Z(k) = 1 - \sum_q B_q e^{\imath k l_q}. \tag{8.3.9}$$

The sum is over all lengths l_q which can be written as sums over periodic orbits,

$$l_q = \sum_i l_{p_i}. \tag{8.3.10}$$

These combinations of periodic orbits are called *pseudo-orbits*. The new prefactors B_q contain all phase and stability factors occuring in the expansion of the product.

The crucial point is that the contributions of orbits and pseudo-orbits of the same lengths have a tendency to cancel. We therefore obtain a rapidly converging sum, if the sum (8.3.9) is ordered according to the lengths of the pseudo-orbits. The method is based on the principle that long orbits can be decomposed into a small number of fundamental cycles acting as building blocks [Art90]. Assume that there are only two fundamental cycles labelled 0 and 1. Then longer composed orbits may be labelled by 01, 001, 011, ... Note that a cyclic permutation of the labels yields the same periodic orbit. It is therefore not necessary to consider the orbits labelled by 10, 010, ..., etc. The zeta function now reads

$$Z(k) = \prod_m \{[1 - t_0^{(m)}][1 - t_1^{(m)}][1 - t_{01}^{(m)}][1 - t_{001}^{(m)}][1 - t_{011}^{(m)}] \cdots\}, \tag{8.3.11}$$

where the $t_p^{(m)}$ are abbreviations for the exponentials entering the product (8.3.8) for the zeta function. Expanding the products we get

$$\begin{aligned} Z(k) = \prod_m \{1 - t_0^{(m)} - t_1^{(m)} - [t_{01}^{(m)} - t_0^{(m)} t_1^{(m)}] \\ - [t_{001}^{(m)} - t_0^{(m)} t_{01}^{(m)}] - [t_{011}^{(m)} - t_1^{(m)} t_{01}^{(m)}] + \cdots\}. \end{aligned} \tag{8.3.12}$$

The dominant contribution comes from the two fundamental cycles, whereas the terms in the subsequent brackets yield rapidly decreasing corrections.

It is essential for the applicability of this technique that there is a symbolic dynamics, meaning that every orbit can be labelled by a symbolic sequence as in the above example. It is further important that there is no or only little pruning, i.e. for every symbolic sequence there should also be a corresponding orbit. The non-Euclidean billiards on a metric with constant negative curvature are ideal examples allowing a symbolic dynamics with no pruning at all, as we shall see later. The importance of the pruning restriction is evident. If the orbit 01 had not existed in the above example, then the term $-t_0^{(m)}t_1^{(m)}$ would not have been cancelled by the term $t_{01}^{(m)}$. The number of possible applications is considerably reduced by this restriction. But whenever the method can be applied, it yields excellent results [Sie91,Tan91]. In an open system consisting of three hard disks the poles of the scattering matrix can also be reproduced using this technique [Cvi89].

To demonstrate the potential of the method, Fig. 8.10 shows the result for the anisotropic Kepler problem [Tan91], a system studied by Gutzwiller in his early work [Gut67]. This system is especially well suited for the present purposes as there is a one-to-one correspondence between the orbits and a binary symbolic dynamics (see Chapter 11 of Ref. [Gut90]). In the figure the so-called *functional determinant* $D(E)$ is plotted, which differs from the dynamical zeta function only by a multiplicative entire function [Kea92]. The zeros of $D(E)$ are therefore identical with the zeros of $Z(E)$. In Fig. 8.10(a) 8 orbits have been considered in the cycle expansion (8.3.12), whereas in

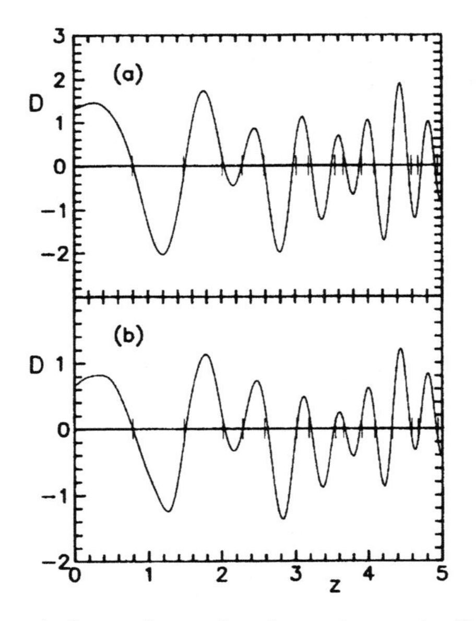

Figure 8.10. Functional determinant for the anisotropic Kepler problem calculated from the cycle expansion (8.3.12) by considering 8 (a) and 71 (b) orbits. The positions of the exact eigenvalues are marked by vertical bars [Tan91] (Copyright 1991 by the American Physical Society).

Fig. 8.10(b) 71 orbits have been taken. In the first case the first 15 eigenvalues are well reproduced, but in the second case agreement is perfect.

The correspondence obtained is even better than we could have expected. In the derivation of the trace formula the stationary phase approximation has been used repeatedly. This is only justified in the semiclassical limit. Nevertheless, in the given example the cycle expansion reproduces even the lowest-lying eigenvalues with a precision within a small percentage.

8.3.2 Riemann zeta function

In the last section we introduced the dynamical zeta function $Z(k)$, whose zeros make up the spectrum of a dynamical system. The notation reminds us of the *Riemann zeta function*, which plays a central role in prime number theory. In fact there are close analogies between the two zeta functions, which will be discussed in the following. Excellent introductions on the Riemann zeta function are given in the books of Titmarsh [Tit51] and Edwards [Edw74]. Therefore only facts important for the following discussion will be recapitulated. The Riemann zeta function is defined by

$$\zeta(z) = \sum_{n=1}^{\infty} \frac{1}{n^z}. \tag{8.3.13}$$

The sum represents a so-called Dirichlet series and converges only for $\mathrm{Re}(z) > 1$, but it is possible to continue the function analytically to the whole complex plane. The sum can be written in the form of the Euler product

$$\zeta(z) = \prod_p \left(1 - \frac{1}{p^z}\right)^{-1}, \tag{8.3.14}$$

where the product is over all prime numbers. To demonstrate the equivalence of both expressions, we write the denominators in the Euler product as geometrical series, and expand the product.

From the analytical continuation of expression (8.3.13) it is rather easy to show that $\zeta(z)$ has zeros for $z = -2n$, where n is an integer. These are the so-called trivial zeros. In addition there is an infinite set of nontrivial zeros. According to the famous *Riemann conjecture* all of them should lie on the line $z = \frac{1}{2} + \imath y$. The hypothesis is as yet unproved, but numerically supported by millions of zeros [Odl87, Odl89].

These nontrivial zeros have a number of unexpected properties making them interesting for quantum chaotic studies. Figure 8.11 shows the two-point correlation function $R_2(x)$ (see Section 3.2.3) for the imaginary parts of the nontrivial zeros of the zeta function. All zeros z_n with n in the range

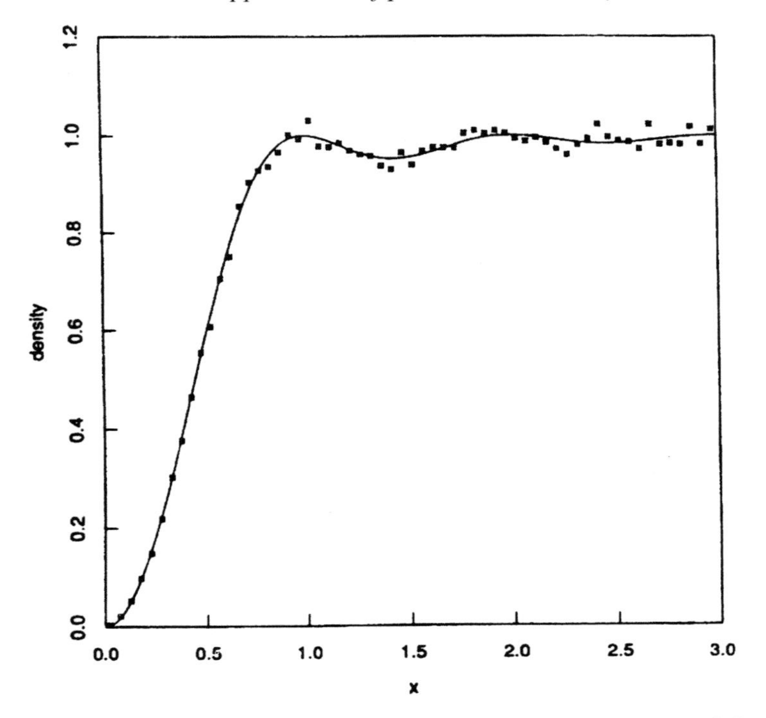

Figure 8.11. Two-point correlation function for the imaginary parts of the nontrivial zeros of the Riemann zeta function on the line $\frac{1}{2} + \imath y$. All zeros z_n with n in the range $10^{12} < n < 10^{12} + 10^5$ have been taken into account. The solid line corresponds to the GUE two-point correlation function [Odl87].

$10^{12} < n < 10^{12} + 10^5$ have been considered [Odl87]. Most surprisingly the data closely follow the two-point correlation function calculated for the GUE. In an unpublished report Odlyzko [Odl89] has even calculated 79×10^6 (!) zeros close to $n = 10^{20}$. The corresponding plot of the two-point correlation function is reprinted in Chapter 1.8 of Mehta's book [Meh91] and is in perfect agreement with the GUE prediction. The imaginary parts of the nontrivial zeros behave, as it seems, exactly as the eigenvalues of a dynamical system without time-reversal invariance. This motivates interest in the Riemann zeta function in the present context.

There is a close connection between the two-point correlation function and the so-called *Hardy–Littlewood conjecture* [Har22] on prime numbers. The simplest form of this conjecture states that the density of pairs of prime numbers $\rho_h(n)$ separated by h should decrease asymptotically with n according to

$$\rho_h(n) \simeq \frac{\alpha(h)}{\ln^2 n}, \tag{8.3.15}$$

where $\alpha(h)$ is a number theoretical function varying erratically with h. A short treatise on the Hardy–Littlewood conjecture, very readable even for nonspecialists, can be found in Ref. [Guy94] (see also Section 22.20 of Ref. [Har60]). On the assumption that the conjecture is true, Bogomolny and Lebœuf could show that the two-point correlation function of the imaginary parts of the nontrivial zeros asymptotically approaches the corresponding GUE behaviour [Bog94]. Subsequently Bogomolny and Keating generalized this result to all n-point correlation functions [Bog95,Bog96b], which is truly remarkable. The nontrivial zeros behave in *every* respect as the eigenvalues of a dynamical system with broken time-reversal symmetry (provided that the Hardy–Littlewood conjecture is true!).

But we have learnt that the universal correlations break down for distances smaller than $\Delta E = \hbar / T_{min}$ where T_{min} is the period of the shortest periodic orbit. The question arises whether there are periodic orbits in this hypothetical dynamical system. The answer is positive. It is even more remarkable that we can explicitly determine the periodic orbits. The respective key is supplied by the Euler product (8.3.14). It already bears a strong resemblance to the dynamical zeta function introduced in Section 8.3.1. We can accentuate the correspondence even more [Ber86] by putting

$$p = e^{l_p}, \tag{8.3.16}$$

and by replacing z by $\frac{1}{2} - \imath k$ in the argument of the zeta function. Then Eq. (8.3.14) reads

$$\zeta\left(\frac{1}{2} - \imath k\right) = \prod_p \left[1 - \exp\left(\imath k l_p - \frac{1}{2} l_p\right)\right]^{-1}. \tag{8.3.17}$$

This is already very close to expression (8.3.8) for the dynamical zeta function, if we interpret the logarithms of the prime numbers as the lengths of the periodic orbits. The instability exponents α_p are now given by the lengths l_p of the orbits. We can even reproduce the sum over m entering into Eq. (8.3.8) by defining

$$Z(k) = \prod_{m=0}^{\infty} \zeta\left(m + \frac{1}{2} - \imath k\right), \tag{8.3.18}$$

whence follows

$$Z(k) = \prod_{p,m} \left\{1 - \exp\left[\imath k l_p - \left(m + \frac{1}{2}\right) l_p\right]\right\}^{-1}. \tag{8.3.19}$$

Apart from the exponent -1 this is identical with expression (8.3.8) derived in the last section for the dynamical zeta function. But this exponent is irrelevant, at least for the calculation of the spectrum, where only the logarithm of $Z(k)$ is used (see Eq. (8.3.7)).

The logarithm l_n of a natural number n with the prime number decomposition $n = \sum_i p_i^{\alpha_i}$ may be written as

$$l_n = \ln n = \sum_i \alpha_i \ln p_i = \sum_i \alpha_i l_{p_i}. \tag{8.3.20}$$

The logarithms of the natural numbers therefore correspond to the pseudo-orbits of the hypothetical dynamical system. The Dirichlet series (8.3.13) may then be written as

$$\zeta\left(\frac{1}{2} - \imath k\right) = 1 + \sum_{n=1}^{\infty} e^{-l_n/2} e^{\imath k l_n}. \tag{8.3.21}$$

This is completely analogous to the cycle expansion introduced in Section 8.3.1 (see Eq. (8.3.9)).

A final correspondence between the hypothetical dynamical system and the Riemann zeta function can be derived from the *Hadamard prime number theorem*. It states that the number of primes $\pi(p)$ smaller than a number p is asymptotically given by

$$\pi(p) \simeq \frac{p}{\ln p}. \tag{8.3.22}$$

Details can be found in every book on number theory, such as Ref. [Har60]. Since the logarithms of the primes correspond to the primitive periodic orbits of the hypothetical dynamical system, the prime number theorem can be translated into the statement that the number of primitive periodic orbits $N(l)$ smaller than a given length l grows according to

$$N(l) \simeq \frac{e^l}{l}. \tag{8.3.23}$$

We have found again the exponential proliferation of periodic orbits in a chaotic system (7.3.21) with a topological entropy given by $h = 1$.

The close analogy between the Riemann zeta function and the dynamical zeta function has led to a number of fruitful mutual applications. First of all, powerful resummation techniques have been developed to calculate the Riemann zeta function on the critical line $\frac{1}{2} + \imath k$ [Ber92]. Such resummation schemes are mandatory for the calculation of the the nontrivial zeros, as the the original Drichlet series (8.3.13) is divergent on this line. It is remarkable that the same resummation techniques can be directly used to obtain rapidly converging expressions for the dynamical zeta function as discussed in Section 8.3.1 [Ber92].

Vice versa the analysis of the zeros of the Riemann zeta function can also profit by techniques originally developed in periodic orbit theory. This is illustrated in Fig. 8.12 showing the number variance for the nontrivial zeros [Ber88]. The stars represent the results of a calculation using the same set of zeros which has already entered into the two-point correlation function shown in Fig. 8.11. Obviously the observed oscillations cannot be described by the universal GUE number variance discussed in Section 3.2.4. The deviations from universal behaviour are again caused by the short periodic orbits discussed in Section 8.2.1. Applying the same technique as for spectral rigidity, Berry has calculated the number variance from the periodic orbits, i.e. from the logarithms of the prime numbers. The solid line shown in Fig. 8.12 represents the result of the calculation. The correspondence with the data is perfect.

We now have to leave this fascinating topic which has exhibited surprising parallels between such different disciplines as prime number theory and the quantum mechanics of chaotic systems. As for the prime numbers, this connection is based on longstanding conjectures which have successfully resisted any attempt to prove them, in the case of the Riemann conjecture for more than a century! As for quantum mechanics, a dynamical system with the logarithms of the prime numbers as the primitive periodic orbits has not been found. There is as yet no formula for the reconstruction of a dynamical system

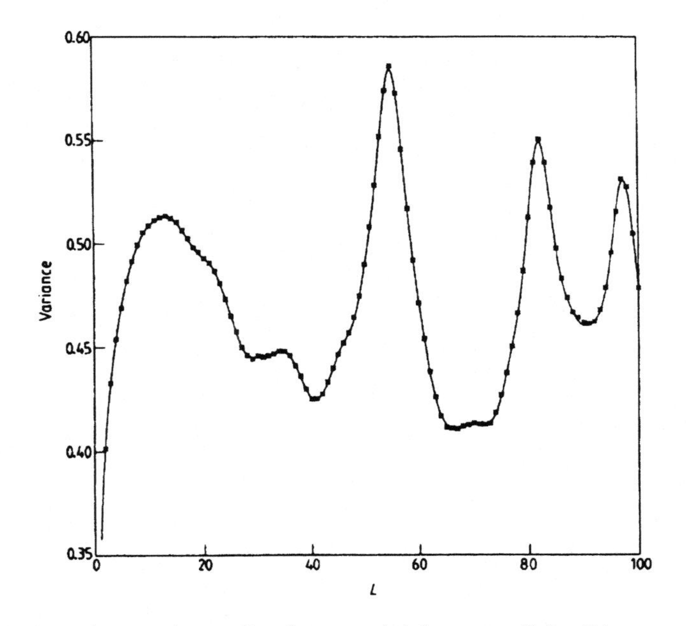

Figure 8.12. Number variance for the nontrivial zeros of the Riemann zeta function. The stars have been obtained from the same data set of zeros used in Fig. 8.11. The solid line is the result of a periodic orbit calculation [Ber88].

from its periodic orbits. Success in this direction would obviously have far-reaching consequences for prime number theory too.

8.4 Surfaces with constant negative curvature

8.4.1 Selberg trace formula

The remaining sections are reserved for discussion of billiards on a metric with constant negative curvature. There are several reasons for the interest in such systems, in spite of the fact that concrete realizations seem hardly imaginable. First, the metric causes neighbouring trajectories to depart exponentially from each other, making such billiards an ideal object for the study of classical chaotic dynamics. Second, the *Selberg trace formula*, expressing the density of states of certain non-Euclidean billiards in terms of their periodic orbits, is exact, in contrast to its Euclidean equivalent, the Gutzwiller trace formula. Last, there are arguments suggesting that the hypothetical dynamical system associated with the Riemann zeta function may be found in billiards or scattering systems on a metric with constant negative curvature.

This section gives a short introduction to non-Euclidean metrics. The derivation of the main relations will only be sketched, and all technical details will be omitted. More details on non-Euclidean billiards can be found in the paper by Balazc and Voros [Bal86], as well as in Chapter 19 of Ref. [Gut90].

Let us start with a somewhat more familiar situation, a metric with constant *positive* curvature. It is realized on the surface of a sphere defined in Cartesian coordinates by the equation

$$x^2 + y^2 + z^2 = R^2, \tag{8.4.1}$$

where R is the radius of the sphere. Expressing x, y, z in polar coordinates

$$x = R\sin\theta\cos\phi, \qquad y = R\sin\theta\sin\phi, \qquad z = R\cos\theta, \tag{8.4.2}$$

the line element on the surface is given by

$$ds^2 = R^2\left(d\theta^2 + \sin^2\theta\, d\phi^2\right). \tag{8.4.3}$$

One important quantity of a metric is its Gaussian curvature K. It is defined by

$$K = \frac{R_{1212}}{|g|}, \tag{8.4.4}$$

where R_{1212} is one component of the Riemann curvature tensor, and $|g|$ is the determinant of the metric tensor (see Section 62 of Ref. [Sok64]). It can be shown that K is an invariant and does not depend on the coordinate system used to describe the metric. For the special case of a diagonal metric tensor with the line element given by

$$ds^2 = g_{11}\, dx_1^2 + g_{22}\, dx_2^2, \tag{8.4.5}$$

Eq. (8.4.4) reads

$$K = -\frac{1}{2\sqrt{|g|}}\left[\frac{\partial}{\partial x_1}\left(\frac{1}{\sqrt{|g|}}\frac{\partial g_{22}}{\partial x_1}\right) + \frac{\partial}{\partial x_2}\left(\frac{1}{\sqrt{|g|}}\frac{\partial g_{11}}{\partial x_1}\right)\right], \tag{8.4.6}$$

where $|g| = g_{11}g_{22}$. Applying Eq. (8.4.6) to the metric defined by Eq. (8.4.3) we find a constant value of $K = 1/R^2$ for the Gaussian curvature of the surface of the sphere.

If we replace the sine in the line element (8.4.3) by the hyperbolic sine,

$$ds^2 = R^2\left(d\lambda^2 + \sinh^2\lambda\, d\phi^2\right), \tag{8.4.7}$$

we obtain a metric with constant *negative* curvature $K = -1/R^2$. For obvious reasons the corresponding surface is called a *pseudosphere*. The pseudosphere cannot be embedded in a three-dimensional Euclidean space as in the case of the ordinary sphere. Embedding in a space with a Minkowski metric, known from the special theory of relativity, can be applied instead [Bal86].

Another realization of the pseudosphere, which is somewhat more practical for calculations, is obtained by substituting λ by

$$r = \frac{\sinh\lambda}{1 + \cosh\lambda}. \tag{8.4.8}$$

In the new variables the line element is given by

$$ds^2 = \frac{4}{(1 - r^2)^2}\left(dr^2 + r^2\, d\phi^2\right), \tag{8.4.9}$$

where we have now put $R = 1$. In the following all distances are therefore measured in units of R. In order to verify that K is indeed independent of the coordinate system, it is a good exercise to check by means of Eq. (8.4.6) that the value of the Gaussian curvature is still $K = -1$. The metric described by Eq. (8.4.9) defines the *Poincaré disk*. It is restricted to $r < 1$. The radius $r = 1$ corresponds to $\lambda \to \infty$ and represents the infinite horizon. We may equivalently express the line element in Cartesian coordinates as

$$ds^2 = \frac{4}{(1 - x^2 - y^2)^2}\left(dx^2 + dy^2\right), \tag{8.4.10}$$

demonstrating that the metric is locally Euclidean. The angles for the metric are identical to the Euclidean ones, merely the distances are distorted. It is practical to interpret the components x and y as real and imaginary parts of the complex number $z = x + \imath y$, and the Poincaré disk as the interior of the unit circle of the complex plain. This allows us to take advantage of the conformal mapping properties of the complex functions. The *Möbius transformation*

$$z' = \frac{\alpha z + \beta}{\beta^* z + \alpha^*} \tag{8.4.11}$$

is of particular importance. α and β are arbitrary complex numbers normalized to $|\alpha|^2 - |\beta|^2 = 1$. The Möbius transformation maps the unit circle onto itself. It it always possible to find a Möbius transformation, by which a given point z in the interior of the unit circle is mapped to $z' = 0$ showing that all points of the Poincaré disk are equivalent.

Integrating the line element (8.4.10) from one point z to another point z' we get for the non-Euclidean distance $d(z, z')$ of the two points

$$\cosh d(z, z') = 1 + \frac{2|z - z'|^2}{(1 - |z|^2)(1 - |z'|^2)}. \tag{8.4.12}$$

In non-Euclidean metrics the propagation of waves is described by a modified Helmholtz equation which for the metric (8.4.10) reads

$$\left[\frac{(1 - x^2 - y^2)^2}{4}\Delta + k^2\right]\psi = 0. \tag{8.4.13}$$

The product of the metric prefactor and the Laplace operator is called the *Laplace–Beltrami operator.* Figure 8.13(a) shows the propagation of waves on the Poincaré metric. The circular trajectories starting from the point left to the centre are the geodesics. They correspond to the straight lines in the Euclidean metric. The circles perpendicular to the geodesics are the fronts of equal phases. The wave fronts seem to be squeezed when approaching the border of the unit circle, but this is only an artifact of the embedding. The non-Euclidean distances between the wave fronts do not change during propagation.

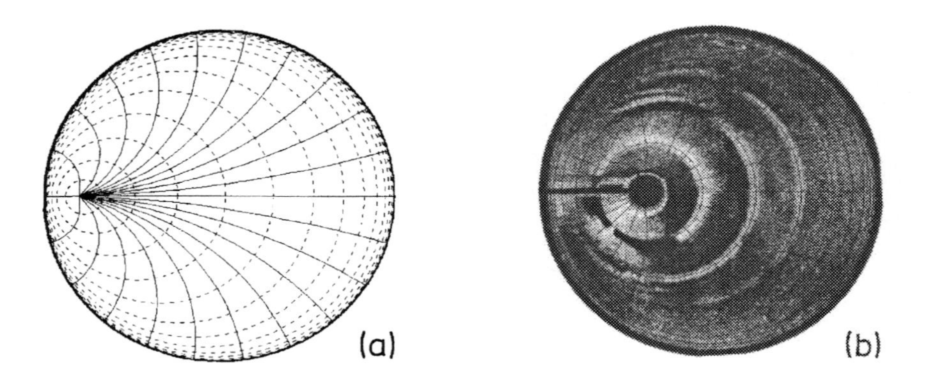

Figure 8.13. (a) Geodesics (solid lines) and fronts of equal phase (dashed lines) in the Poincaré disk for a particle starting from a point left of the centre. (b) Experimental realization of the Poincaré metric in a water vessel with an appropriately adjusted depth profile. The surface waves were periodically excited by a dipper [Sch96].

We may also consider the modified Helmholtz equation (8.4.13) from another point of view. Dividing the equation by the metric prefactor, we obtain an ordinary wave equation which describes wave propagation in a Euclidean space, but now with a position-dependent velocity $v = (1 - r^2)/2k$. This allows an experimental realization of the Poincaré disk. A demonstration with water surface waves has been achieved by Schanze [Sch96]. In the experiments he used the fact that surface gravity waves propagate with the velocity $v = \sqrt{gh}$, where h is the depth of the vessel (see Section 2.1.2). Wave propagation on the Poincaré disk was simulated by a suitable modelling of the depth profile. A typical result is shown in Fig. 8.13(b). Because of the rather strong damping of the waves such a set-up is not suited for quantitative studies.

The pseudosphere is unbounded and therefore does not allow chaotic dynamics. A free particle will move on a geodesic and eventually escape to infinity. To obtain chaotic dynamics we have to make the system *compact*. This is illustrated for the Euclidean plane. It can be tessellated in a number of different ways, in the simplest case by means of rectangles. A compact surface is obtained by identifying the opposite sides of the rectangle. Whenever a particle leaves the fundamental domain on one side, it reenters on the opposite side at the same angle. The resulting surface has the topology of a torus, but is still flat. We may alternatively describe the surface in terms of a billiard with periodic boundary conditions.

To apply the same procedure to the pseudosphere we first need a tessellation. For surfaces with constant curvature the curvature radius defines a length scale, in contrast to the Euclidean metric which contains no characteristic length. As a consequence the variety of possible tessellations is much larger in non-Euclidean metrics. Tessellations of the Poincaré disk by squares and hexagons were repeatedly used by the Dutch artist M. Escher for his graphic arts. A tessellation into octagons, not possible for the Euclidean plane, is shown in Fig. 8.14(a) [Bal86]. All octagons, apart from the central one, are strongly distorted by the embedding, but nevertheless they all have the same shape and area. The figure demonstrates how the dynamics becomes chaotic, as soon as the pseudosphere has been made compact. In Fig. 8.14(a) a trajectory is shown which ignores the tessellation. It runs unspectacularly to the left margin with a minute angular deviation of 10^{-3}, too small to be seen in the figure. In Figs. 8.14(b) to (e) the same trajectory is shown, but now the motion is restricted to the fundamental octagon by identifying opposite sides. The seeming concentration of trajectories in the corners is once more an artifact resulting from distortion by the embedding. We may ask why such a complicated fundamental domain as the regular octagon has been used to make the pseudosphere compact. The answer is simple: there is no easier way to achieve this. This is a

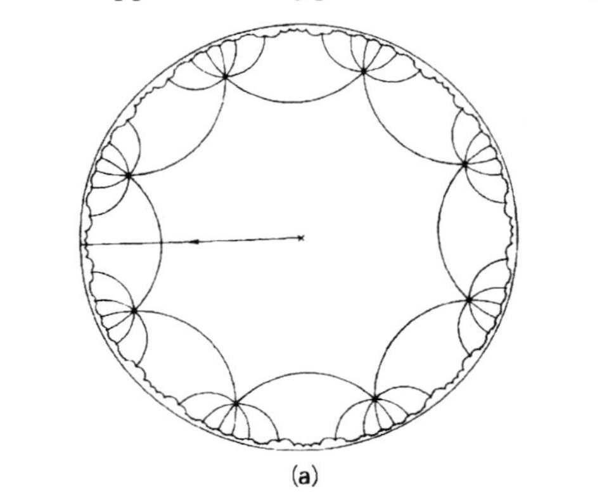

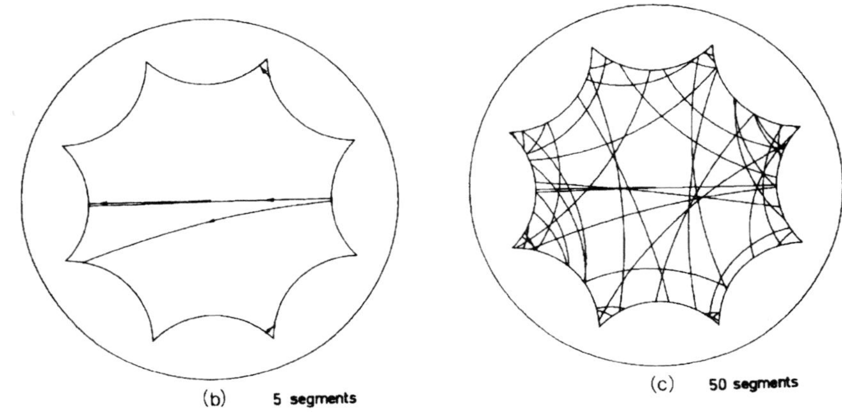

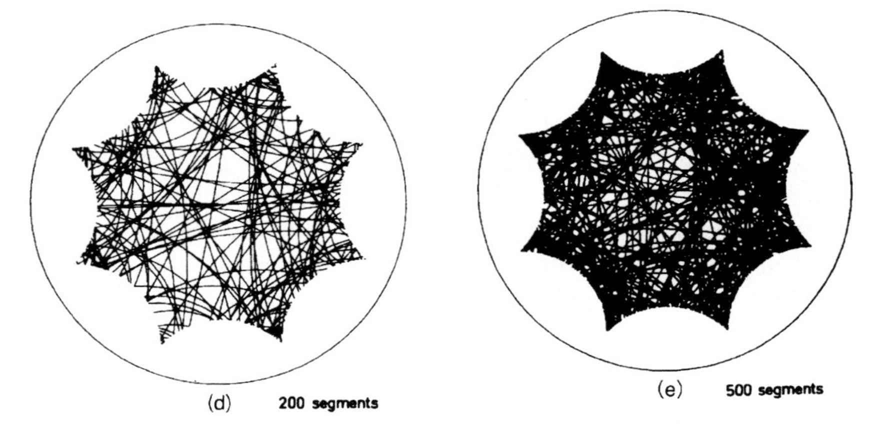

Figure 8.14. Chaotic dynamics of a particle on the Poincaré disk in the regular octagon with periodic boundary conditions. The particle moves to the left with an angular deviation of 10^{-3} from the horizontal. In (a) the particle runs on a geodesic towards the boundary ignoring the tessellations. In figures (b) to (e) the trajectory reenters the fundamental domain from the opposite side whenever it crosses a boundary [Bal86] (Copyright 1986 by the American Physical Society).

consequence of the *Gauss–Bonnet theorem*, relating the angle sum in a polygon with the Gaussian curvature via

$$\int K\,dA = (2-n)\pi + \sum_i \alpha_i, \tag{8.4.14}$$

where the integral is over the area of the polygon. The α_i correspond to the inner angles, and n is the number of corners (see Section 74 of Ref. [Sok64]). For a compact surface on a metric with constant curvature K we obtain from the Gauss–Bonnet theorem the relation

$$-KA = 4\pi(g-1), \tag{8.4.15}$$

between the area A of the surface and its genus g [Bal86]. The genus counts the number of holes in the topology of the surface. Let us take as an example a surface with constant positive curvature $K = 1$. Equation (8.4.15) shows that a compact surface can be obtained with $g = 0$. The relation further yields that the area of this surface is given by $A = 4\pi$. This is exactly the surface of the unit sphere. For the Euclidean plane with $K = 0$ it follows that the genus of the compact surface must be $g = 1$, corresponding to a torus, in accordance with the example given above. The area may be arbitrary in this case. Finally, for a constant negative curvature $K = -1$, g must be at least 2, but larger values are possible. In fact it is easily seen that the octagon with identified opposite sides has the topology of a two-handled sphere (readers having problems in imagining this may consult the references given above). According to Eq. (8.4.15) the area is now given by $A = 4\pi$, as for the ordinary sphere.

The motion of a classical particle on compact surfaces with constant negative curvature shows the strongest chaotic behaviour known. Every trajectory can be labelled by a random sequence of symbols, and arbitrary sequences of symbols are realized by trajectories. This is the so-called *Bernoulli property.* For the octagon the symbolic alphabet contains just four letters denoting the pairs of opposite sides. If the side pairs are labelled by 1, 2, 3, 4, starting with the top-down pair and continuing clockwise, the trajectory shown in Fig. 8.14 starts with the sequence 3321.

We have already mentioned that for a large class of billiards, including the regular octagon, there is an expansion of the spectra in terms of periodic orbits, which is exact in contrast to the Gutzwiller formula. We can only sketch the essential idea of the derivation. Details can be found in the references given above. We start with the Green function $G_0(z, z', E)$ for the free particle on the pseudosphere. $G_0(z, z', E)$ is obtained as the solution of the inhomogeneous wave equation

$$\left[\frac{(1-x^2-y^2)^2}{4}\Delta + E\right] G_0(z, z', E) = \delta(x-x')\delta(y-y'), \tag{8.4.16}$$

which is given by

$$G_0(z, z', E) = -\frac{1}{2\pi} Q_{-1/2+\imath k}[\cosh d(z, z')], \tag{8.4.17}$$

where $d(z, z')$ is the non-Euclidean distance between z and z' (see Eq. (8.4.12)). $Q_l(x)$ is the Legendre function of the second kind. The wave number k is related to the energy via

$$k = \sqrt{E - \frac{1}{4}}. \tag{8.4.18}$$

$Q_{-1/2+\imath k}(\cosh x)$ asymptotically behaves as $e^{-\imath kx}$ which justifies the interpretation of k as a wave number. The most important properties of the less known function $Q_l(x)$ are listed in Appendix G of Ref. [Bal86] (see also Ref. [Aur93]).

The Green function $G(z, z', E)$ for the octagon with periodic boundary conditions is obtained from the free particle Green function $G_0(z, z', E)$ by the method of images as

$$G(z, z', E) = \sum_{g \in G} G_0[z, g(z'), E]. \tag{8.4.19}$$

Both z and z' are taken from the fundamental domain. The sum is over all images of the fundamental octagon. Remember that for the Euclidean case the Green function of the rectangle can be written in complete analogy as a sum over images of the free Green function, in this case given by a Hankel function (see Chapter 7.2 of Ref. [Mor53]).

The images $g(z')$ of z' are obtained by Möbius transformations (8.4.11) as

$$g(z') = \frac{\alpha_g z' + \beta_g}{\beta_g^* z' + \alpha_g^*}. \tag{8.4.20}$$

The complete set of these transformations forms the group G. It can be constructed from the generators T_1, T_2, T_3, T_4 and their inverses, where the T_i are translational operators, also called *boosts*, which map the fundamental octagon to its neighbours adjacent to the ith side. The eight generators are not completely independent, since an eight-fold repeated mapping of the fundamental octagon around one corner reproduces the original octagon. Each group element can be written in exactly one way as a product of generators. Each finite product of generators defines a symbolic sequence for a periodic orbit. For example, the element $T_1T_2T_3$, can be associated with the periodic orbit labelled by the symbolic sequence 123123.... This shows that sum (8.4.19) can be alternatively interpreted as a sum over periodic orbits.

The *primitive* elements play a special role. A group element is called primitive if it is not the unit element or if it cannot be expressed in terms of powers of other group elements. In view of the just established correspondence the primitive group elements correspond to the primitive periodic orbits. Not all primitive elements, however, belong to different periodic orbits. Let us consider once more the orbit 123123.... It may be associated with each of the three primitive elements $p = T_1 T_2 T_3$, $p' = T_2 T_3 T_1$, and $p'' = T_3 T_1 T_2$, since the choice of the starting element in the symbolic dynamics is arbitrary.

All primitive group elements p and p', which are related to each other via $p' = gpg^{-1}$, where g is an arbitrary group element, belong to the same *conjugacy class* $[p]$. All elements of one conjugacy class belong to the same orbit according to the above reasoning. We have thus established a one-to-one correspondence between the conjugacy classes of the group elements and the periodic orbits.

In the next step we rewrite the sum (8.4.19) as a sum over the conjugacy classes $[p^n]$, where p runs over the primitive elements and n over its repetitions, as well as over the elements of each conjugacy class. The sum over the elements of the conjugacy classes can be calculated explicitly. The result is a representation of the Green function in terms of a sum over the conjugacy classes of the group elements, i.e. the periodic orbits. Finally we calculate the spectrum in the usual way from the imaginary part of the trace of the Green function. The result is the *Selberg trace formula*

$$\begin{aligned}\overline{\rho}(k) &= \sum_n \delta(k - k_n) \\ &= \frac{A}{2\pi} k \tanh \pi k + \frac{1}{2\pi} \sum_{[p]} \sum_{n=1}^{\infty} \frac{l_p}{\sinh \dfrac{nl_p}{2}} \cos nkl_p. \qquad (8.4.21)\end{aligned}$$

The first sum is over the conjugacy classes of the primitive elements p, and the second is over the repetitions.

The Selberg trace formula holds exactly, in contrast to the Gutzwiller trace formula, since a semiclassical approximation has nowhere been applied. It suffers, however, from the same divergences as its Euclidean equivalent. For this reason smoothing functions have been used to ensure convergence in the derivations given in Refs. [Bal86, Gut90]. A discussion of the Selberg trace formula from the mathematical point of view can be found in the monographs by Hejhal [Hej76, Hej83].

There is a striking similarity between the Selberg trace formula and the Gutzwiller trace formula (compare Eqs. (8.1.3) to (8.1.5)). In the limit of large k values the two trace formulas become indistinguishable. We must not be surprised at that fact, since the metric is locally Euclidean, and in the short wavelength limit the curvature of the metric is not noticed by the wave functions. As the area of the octagon is 4π, we could have replaced the prefactor $A/2\pi$ of the smooth part by 2. This has not been done, to make the correspondence between the two trace formulas more evident. The instability exponent α_p which cannot be universally specified in ordinary chaotic systems, is given by the length l_p of the primitive orbit in the octagon. In Section 8.3.2 we have found the same situation in the periodic orbit interpretation of the Riemann zeta function (see Eq. (8.3.19)).

8.4.2 Non-Euclidian billiards

In view of the close relation between the Selberg and the Gutzwiller trace formulas it is not surprising that the spectra of Euclidean and non-Euclidean billiards show similar phenomenology. Steiner and coworkers have studied the spectrum of an octagon on the pseudosphere and found complete agreement with random matrix predictions for level spacing distribution, number variance, spectral rigidity, and other spectral correlations [Aur90]. As the regular octagon has a large number of symmetries, the authors studied a desymmetrized variant of the octagon. They further developed a non-Euclidean equivalent of Bogomolny's scar formula [Aur91], as well as a semiclassical theory of the number variance [Aur95]. The calculation of the Gaussian smoothed density of states of the octagon billiard in terms of periodic orbits [Aur88] has already been discussed in Section 8.3.1.

There are, however, striking exceptions from the general behaviour, which seem to contradict our entire previous understanding of level repulsion in chaotic systems. In 1986 Bohigas, Giannoni, and Schmitt [Boh86] discovered clear deviations from the Wigner behaviour expected in the level spacing distribution of a triangle on the pseudosphere with angles $\pi/8$, $\pi/2$ and $\pi/3$. This triangle can be completed by a repeated reflection to the regular octagon [Bal86], thus tessellating the plane. The authors speculated that the deviations from the usual level spacing distribution might have been caused by this fact [Sch89]. But the spectrum of the desymmetrized octagon discussed above is a counter example. It shows GOE behaviour, although the octagon tessellates the pseudosphere. Meanwhile it has been recognized that all triangles with this unusual level spacing distribution are associated with the so-called *arithmetical groups*, and that number theoretical degeneracies in the length spectrum are responsible for the deviations from universal behaviour.

We shall discuss only one special case of the arithmetical groups, namely the *modular group*. It is associated with the triangle with angles $\pi/4$, $\pi/4$, and π/∞, also called *Artin's billiard*. Triangles where one of the angles is zero are obtained whenever one of the corners is located on the unit circle.

Figure 8.15 shows the level spacing distribution in Artin's billiard as calculated by Bolte *et al.* [Bol92]. Though the underlying classical dynamics is completely chaotic, Poissonian behaviour is found, apart from minor deviations at small distances. Essentially the same results have been obtained independently by Bogomolny *et al.* [Bog92]. For the spectral rigidity Bogomolny *et al.* found the behaviour known from the rectangular billiards. For small L values $\Delta_3(L)$ increases linearly with L, but for larger L values the rigidity saturates. For the purpose of comparison the authors also calculated the spectrum of a triangle with angles $\pi/2$, $\pi/5$, π/∞, which is not associated with one of the arithmetic groups, and found complete agreement with the GOE predictions.

For the following discussion we switch to an alternative embedding of the pseudosphere, the Poincaré half plane. It is obtained from the Poincaré disk by means of the conformal mapping

$$z' = \imath \frac{-z+1}{z+1}, \tag{8.4.22}$$

by which the interior of the unit circle is mapped onto the upper half of the complex plane. The line element is now given by

$$ds^2 = \frac{1}{y^2}\left(dx^2 + dy^2\right), \tag{8.4.23}$$

whence follows for the non-Euclidean distance between the points z and z'

$$\cosh d(z, z') = 1 + \frac{(x-x')^2 + (y-y')^2}{2yy'}. \tag{8.4.24}$$

In the new metric the Schrödinger equation reads

$$\left(y^2\Delta + E\right)\psi = 0. \tag{8.4.25}$$

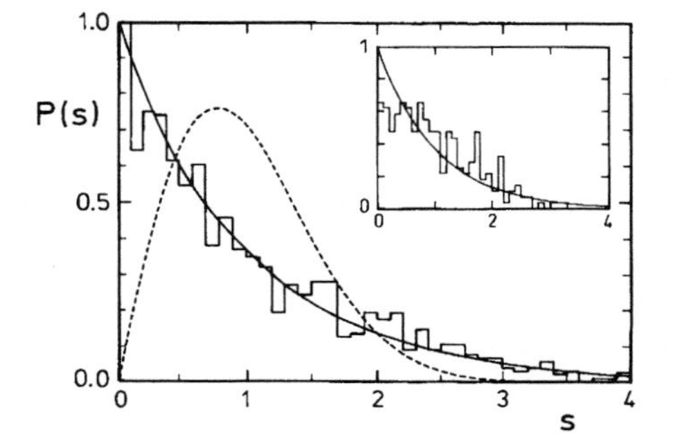

Figure 8.15. Level spacing distribution in Artin's billiard with periodic boundary conditions in the momentum range $250 \leqslant p_n \leqslant 300$. The inset shows the same distribution in the low momentum range $p \leqslant 100$ [Bol92] (Coyright 1992 by the American Physical Society).

In the Poincaré half plane Artin's billiard is bounded from below by the unit circle, and from the right and the left sides by the straight lines $z = -\frac{1}{2} + \imath y$ and $z = \frac{1}{2} + \imath y$, respectively. All copies of the fundamental domain entering into the tessellation of the pseudosphere are obtained by the map

$$z' = \frac{az + b}{cz + d}, \tag{8.4.26}$$

where a, b, c, d are *integers* obeying $ad - bc = 1$. The set of all these mappings forms the modular group. It is easily seen that all group elements can be composed of the two generators

$$T: \quad z' = z + 1, \tag{8.4.27}$$

$$S: \quad z' = -\frac{1}{z}. \tag{8.4.28}$$

T is the elementary translation, and S is the inversion at the unit circle. Once more there is a one-to-one correspondence between the conjugacy classes and the periodic orbits, allowing us to derive a trace formula for this system as well [Bog93].

Number theoretical considerations show that the mean number $\langle n(l) \rangle$ of orbits of the same length increases according to

$$\langle n(l) \rangle = 2\frac{e^{l/2}}{l}. \tag{8.4.29}$$

This exponentially increasing degeneracy in the length spectrum is responsible for the deviations from universal behaviour. As a consequence the eigenvalues become uncorrelated, and a Poisson level spacing distribution is observed. Details can be found in Ref. [Bog96c].

A special polygon where all corners are infinite, and consequently all angles are zero, is the singular square. Its four corners are located on the unit circle at the positions $1, \imath, -1, -\imath$. The circumference of the singular square is infinite, but its area is finite and given by $A = 2\pi$. This is a direct consequence of the Gauss–Bonnet theorem (8.4.14). The singular square is discussed in detail in Chapter 19 of Ref. [Gut90].

Again it is more practical to use embedding in the Poincaré half plane. Now three of the corners are on the real axis at $-1, 0, 1$, while the fourth corner is at $\imath\infty$ (see Fig. 8.16). Figure 8.16 suggests that the singular square should be treated as a scattering system. Now let us see what will happen if a plane wave enters from above, and is scattered back from the circular arcs forming the two lower borders.

The Schrödinger equation (8.4.25) has two particularly simple solutions

$$\psi_-(x, y) = y^{1/2 - \imath k}, \quad \psi_+(x, y) = y^{1/2 + \imath k}, \tag{8.4.30}$$

where again the wave number is given by $k = \sqrt{E - 1/4}$ (see Eq. (8.4.18)). The two solutions (8.4.30) correspond asymptotically to plane waves entering

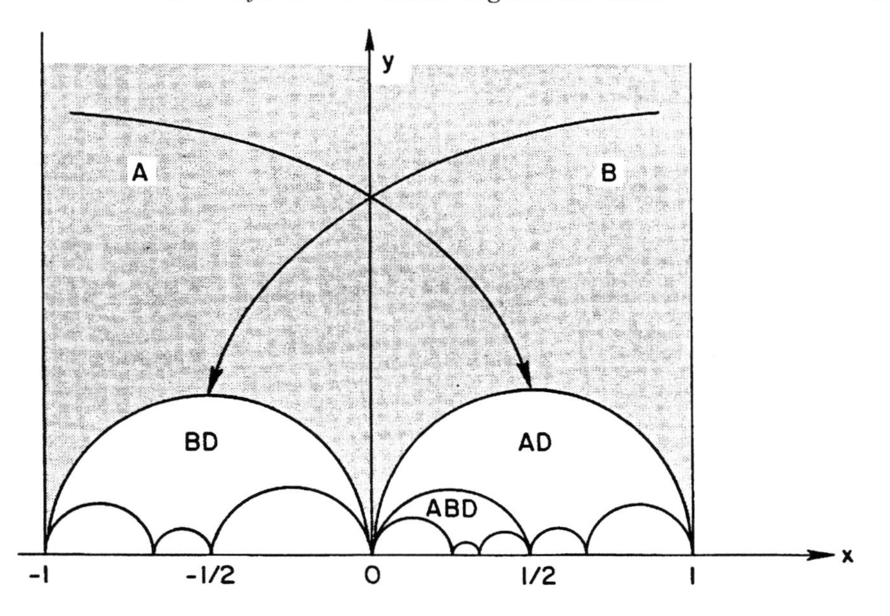

Figure 8.16. The singular square (grey) and a number of its copies, embedded in the Poincaré half plane [Gut83] (with kind permission from Elsevier Science).

and leaving along the imaginary axis, as can be easily derived from the non-Euclidean distance relation (8.4.24).

The solution of the scattering problem is obtained by a superposition of the incoming wave function $\psi_-(x, y)$ with all its images from the copies of the fundamental domain. The images are obtained once more by means of Möbius transformations. It is left as an exercise to readers not familiar with conformal mappings to find the Möbius transformations which are necessary to map the fundamental domain to its adjacent images below the two semi-circles. These two mappings, together with their inverses, are the generators of the associated group G.

For an arbitrary Möbius transformation the image y' of y is obtained from Eq. (8.4.26) by taking on both sides the imaginary part,

$$y' = \frac{y}{|cz + d|^2}. \tag{8.4.31}$$

From this we get for the solution of the scattering problem

$$\psi(x, y) = y^{1/2-\imath k} + \sum_{g \in G}{}' \left(\frac{y}{|c_g z + d_g|^2} \right)^{1/2-\imath k}, \tag{8.4.32}$$

where the sum is over all group elements with the exception of the identity element.

It is evident from Fig. 8.16 that $\psi(x, y)$ is periodic in x with a period of 2. It is shown in Ref. [Gut83] that all Fourier components of $\psi(x, y)$, apart from the term independent of x, exponentially decrease for large y. The asymptotic scattering solution is therefore obtained by taking the average of $\psi(x, y)$ over x. We skip the technical details, and immediately turn to the final solution, which reads

$$\langle \psi(x, y) \rangle_x = y^{1/2-ik} + e^{i\phi(k)} y^{1/2+ik}, \tag{8.4.33}$$

where the phase factor is given by

$$e^{i\phi(k)} = \frac{Z(1+2ik)}{Z(1-2ik)} \tag{8.4.34}$$

with

$$Z(z) = \pi^{-z/2} \Gamma\left(\frac{z}{2}\right) \zeta(z). \tag{8.4.35}$$

Thus the singular square behaves exactly like a scattering system with one scattering channel, where the scattering phase $\phi(k)$ is related to the Riemann zeta function via Eqs. (8.4.34) and (8.4.35). The arguments of the zeta functions are located on the line parallel to the critical line $\frac{1}{2} + ik$ at a distance of $\frac{1}{2}$. The nontrivial zeros of the zeta function therefore play the role of the poles of the scattering matrix. This is demonstrated in Fig. 8.17 where the

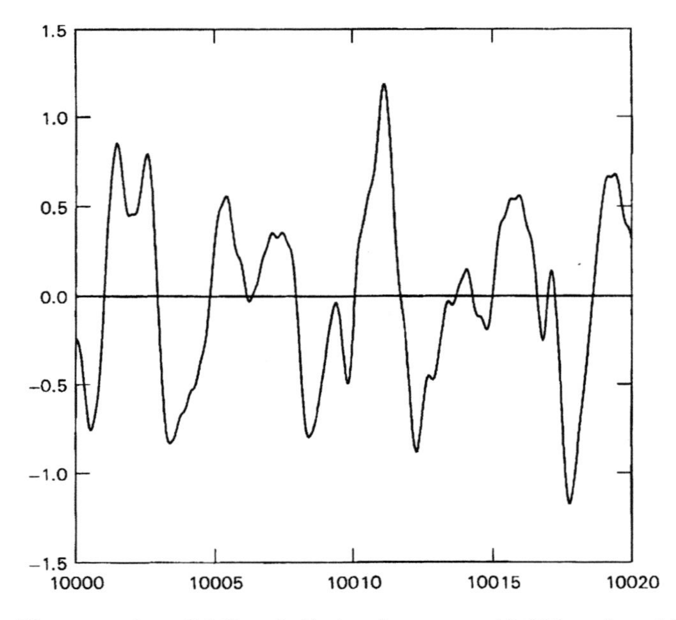

Figure 8.17. Phase angle of $\zeta(1 + 2ik)$ in the range $10\,000 \leqslant k \leqslant 10\,020$ [Gut83] (with kind permission from Elsevier Science).

phase angle of $\zeta(1 + 2\imath k)$ is shown over some k range [Gut83]. The similarity with the Ericson fluctuations and the universal conductance fluctuations, discussed in Chapter 6, is obvious.

For the case of the singular square we may get the real impression that we are close to the solution of the problem of finding a dynamical system having the nontrivial zeros of the Riemann zeta function as eigenvalues. But, as already mentioned, the problem is as yet unsolved.

References

Chapter 1

[Bay89] J.E. Bayfield, G. Casati, I. Guarneri, and D.W. Sokol. Localization of classically chaotic diffusion for hydrogen atoms in microwave fields. *Phys. Rev. Lett.* **63**, 364 (1989).

[Ber87] M.V. Berry. The Bakerian lecture, 1987: Quantum chaology. *Proc. R. Soc. Lond. A* **413**, 183 (1987).

[Ber89] M.V. Berry. Some quantum-to-classical asymptotics. In M.-J. Giannoni, A. Voros, and J. Zinn-Justin, editors, *Chaos and Quantum Physics* page 251. Les Houches Session LII North-Holland 1989.

[Cas79] G. Casati, B.V. Chirikov, F.M. Izraelev, and J. Ford. Stochastic behavior of a quantum pendulum under a periodic perturbation. In G. Casati and J. Ford, editors, *Stochastic Behaviour in Classical and Quantum Hamiltonian Systems* Lecture Notes in Physics 93 page 334. Springer-Verlag Berlin 1979.

[Chi79] B.V. Chirikov. A universal instability of many-dimensional oscillator systems. *Phys. Rep.* **52**, 263 (1979).

[Gal88] E.J. Galvez, B.E. Sauer, L. Moorman, P.M. Koch, and D. Richards. Microwave ionization of H atoms: Breakdown of classical dynamics for high frequencies. *Phys. Rev. Lett.* **61**, 2011 (1988).

[Gei86] T. Geisel, G. Radons, and J. Rubner. Kolmogorov–Arnol'd–Moser barriers in the quantum dynamics of chaotic systems. *Phys Rev. Lett.* **57**, 2883 (1986).

[Gia89] M.-J. Giannoni, A. Voros, and J. Zinn-Justin, editors. *Chaos and Quantum Physics*. Les Houches Session LII North-Holland 1989.

[Moo94] F.L. Moore, J.C. Robinson, C. Bharucha, P.E. Williams, and M.G. Raizen. Observation of dynamical localization in atomic momentum transfer: A new testing ground for quantum chaos. *Phys. Rev. Lett.* **73**, 2974 (1994).

[Ott93] E. Ott. *Chaos in Dynamical Systems.* Cambridge University Press 1993.

[Por65] C.E. Porter. *Statistical Theory of Spectra: Fluctuations*. Academic Press New York 1965.

[Sch84] H.G. Schuster. *Deterministic Chaos: An Introduction.* Physik Verlag Weinheim 1984.

[Ste95] J. Stein, H.-J. Stöckmann, and U. Stoffregen. Microwave studies of billiard Green functions and propagators. *Phys. Rev. Lett.* **75**, 53 (1995).

Chapter 2

[Alt95] H. Alt, H.-D. Gräf, H.L. Harney, R. Hofferbert, H. Lengeler, A. Richter, P. Schardt, and H.A. Weidenmüller. Gaussian orthogonal ensemble statistics in a microwave stadium billiard with chaotic dynamics: Porter–Thomas distribution and algebraic decay of time correlations. *Phys. Rev. Lett.* **74**, 62 (1995).

[Alt96] H. Alt, H.-D. Gräf, R. Hofferbert, C. Rangacharyulu, H. Rehfeld, A. Richter, P. Schardt, and A. Wirzba. Chaotic dynamics in a three-dimensional superconducting microwave billiard. *Phys. Rev. E* **54**, 2303 (1996).

[Alt97] H. Alt, C. Dembowski, H.-D. Gräf, R. Hofferbert, H. Rehfeld, A. Richter, R. Schuhmann, and T. Weiland. Wave dynamical chaos in a superconducting three-dimensional Sinai billiard. *Phys. Rev. Lett.* **79**, 1026 (1997).

[Bal70] R. Balian and C. Bloch. Distribution of eigenfrequencies for the wave equation in a finite domain. I. Three-dimensional problem with smooth boundary surface. *Ann. Phys.* **60**, 401 (1970).

[Bal71] R. Balian and C. Bloch. Distribution of eigenfrequencies for the wave equation in a finite domain. II. Electromagnetic field. Riemannian spaces. *Ann. Phys.* **64**, 271 (1971).

[Bal76] H.P. Baltes and E.R. Hilf. *Spectra of Finite Systems.* BI-Wissenschaftsverlag Mannheim 1976.

[Bay74] J.E. Bayfield and P.M. Koch. Multiphoton ionization of highly excited hydrogen atoms. *Phys. Rev. Lett.* **33**, 258 (1974).

[Ber94] M.J. Berry, J.H. Baskey, R.M. Westervelt, and A.C. Gossard. Coherent electronic backscattering in ballistic microstructures. *Phys. Rev. B* **50**, 8857 (1994).

[Blü92] R. Blümel, I.H. Davidson, W.P. Reinhardt, H. Lin, and M. Sharnoff. Quasilinear ridge structures in water surface waves. *Phys. Rev. A* **45**, 2641 (1992).

[Boh91] O. Bohigas, O. Legrand, C. Schmit, and D. Sornette. Comment on spectral statistics in elastodynamics. *J. Acoust. Soc. Am.* **89**, 1456 (1991).

[Cha94] A.M. Chang, H.U. Baranger, L.N. Pfeiffer, and K.W. West. Weak localization in chaotic versus nonchaotic cavities: A striking difference in the line shape. *Phys. Rev. Lett.* **73**, 2111 (1994).

[Cha95] I.H. Chan, R.M. Clarke, C.M. Marcus, K. Campman, and A.C. Gossard. Ballistic conductance fluctuations in shape space. *Phys. Rev. Lett.* **74**, 3876 (1995).

[Chi96] P.A. Chinnery and V.F. Humphrey. Experimental visualization of acoustic resonances within a stadium-shaped cavity. *Phys. Rev. E* **53**, 272 (1996).

[Chi97] P.A. Chinnery, V.F. Humphrey, and C. Beckett. The schlieren image of two-dimensional ultrasonic fields and cavity resonances. *J. Acoust. Soc. Am.* **101**, 250 (1997).

[Cla95] R.M. Clarke, I.H. Chan, C.M. Marcus, C.I. Duruöz, J.S. Harris, Jr., K. Campman, and A.C. Gossard. Temperature dependence of phase breaking in ballistic quantum dots. *Phys. Rev. B* **52**, 2656 (1995).

[Cro93a] M.F. Crommie, C.P. Lutz, and D.M. Eigler. Confinement of electrons to quantum corrals on a metal surface. *Science* **262**, 218 (1993).

[Cro93b] M.F. Crommie, C.P. Lutz, and D.M. Eigler. Imaging standing waves in a two-dimensional electron gas. *Nature* **363**, 524 (1993).

[Cro95] M.F. Crommie, C.P. Lutz, D.M. Eigler, and E.J. Heller. Quantum corrals. *Physica D* **83**, 98 (1995).

[Dal91] A.D. Dalmédico. Sophie Germain. *Scientific American* **December**, 77 (1991).

[Dat95] S. Datta. *Electronic Trasnport in Mesoscopic Systems*. Cambridge University Press 1995.

[Deu95] S. Deus, P.M. Koch, and L. Sirko. Statistical properties of the eigenfrequency distribution of three-dimensional microwave cavities. *Phys. Rev. E* **52**, 1146 (1995).

[Dör98] U. Dörr, H.-J. Stöckmann, M. Barth, and U. Kuhl. Scarred and chaotic field distributions in three-dimensional Sinai-microwave resonators. *Phys. Rev. Lett.* **80**, 1030 (1998).

[Dor90] E. Doron, U. Smilansky, and A. Frenkel. Experimental demonstration of chaotic scattering of microwaves. *Phys. Rev. Lett.* **65**, 3072 (1990).

[Ell95] C. Ellegaard, T. Guhr, K. Lindemann, H.Q. Lorensen, J. Nygård, and M. Oxborrow. Spectral statistics of acoustic resonances in aluminium blocks. *Phys. Rev. Lett.* **75**, 1546 (1995).

[Ell96] C. Ellegaard, T. Guhr, K. Lindemann, J. Nygård, and M. Oxborrow. Symmetry breaking and spectral statistics of acoustic resonances in quartz blocks. *Phys. Rev. Lett.* **77**, 4918 (1996).

[Fle91] N.H. Fletcher and T.D. Rossing. *The Physics of Musical Instruments*. Springer-Verlag New York 1991.

[Fle92] R. Fleischmann, T. Geisel, and R. Ketzmerick. Magnetoresistance due to chaos and nonlinear resonances in lateral surface superlattices. *Phys. Rev. Lett.* **68**, 1367 (1992).

[Fro94] T.M. Fromhold, L. Eaves, F.W. Sheard, M.L. Leadbeater, T.J. Foster, and P.C. Main. Magnetotunneling spectroscopy of a quantum well in the regime of classical chaos. *Phys. Rev. Lett.* **72**, 2608 (1994).

[Fro95] T.M. Fromhold, P.B. Wilkinson, F.W. Sheard, L. Eaves, J. Miao, and G. Edwards. Manifestation of classical chaos in the energy level spectrum of a quantum well. *Phys. Rev. Lett.* **75**, 1142 (1995).

[Gor92] C. Gordon, D.L. Webb, and S. Wolpert. One cannot hear the shape of a drum. *Bull. Am. Math. Soc.* **27**, 134 (1992).

[Grä92] H.-D. Gräf, H.L. Harney, H. Lengeler, C.H. Lewenkopf, C. Rangacharyulu, A. Richter, P. Schardt, and H.A. Weidenmüller. Distribution of eigenmodes in a superconducting stadium billiard with chaotic dynamics. *Phys. Rev. Lett.* **69**, 1296 (1992).

[Guh98] T. Guhr, A. Müller-Groeling, and H.A. Weidenmüller. Random matrix theories in quantum physics: common concepts. *Phys. Rep.* **299**, 189 (1998).

[Gut90] M.C. Gutzwiller. *Chaos in Classical and Quantum Mechanics*. Springer-Verlag New York 1990.

[Haa91] F. Haake, G. Lenz, P. Šeba, J. Stein, H.-J. Stöckmann, and K. Życzkowski. Manifestation of wave chaos in pseudointegrable microwave resonators. *Phys. Rev. A* **44**, R6161 (1991).

[Hel84] E.J. Heller. Bound-state eigenfunctions of classically chaotic Hamiltonian systems: Scars of periodic orbits. *Phys. Rev. Lett.* **53**, 1515 (1984).

[Hel94] E.J. Heller, M.F. Crommie, C.P. Lutz, and D.M. Eigler. Scattering and absorption of surface electron waves in quantum corrals. *Nature* **369**, 464 (1994).

[Hol86] A. Holle, G. Wiebusch, J. Main, B. Hager, H. Rottke, and K.H. Welge. Diamagnetism of hydrogen atom in the quasi-Landau regime. *Phys. Rev. Lett.* **56**, 2594 (1986).

[Hut81] C.M. Hutchins. The acoustics of violin plates. *Scientific American* **December**, 127 (1981).

[Jac62] J.D. Jackson. *Classical Electrodynamics*. John Wiley & Sons New York 1962.

[Kac66] M. Kac. Can one hear the shape of a drum? *Am. Math. Monthly* **73 part II**, 1 (1966).
[Kol94] M. Kollmann, J. Stein, U. Stoffregen, H.-J. Stöckmann, and B. Eckhardt. Periodic orbit analysis of billiard level dynamics. *Phys. Rev. E* **49**, R1 (1994).
[Krü94] C.A. Krülle, T. Doderer, D. Quenter, R.P. Hübener, R. Pöpel, and J. Niemeyer. Standing wave patterns of microwaves propagating in Josephson tunnel junctions with regular and chaotic billiard geometries. *Physica D* **78**, 214 (1994).
[Kud94] A. Kudrolli, S. Sridhar, A. Pandy, and R. Ramaswamy. Signatures of chaos in quantum billliards: Microwave experiments. *Phys. Rev. E* **49**, R11 (1994).
[Kud95] A. Kudrolli, V. Kidambi, and S. Sridhar. Experimental studies of chaos and localization in quantum wave functions. *Phys. Rev. Lett.* **75**, 822 (1995).
[Lan59] L.D. Landau and E.M. Lifshitz. *Theory of Elasticity (Volume 7 of Course of Theoretical Physics)*. Pergamon Press London 1959.
[Lau94] H.-M. Lauber. *Experimenteller Nachweis geometrischer Phasen und Untersuchungen der Wellenmechanik nichtintegrabler Systeme mit Mikrowellenresonatoren.* Dissertation Ruprecht-Karls-Universität Heidelberg 1994.
[Lax62] B. Lax and K.J. Button. *Microwave Ferrites and Ferrimagnetics*. McGraw-Hill New York 1962.
[Lee85] P.A. Lee and T.V. Ramakrishnan. Disordered electronic systems. *Rev. Mod. Phys.* **57**, 287 (1985).
[Lin86] P.E. Lindelof, J. Nørregaard, and J. Hanberg. New light on the scattering mechanisms in Si inversion layers by weak localization experiments. *Phys. Scr.* **T14**, 17 (1986).
[Mai52] L.C. Maier and J.C. Slater. Field strength measurements in resonant cavities. *J. Appl. Phys.* **23**, 68 (1952).
[Mai86] J. Main, G. Wiebusch, A. Holle, and K.H. Welge. New quasi-Landau structure of highly excited atoms: The hydrogen atom. *Phys. Rev. Lett.* **75**, 2789 (1986).
[Mar92] C.M. Marcus, A.J. Rimberg, R.M. Westervelt, P.F. Hopkins, and A.C. Gossard. Conductance fluctuations and chaotic scattering in ballistic microstructures. *Phys. Rev. Lett.* **69**, 506 (1992).
[Mar93] C.M. Marcus, R.M. Westervelt, P.F. Hopkins, and A.C. Gossard. Phase breaking in ballistic quantum dots: Experiment and analysis based on chaotic scattering. *Phys. Rev. B* **48**, 2460 (1993).
[McD79] S.W. McDonald and A.N. Kaufman. Spectrum and eigenfunctions for a Hamiltonian with stochastic trajectories. *Phys. Rev. Lett.* **42**, 1189 (1979).
[Mel88] F. Melde. *Chladnis's Leben and Wirken.* N.G. Elwert'sche Verlagsbuchhandlung Marburg 1888.
[Mel91] F. Melde. Akustik. In A. Winkelmann, editor, *Handbuch der Physik* page 683. Trewendt-Verlag Breslau 1891.
[Mil68] L.M. Milne-Thomson. *Theoretical Hydrodynamics*. Macmillan & Co London fifth edition 1968.
[O'Con87] P. O'Connor, J. Gehlen, and E.J. Heller. Properties of random superpositions of plane waves. *Phys. Rev. Lett.* **58**, 1296 (1987).
[Por65] C.E. Porter. *Statistical Theory of Spectra: Fluctuations.* Academic Press New York 1965.
[Ric98] A. Richter. Playing billiards with microwaves – quantum manifestations of classical chaos. In D.A. Hejhal, J. Friedman, M.C. Gutzwiller, and A.M. Odlyzko, editors, *Emerging Applications of Number Theory* IMA volume 109 page 479 New York 1998. Springer-Verlag.

[Sch87] M.R. Schröder. Normal frequency and excitation statistics in rooms: Model experiments with electric waves. *J. Audio Eng. Soc.* **35**, 307 (1987).

[Sch96] H. Schanze. *Realisierung von nichteuklidischen Billards durch Wellenwannen*. Diplomarbeit Philipps-Universität Marburg 1996.

[Sch97] K. Schaadt. *The Quantum Chaology of Acoustic Resonators*. Master of Science thesis, The University of Copenhagen. The Niels Bohr Institute 1997.

[Šeb90] P. Šeba. Wave chaos in singular quantum billiard. *Phys. Rev. Lett.* **65**, 1855 (1990).

[Sin63] Y.G. Sinai. On the foundations of the ergodic hypothesis for a dynamical system of statistical mechanics. *Sov. Math. Dokl.* **4**, 1818 (1963).

[Sir97] L. Sirko, P.M. Koch, and R. Blümel. Experimental identification of non-Newtonian orbits produced by ray splitting in a dielectric-loaded microwave cavity. *Phys. Rev. Lett.* **78**, 2940 (1997).

[So95] P. So, S.M. Anlage, E. Ott, and R.N. Oerter. Wave chaos experiments with and without time reversal symmetry: GUE and GOE statistics. *Phys. Rev. Lett.* **74**, 2662 (1995).

[Soo60] R.F. Soohoo. *Theory and Application of Ferrites*. Prentice-Hall Englewood Cliffs, New Jersey 1960.

[Sri91] S. Sridhar. Experimental observation of scarred eigenfunctions of chaotic microwave cavities. *Phys. Rev. Lett.* **67**, 785 (1991).

[Sri92] S. Sridhar and E.J. Heller. Physical and numerical experiments on the wave mechanics of classically chaotic systems. *Phys. Rev. A* **46**, R1728 (1992).

[Sri94] S. Sridhar and A. Kudrolli. Experiments on not 'hearing the shape' of drums. *Phys. Rev. Lett.* **72**, 2175 (1994).

[Ste92] J. Stein and H.-J. Stöckmann. Experimental determination of billiard wave functions. *Phys. Rev. Lett.* **68**, 2867 (1992).

[Ste95] J. Stein, H.-J. Stöckmann, and U. Stoffregen. Microwave studies of billiard Green functions and propagators. *Phys. Rev. Lett.* **75**, 53 (1995).

[Stö90] H.-J. Stöckmann and J. Stein. 'Quantum' chaos in billiards studied by microwave absorption. *Phys. Rev. Lett.* **64**, 2215 (1990).

[Stö95a] H.-J. Stöckmann, J. Stein, and M. Kollmann. Microwave studies in irregularly shaped billiards. In C. Casati and B. Chirikov, editors, *Quantum Chaos Between Order and Disorder* page 661. Cambridge University Press 1995.

[Sto95b] U. Stoffregen, J. Stein. H.-J. Stöckmann, M. Kuś, and F. Haake. Microwave billiards with broken time reversal symmetry. *Phys. Rev. Lett.* **74**, 2666 (1995).

[Wea89] R.L. Weaver. Spectral statistics in elastodynamics. *J. Acoust. Soc. Am.* **85**, 1005 (1989).

[Wei91] D. Weiss, M.L. Roukes, A. Menschig, P. Grambow, K. von Klitzing, and G. Weimann. Electronic pinball and commensurate orbits in a periodic array of scatterers. *Phys. Rev. Lett.* **66**, 2790 (1991).

[Wei93] D. Weiss, K. Richter, A. Menschig, R. Bergmann, H. Schweizer, K. von Klitzing, and G. Weimann. Quantized periodic orbits in large antidot arrays. *Phys. Rev. Lett.* **70**, 4118 (1993).

[Wei95] D. Weiss and K. Richter. Complex and quantized electron motion in antidot arrays. *Physica D* **83**, 290 (1995).

[Wil96] P.B. Wilkinson, T.M. Fromhold, L. Eaves, F. W. Sheard, N. Miura, and T. Takamasu. Observation of 'scarred' wavefunctions in a quantum well with chaotic electron dynamics. *Nature* **380**, 608 (1996).

Chapter 3

[Abr70] M. Abramowitz and I.A. Stegun. *Handbook of Mathematical Functions.* Dover Publications New York 1970.

[Alb91] S. Albeverio and P. Šeba. Wave chaos in quantum systems with point interaction. *J. Stat. Phys.* **64**, 369 (1991).

[Alt97] H. Alt, H.-D. Gräf, T. Guhr, H.L. Harney, R. Hofferbert, H. Rehfeld, A. Richter, and P. Schardt. Correlation-hole method for the spectra of superconducting microwave billiards. *Phys. Rev. E* **55**, 6674 (1997).

[And96] A.V. Andreev, O. Agam, B.D. Simons, and B.L. Altshuler. Quantum chaos, irreversible classical dynamics, and random matrix theory. *Phys. Rev. Lett.* **76**, 3947 (1996).

[Bal68] R. Balian. Random matrices and information theory. *Il Nuovo Cimento* **LVII B, N.1**, 183 (1968).

[Ber84] M.V. Berry and M. Robnik. Semiclassical level spacings when regular and chaotic orbits coexist. *J. Phys. A* **17**, 2413 (1984).

[Boh83] O. Bohigas, R.U. Haq, and A. Pandey. Fluctuation properties of nuclear energy levels and widths: comparison of theory with experiments. In K.H. Böckhoff, editor, *Nuclear Data for Science and Technology* page 809. Reidel Dordrecht 1983.

[Boh84] O. Bohigas, M.J. Giannoni, and C. Schmit. Characterization of chaotic spectra and universality of level fluctuation laws. *Phys. Rev. Lett.* **52**, 1 (1984).

[Boh85] O. Bohigas, R.U. Haq, and A. Pandey. Higher-order correlations in spectra of complex systems. *Phys. Rev. Lett.* **54**, 1645 (1985).

[Boh89] O. Bohigas. Random matrix theories and chaotic dynamics. In M.-J. Giannoni, A. Voros, and J. Zinn-Justin, editors, *Chaos and Quantum Physics* page 87. Les Houches Session LII North-Holland 1989.

[Bro73] T.A. Brody. A statistical measure for the repulsion of energy levels. *Lettere Al Nuovo Cimento* **7**, 482 (1973).

[Bro81] T.A. Brody, J. Flores, J.B. French, P.A. Mello, A. Pandey, and S.S.M. Wong. Random-matrix physics: spectrum and strength fluctuations. *Rev. Mod. Phys.* **53**, 385 (1981).

[Cas85] G. Casati, B.V. Chirikov, and I. Guarneri. Energy-level statistics of integrable quantum systems. *Phys. Rev. Lett.* **54**, 1350 (1985).

[Cas91] G. Casati, F. Izrailev, and L. Molinari. Scaling properties of the eigenvalue spacing distribution for band random matrices. *J. Phys. A* **24**, 4755 (1991).

[Cau89] E. Caurier and B. Grammaticos. Extreme level repulsion for chaotic quantum Hamiltonians. *Phys. Lett. A* **136**, 387 (1989).

[Con97] R.D. Connors and J.P. Keating. Two-point spectral correlations for the square billiard. *J. Phys. A* **30**, 1817 (1997).

[Deu95] S. Deus, P.M. Koch, and L. Sirko. Statistical properties of the eigenfrequency distribution of three-dimensional microwave cavities. *Phys. Rev. E* **52**, 1146 (1995).

[Edw76] S.F. Edwards and R.C. Jones. The eigenvalue spectrum of a large symmetric random matrix. *J. Phys. A* **9**, 1595 (1976).

[Efe83] K.B. Efetov. Supersymmetry and theory of disordered metals. *Adv. Phys.* **32**, 53 (1983).

[Efe97] K. Efetov. *Supersymmetry in Disorder and Chaos*. Cambridge University Press 1997.

[Ell96] C. Ellegaard, T. Guhr, K. Lindemann, J. Nygård, and M. Oxborrow. Symmetry breaking and spectral statistics of acoustic resonances in quartz blcoks. *Phys. Rev. Lett.* **77**, 4918 (1996).

[Guh90] T. Guhr and H.A. Weidenmüller. Isospin mixing and spectal fluctuation properties. *Ann. Phys.* **199**, 412 (1990).

[Guh98] T. Guhr, A. Müller-Groeling, and H.A. Weidenmüller. Random matrix theories in quantum physics: common concepts. *Phys. Rep.* **299**, 189 (1998).

[Haa91a] F. Haake. *Quantum Signatures of Chaos*. Springer-Verlag Berlin 1991.

[Haa91b] F. Haake, G. Lenz, P. Šeba, J. Stein, J.-J. Stöckmann, and K. Życzkowski. Manifestation of wave chaos in pseudointegrable microwave resonators. *Phys. Rev. A.* **44**, R6161 (1991).

[Haq82] R.U. Haq, A. Pandey, and O. Bohigas. Fluctuation properties of nuclear energy levels: Do theory and experiment agree? *Phys Rev. Lett.* **48**, 1086 (1982).

[Hön89] A. Hönig and D. Wintgen. Spectral properties of strongly perturbed Coulomb systems: Fluctuation properties. *Phys. Rev. A* **39**, 5642 (1989).

[Izr88] F.M. Izrailev. Quantum localization and statistics of quasienergy spectrum in a classically chaotic system. *Phys. Lett. A* **134**, 13 (1988).

[Izr90] F.M. Izrailev. Simple models of quantum chaos: Spectrum and eigenfunctions. *Phys. Rep.* **196**, 299 (1990).

[Kud94] A. Kudrolli, S. Sridhar, A. Pandey, and R. Ramaswamy. Signatures of chaos in quantum billiards: Microwave experiments. *Phys. Rev. E* **49**, R11 (1994).

[Leg92] O. Legrand, C. Schmit, and D. Sornette. Quantum chaos methods applied to high-frequency plate vibrations. *Europhys. Lett* **18**, 101 (1992).

[Len91] G. Lenz and F. Haake. Reliability of small matrices for large spectra with nonuniversal fluctuation. *Phys. Rev. Lett.* **67**, 1 (1991).

[Lev86] L. Leviandier, M. Lombardi, R. Jost, and J.P. Pique. Fourier transform: A tool to measure statistical level properties in very complex spectra. *Phys. Rev. Lett.* **56**, 2449 (1986).

[Ley97] F. Leyvraz and T.H. Seligman. Comment on 'quantum chaos, irreversible classical dynamics, and random matrix theory'. *Phys. Rev. Lett.* **79**, 1778 (1997).

[Lom94] M. Lombardi, O. Bohigas, and T.H. Seligman. New evidence of GOE statistics for compound nuclear resonances. *Phys. Lett. B* **324**, 263 (1994).

[Mag66] W. Magnus, F. Oberhettinger, and R.P. Soni. *Formulas and Theorems for the Special Functions of Mathematical Physics*. Springer-Verlag New York 1966.

[Meh91] M.L. Mehta. *Random Matrices.* Academic Press San Diego second edition 1991.

[Mit88] G.E. Mitchell, E.G. Bilpuch, P.M. Endt, and J.F. Shriner. Broken symmetries and chaotic behaviour in ^{26}Al. *Phys. Rev. Lett.* **61**, 1473 (1988).

[Mor53] P.M. Morse and H. Feshbach. *Methods of Theoretical Physics.* McGraw-Hill New York 1953.

[Muz95] B.A. Muzykantskiĭ and D.E. Khmelnitskiĭ. Effective action in the theory of quasi-ballistic disordered conductors. *JETP Lett.* **62**, 76 (1995).

[Ott93] E. Ott. *Chaos in Dynamical Systems.* Cambridge University Press 1993.

[Oxb95] M. Oxborrow and C. Ellegaard. Quantum chaology in quartz. In *Proceedings of the 3rd Experimental Chaos Conference, Edinburgh, Scotland, UK* 1995.

[Piq87] J.P. Pique, Y. Chen, R.W. Field, and J.L. Kinsey. Chaos and dynamics on 0.5–300-ps time scales in vibrationally excited actylene: Fourier transform of stimulated-emission pumping spectrum. *Phys. Rev. Lett.* **58**, 475 (1987).

[Plu95] Z. Pluhař, H.A. Weidenmüller, J.A. Zuk, C.H. Lewenkopf, and F.J. Wegner.

Crossover from orthogonal to unitary symmetry for ballistic electron transport in chaotic microstructures. *Ann. Phys.* **243**, 1 (1995).
[Por65] C.E. Porter. *Statistical Theory of Spectra: Fluctuations.* Academic Press New York 1965.
[Pro93] T. Prosen and M. Robnik. Energy level statistics in the transition region between integrability and chaos. *J. Phys. A* **26**, 2371 (1993).
[Pro94] T. Prosen and M. Robnik. Numerical demonstration of the Berry–Robnik level spacing distribution. *J. Phys. A* **27**, L459 (1994).
[Ric81] P.J. Richens and M.V. Berry. Pseudointegrable systems in classical and quantum mechanics. *Physica D* **2**, 495 (1981).
[Rob83] M. Robnik. Classical dynamics of a family of billiards with analytic boundaries. *J. Phys. A* **16**, 3971 (1983).
[Rob98] M. Robnik and G. Veble. On spectral statistics of classically integrable systems. *J. Phys. A* **31**, 4669 (1998).
[Šeb90] P. Šeba. Wave chaos in singular quantum billiard. *Phys. Rev. Lett.* **64**, 1855 (1990).
[Shi94] T. Shigehara. Conditions for the appearance of wave chaos in quantum singular systems with a pointlike scatterer. *Phys. Rev. E* **50**, 4357 (1994).
[Shi95] Y. Shimizu and A. Shudo. Polygonal billiards: Correspondence between classical trajectories and quantum eigenstates. *Chaos, Solitons & Fractals* **5**, 1337 (1995).
[Sil95] H. Silberbauer, P Rotter, U. Rössler, and M. Suhrke. Quantum chaos in magnetic band structures. *Europhys. Lett.* **31**, 393 (1995).
[So95] P. So, S.M. Anlage, E. Ott, and R.N. Oerter. Wave chaos experiments with and without time reversal symmetry: GUE and GOE statistics. *Phys. Rev. Lett.* **74**, 2662 (1995).
[Sto95] U. Stoffregen, J. Stein, H.-J. Stöckmann, M. Kuś, and F. Haake. Microwave billiards with broken time reversal symmetry. *Phys. Rev. Lett.* **74**, 2666 (1995).
[Ver85a] J.J.M. Verbaarschot, H.A. Weidenmüller, and M.R. Zirnbauer. Grassmann integration in stochastic quantum physics: The case of compound-nucleus scattering. *Phys. Rep.* **129**, 367 (1985).
[Ver85b] J.J.M. Verbaarschot, and M.R. Zirnbauer. Critique of the replica trick. *J. Phys. A* **18**, 1093 (1985).
[Zim88] T. Zimmermann, H. Köppel, L.S. Cederbaum, G. Persch, and W. Demtröder. Confirmation of random-matrix fluctuations in molecular spectra. *Phys. Rev. Lett.* **61**, 3 (1988).
[Zir99] M.R. Zirnbauer. Pair correlations of quantum chaotic maps from supersymmetry. In I.V. Lerner, J.P. Keating, D.E. Khmelnitskii, editors. *Supersymmetry and Trace Formulae: Chaos and Disorder* page 153, Plenum Press 1999.

Chapter 4

[Alt96] A. Altland and M.R. Zirnbauer. Field theory of the quantum kicked rotor. *Phys. Rev. Lett.* **77**, 4536 (1996).
[And58] P.W. Anderson. Absence of diffusion in certain random lattices. *Phys. Rev.* **109**, 1492 (1958).
[And78] P.W. Anderson. Local moments and localized states. *Rev. Mod. Phys.* **50**, 191 (1978).

[Bar95] P.J. Bardroff, I. Bialynicki-Birula, D.S. Krähmer, G. Kurizki, E. Mayr, P. Stifter, and W.P. Schleich. Dynamical localization: Classical vs quantum oscillations in momentum spread of cold atoms. *Phys. Rev. Lett.* **74**, 3959 (1995).

[Bay74] J.E. Bayfield and P.M. Koch. Multiphoton ionization of highly excited hydrogen atoms. *Phys. Rev. Lett.* **33**, 258 (1974).

[Bay89] J.E. Bayfield, G. Casati, I. Guarneri, and D.W. Sokol. Localization of classically chaotic diffusion for hydrogen atoms in microwave fields. *Phys. Rev. Lett.* **63**, 364 (1989).

[Bel96] M.R.W. Bellermann, P.M. Koch, D.R. Mariani, and D. Richards. Polarization independence of microwave 'ionization' thresholds of excited hydrogen atoms near the principal resonance. *Phys. Rev. Lett.* **76**, 892 (1996).

[Ben95] O. Benson, G. Raithel, and H. Walther. Rubidium Rydberg atoms in strong fields. In G. Casati and B. Chirikov, editors, *Quantum Chaos* page 247. Cambridge University Press 1995.

[Blü97] R. Blümel and W.P. Reinhardt. *Chaos in Atomic Physics.* Cambridge University Press 1997.

[Buc95a] A. Buchleitner and D. Delande. Nondispersive electronic wave packets in multiphoton processes. *Phys. Rev. Lett.* **75**, 1487 (1995).

[Buc95b] A. Buchleitner and D. Delande. Spectral aspects of the microwave ionization of atomic Rydberg states. *Chaos, Solitons & Fractals* **5**, 1125 (1995).

[Cas79] G. Casati, B.V. Chirikov, F.M. Izraelev, and J. Ford. Stochastic behavior of a quantum pendulum under a periodic perturbation. In G. Casati and J. Ford, editors, *Stochastic Behaviour in Classical and Quantum Hamiltonian Systems,* Lecture Notes in Physics 93 page 334. Springer-Verlag Berlin 1979.

[Cas87a] G. Casati, B.V. Chirikov, D.L. Shepelyansky, and I. Guarneri. Relevance of classical chaos in quantum mechanics: The hydrogen atom in a monochromatic field. *Phys. Rep.* **154**, 77 (1987).

[Cas87b] G. Casati, I. Guarneri, and D.L. Shepelyansky. Exponential photonic localization for the hydrogen atom in a monochromatic field. *Phys. Rev. A* **36**, 3501 (1987).

[Cas90] G. Casati, I. Guarneri, and D.L. Shepelyanski. Classical chaos, quantum localization and fluctuations: A unified view. *Physica A* **163**, 205 (1990).

[Cas91] G. Casati, F. Izrailev, and L. Molinari. Scaling properties of the eigenvalue spacing distribution for band random matrices. *J. Phys. A* **24**, 4755 (1991).

[Cas95] G. Casati and B. Chirikov. The legacy of chaos in quantum mechanics. In G. Casati and B. Chirikov, editors. *Quantum Chaos* page 3. Cambridge University Press 1995.

[Cas98] G. Casati, F.M. Izrailev, and V.V. Sokolov. Comment on 'dynamical theory of quantum chaos or a hidden random matrix ensemble?'. *Phys. Rev. Lett.* **80**, 640 (1988).

[Chi79] B.V. Chirikov. A universal instability of many-dimensional oscillator systems. *Phys. Rep.* **52**, 263 (1979).

[Chi81] B.V. Chirikov, F.M. Izrailev, and D.L. Shepelyansky. Dynamical stochasticity in classical and quantum mechanics. *Sov. Sci. Rev. C* **2**, 209 (1981).

[Chi88] B.V. Chirikov, F.M. Izrailev, and D.L. Shepelyansky. Quantum chaos: localization vs. ergodicity. *Physica D* **33**, 77 (1988).

[Chu91] S. Chu. Laser manipulation of atoms and particles. *Science* **253**, 861 (1991).

[Fey57] R.P. Feynman, F.L. Vernon, and R.W. Hellworth. Geometrical representation of the Schrödinger equation for solving maser problems. *J. Appl. Phys.* **28**, 49 (1957).

[Fis82] S. Fishman, D.R. Grempel, and R.E. Prange. Chaos, quantum recurrences and Anderson localization. *Phys. Rev. Lett.* **49**, 509 (1982).

[Fis89] S. Fishman, R.E. Prange, and M. Griniasty. Scaling theory for the localization length of the kicked rotor. *Phys. Rev. A* **39**, 1628 (1989).

[Fle95] R. Fleischmann, T. Geisel, R. Ketzmerick, and G. Petschel. Quantum diffusion, fractal spectra, and chaos in semiconductor microstructures. *Physica D* **86**, 171 (1995).

[Gal88] E.J. Galvez, B.E. Sauer, L. Moorman, P.M. Koch, and D. Richards. Microwave ionization of H atoms: Breakdown of classical dynamics for high frequencies. *Phys. Rev. Lett.* **61**, 2011 (1988).

[Gei86] T. Geisel, G. Radons, and J. Rubner. Kolmogorov–Arnol'd–Moser barriers in the quantum dynamics of chaotic systems. *Phys. Rev. Lett.* **57**, 2883 (1986).

[Gei91] T. Geisel, R. Ketzmerick, and G. Petschel. New class of level statistics in quantum systems with unbounded diffusion. *Phys. Rev. Lett.* **66**, 1651 (1991).

[Gei95] T. Geisel, R. Ketzmerick, and G. Petschel. Unbounded quantum diffusion and fractal spectra. In G. Casati and B. Chirikov, editors, *Quantum Chaos* page 633. Cambridge University Press 1995.

[Gra92] R. Graham, M. Schlautmann, and P. Zoller. Dynamical localization of atomic-beam deflection by a modulated standing light wave. *Phys. Rev. A* **45**, R19 (1992).

[Gre84] D.R. Grempel, R.E. Prange, and S. Fishman. Quantum dynamics of a nonintegrable system. *Phys. Rev. A* **29**, 1639 (1984).

[Gri88] M. Griniasty and S. Fishman. Localization by pseudorandom potentials in one dimension. *Phys. Rev. Lett.* **60**, 1334 (1988).

[Haa87] F. Haake, M. Kuś, and R. Scharf. Classical and quantum chaos for a kicked top. *Z. Phys. B* **65**, 381 (1987).

[Haa91] F. Haake. *Quantum Signatures of Chaos*. Springer-Verlag Berlin 1991.

[Hof76] D.R. Hofstadter. Energy levels and wave functions of Bloch electrons in rational and irrational magnetic fields. *Phys. Rev. B* **14**, 2239 (1976).

[Ish73] K. Ishii. Localization of eigenstates and transport phenomena in the one-dimensional disordered system. *Prog. Theor. Phys. Suppl.* **53** 77 (1973).

[Izr90] F.M. Izrailev. Simple models of quantum chaos: Spectrum and eigenfunctions, *Phys. Rep.* **196**, 299 (1990).

[Jen91] R.V. Jensen, S.M. Susskind, and M.M. Sanders. Chaotic ionization of highly excited hydrogen atoms: Comparison of classical and quantum theory with experiment. *Phys. Rep.* **201**, 1 (1991).

[Ket97] R. Ketzmerick, K. Kruse, S. Kraut, and T. Geisel. What determines the spreading of a wave packet? *Phys. Rev. Lett.* **79**, 1959 (1997).

[Koc95a] P.M. Koch. Microwave 'ionization' of excited hydrogen atoms: How nonclassical local stability brought about by scarred matrix states is affected by broadband noise and by varying the pulse envelope. *Physica D* **83**, 178 (1995).

[Koc95b] P.M. Koch and K.A.H. van Leeuwen. The importance of resonances in microwave 'ionization' of excited hyrdrogen atoms. *Phys. Rep.* **255**, 289 (1995).

[Kra93] B. Kramer and A. MacKinnon. Localization: theory and experiment. *Rep. Prog. Phys.* **56**, 1469 (1993).

[Kuh98] U. Kuhl and H.-J. Stöckmann. Microwave realization of the Hofstadter butterfly. *Phys. Rev. Lett.* **80**, 3232 (1998).

[Lee85] P.A. Lee and T.V. Ramakrishnan. Disordered electronic systems. *Rev. Mod. Phys.* **57**, 287 (1985).

[Meh91] M.L. Mehta. *Random Matrices*. Academic Press San Diego second edition 1991.
[Moo94] F.L. Moore, J.C. Robinson, C. Bharucha, P.E. Williams, and M.G. Raizen. Observation of dynamical localization in atomic momentum transfer: A new testing ground for quantum chaos. *Phys. Rev. Lett.* **73**, 3974 (1994).
[Moo95] F.L. Moore, J.C. Robinson, C.F. Bharucha, B. Sundaram, and M.G. Raizen. Atom optics realization of the quantum δ-kicked rotor. *Phys. Rev. Lett.* **75**, 4598 (1995).
[Ott93] E. Ott. *Chaos in Dyamical Systems*. Cambridge University Press 1993.
[Pan84] B. Pannetier, J. Chaussy, R. Rammal, and J.C. Villegier. Experimental fine tuning of frustration: Two-dimensional superconducting network in a magnetic field. *Phys. Rev. Lett.* **53**, 1845 (1984).
[Rob95] J.C. Robinson, C. Bharucha, F.L. Moore, R. Jahnke, G.A. Georgakis, Q. Niu, M.G. Raizen, and B. Sundaram. Study of quantum dynamics in the transition from classical stability to chaos. *Phys. Rev. Lett.* **74**, 3963 (1995).
[Sch84] H.G. Schuster. *Deterministic Chaos: An Introduction*. Physik Verlag Weinheim 1984.
[Sch88] R. Scharf, B. Dietz, M. Kuś, F. Haake, and M.V. Berry. Kramer's degeneracy and quartic level repulsion. *Europhys. Lett.* **5**, 383 (1988).
[Sch96] T. Schlösser, K. Ensslin, J.P. Kotthaus, and M. Holland. Experimental observation of an artificial bandstructure in lateral superlattices. In T. Martin, G. Montambaux, and J. Trân Thanh Vân, editors, *Correlated Fermions and Transport in Mesoscopic Systems* page 423. Frontieres 1996.
[She86] D.L. Shepelyansky. Localization of quasienergy eigenfunctions in action space. *Phys. Rev. Lett.* **56**, 677 (1986).
[She87] D.L. Shepelyansky. Localization of diffusive excitation in multi-level systems. *Physica D* **28**, 103 (1987).
[Tho72] D.J. Thouless. A relation between the density of states and range of localization for one dimensional random systems. *J. Phys. C* **5**, 77 (1972).
[Tho74] D.J. Thouless. Electrons in disordered systems and the theory of localization. *Phys. Rep.* **13**, 93 (1974).
[Tho88] D.J. Thouless. Localization by a potential with slowly varying period. *Phys. Rev. Lett.* **61**, 2141 (1988).
[Zak95] J. Zakrzewski, D. Delande, and A. Buchleitner. Nonspreading electronic wave packets and conductance fluctuations. *Phys. Rev. Lett.* **75**, 4015 (1995).

Chapter 5

[Aha59] Y. Aharonov and D. Bohm. Significance of electromagnetic potentials in the quantum theory. *Phys. Rev.* **115**, 485 (1959).
[Ber84a] M.V. Berry. Quantal phase factors accompanying adiabatic changes. *Proc. R. Soc. Lond. A* **392**, 45 (1984).
[Ber84b] M.V. Berry and M. Wilkinson. Diabolical points in the spectra of triangles. *Proc. R. Soc. Lond. A* **392**, 15 (1984).
[Cal71] F. Calogero. Solution of the one-dimensional N-body problems with quadratic and/or inversely quadratic pair potentials. *J. Math. Phys.* **12**, 419 (1971).
[Gas90] P. Gaspard, S.A. Rice, H.J. Mikeska, and K. Nakamura. Parametric motion of energy levels: Curvature distribution. *Phys. Rev. A* **42**, 4015 (1990).
[Gau66] M. Gaudin. Une famille à un paramètre d'ensembles unitaires. *Nuclear*

Physics **85**, 545 (1966).
[Haa90] F. Haake and G. Lenz. Classical Hamiltonian dynamics of rescaled quantum levels. *Europhys. Lett.* **13**, 577 (1990).
[Haa91] F. Haake. *Quantum Signatures of Chaos*. Springer-Verlag Berlin 1991.
[Has93] H. Hasegawa and M. Robnik. On the applicability of the energy level dynamics for the Hamiltonian systems in the transition region between integrability and chaos. *Europhys. Lett.* **23**, 171 (1993).
[Hen86] P. Henrici. *Applied and Computational Complex Analysis, Vol. 3.* John Wiley & Sons New York 1986.
[Hop92] J. Hoppe. *Lectures on Integrable Systems.* Springer-Verlag Berlin 1992.
[Kol94] M. Kollmann, J. Stein, U. Stoffregen, H.-J. Stöckmann, and B. Eckhardt. Periodic orbit analysis of billiard level dynamics. *Phys. Rev. E* **49**, R1 (1994).
[Kuś87] M. Kuś, R. Scharf, and F. Haake. Symmetry versus degree of level repulsion for kicked quantum systems. *Z. Phys. B* **66**, 129 (1987).
[Kuś88] M. Kuś. Dynamics and statistics of quasi-energy levels for kicked quantum systems. *Europhys. Lett.* **5**, 1 (1988).
[Lau94] H.-M. Lauber, P. Weidenhammer, and D. Dubbers. Geometric phases and hidden symmetries in simple resonators. *Phys. Rev. Lett.* **72**, 1004 (1994).
[Ma94] J.-Z. Ma and H. Hasegawa. Parametric modelling of Poisson-Gaussian random matrix ensembles. *Z. Phys. B* **93**, 529 (1994).
[Ma95] J.-Z. Ma. Interpolation between the Poisson and circular unitary ensemble: long-range statistics. *Phys. Lett. A* **203**, 312 (1995).
[Mos75] J. Moser. Three integrable Hamiltonian systems connected with isospectral deformations. *Adv. Math.* **16**, 197 (1975).
[Nak86] K. Nakamura and M. Lakshmanan. Complete integrability in a quantum description of chaotic systems. *Phys. Rev. Lett.* **57**, 1661 (1986).
[Nak88] K. Nakamura and H. Thomas. Quantum billiard in a magnetic field: chaos and diamagnetism. *Phys. Rev. Lett.* **61**, 247 (1988).
[Nak93] K. Nakamura. *Quantum Chaos. A New Paradigm of Nonlinear Dynamics.* Cambridge University Press 1993.
[Opp94] F.v. Oppen. Exact distribution of eigenvalue curvatures of chaotic quantum systems. *Phys. Rev. Lett.* **73**, 798 (1994).
[Opp95] F.v. Oppen. Exact distributions of eigenvalue curvatures for time-reversal-invariant chaotic systems. *Phys. Rev. E* **51**, 2647 (1995).
[Pec83] P. Pechukas. Distribution of energy eigenvalues in the irregular spectrum. *Phys. Rev. Lett.* **51**, 943 (1983).
[Sha89] A. Shapere and F. Wilczek. *Geometric Phases in Physics*. World Scientific Singapore 1989.
[Sie95] M. Sieber, H. Primack, U. Smilansky, I. Ussishkin, and H. Schanz. Semiclassical quantization of billiards with mixed boundary conditions. *J. Phys. A* **28**, 5041 (1995).
[Sim93] B.D. Simons and B.L. Altshuler. Universalities in the spectra of disordered and chaotic systems. *Phys. Rev. B* **48**, 5422 (1993).
[Sli80] C.P. Slichter. *Principles of Magnetic Resonance*. Springer-Verlag Berlin 1980.
[Stö97] H.-J. Stöckmann, U. Stoffregen, and M. Kollmann. A relation between billiard geometry and the temperature of its eigenvalue gas. *J. Phys. A* **30**, 129 (1997).
[Tak91] T. Takami. Curvature distribution of stadium billiard. *J. Phys. Soc. Jpn.* **60**, 2489 (1991).

[Tak92] T. Takami and H. Hasegawa. Curvature distribution of chaotic quantum systems: Universality and nonuniversality. *Phys. Rev. Lett.* **68**, 419 (1992).
[Yuk85] T. Yukawa. New approach to the statistical properties of energy levels. *Phys. Rev. Lett.* **54**, 1883 (1985).
[Yuk86] T. Yukawa. Lax form of the quantum mechanical eigenvalue problem. *Phys. Lett. A* **116**, 227 (1986).
[Zak91] J. Zakrzewski and M. Kuś. Distributions of avoided crossings for quantum chaotic systems. *Phys. Rev. Lett.* **67**, 2749 (1991).
[Zak93a] J. Zakrzewski and D. Delande. Parametric motion of energy levels in quantum chaotic systems. I. Curvature distributions. *Phys. Rev. E* **47**, 1650 (1993).
[Zak93b] J. Zakrzewski, D. Delande, and M. Kuś. Parametric motion of energy levels in quantum chaotic systems. II. Avoided-crossing distributions. *Phys. Rev. E* **47**, 1665 (1993).

Chapter 6

[Akk94] E. Akkermans, G. Montambaux, J.-L. Pichard, and J. Zinn-Justin, editors. *Mesoscopic Quantum Physics*. Les Houches Session LXI North-Holland 1994.
[Alt93] H. Alt, P.v. Brentano, H.-D. Gräf, R.-D. Herzberg, M. Philipp, A. Richter, and P. Schardt. Resonances of a superconducting microwave cavity: A test of the Breit-Wigner formula over a large dynamic range. *Nucl. Phys. A* **560**, 293 (1993).
[Alt95] H. Alt, H.-D. Gräf, H.L. Harney, R. Hofferbert, H. Lengeler, A. Richter, P. Schardt, and H.A. Weidenmüller. Gaussian orthogonal ensemble statistics in a microwave stadium billiard with chaotic dynamics: Porter-Thomas distribution and algebraic decay of time correlations. *Phys. Rev. Lett.* **74**, 62 (1995).
[Bal77] R. Balian and B. Duplantier. Electromagnetic waves near perfect conductors. I. Multiple scattering expansion. Distribution of modes. *Ann. Phys.* **104**, 300 (1997).
[Bar99] H.U. Baranger and R.M. Westerveld. Chaos in ballistic nanostructures. In G. Timp, editor, *Nanotechnology* page 537. Springer-Verlag Berlin 1999.
[Bar93] H.U. Baranger, R.A. Jalabert, and A.D. Stone. Weak localization and integrability in ballistic cavities. *Phys. Rev. Lett.* **70**, 3876 (1993).
[Bar94] H.U. Baranger and P.A. Mello. Mesoscopic transport through chaotic cavities: A random S-matrix theory approach. *Phys. Rev. Lett.* **73**, 142 (1994).
[Bee97] C.W.J. Beenakker. Random-matrix theory of quantum transport. *Rev. Mod. Phys.* **69**, 731 (1997).
[Ber77] M.V. Berry. Regular and irregular semiclassical wavefunctions. *J. Phys. A* **10**, 2083 (1977).
[Bla52] J.M. Blatt and V.F. Weisskopf. *Theoretical Nuclear Physics*. John Wiley & Sons New York 1952.
[Blü88] R. Blümel and U. Smilansky. Classical irregular scattering and its quantum-mechanical implications. *Phys. Rev. Lett.* **60**, 477 (1988).
[Blü89] R. Blümel and U. Smilansky. A simple model for chaotic scattering. II. Quantum mechanical theory. *Physica D* **36**, 111 (1989).
[Blü90] R. Blümel and U. Smilansky. Random-matrix description of chaotic scattering: Semiclassical approach. *Phys. Rev. Lett.* **64**, 241 (1990).
[Blü92a] R. Blümel, I.H. Davidson, W.P. Reinhardt, H. Lin, and M. Sharnoff. Quasilinear ridge structures in water surface waves. *Phys. Rev. A* **45**, 2641 (1992).

[Blü92b] R. Blümel, B. Dietz, C. Jung, and U. Smilansky. On the transition to chaotic scattering. *J. Phys. A* **25**, 1483 (1992).
[Cha94] A.M. Chang, H.U. Baranger, L.N. Pfeiffer, and K.W. West. Weak localization in chaotic versus nonchaotic cavities: A striking difference in the line shape. *Phys. Rev. Lett.* **73**, 2111 (1994).
[Chi96] P.A. Chinnery and V.F. Humphrey. Experimental visualization of acoustic resonances within a stadium-shaped cavity. *Phys. Rev. E* **53**, 272 (1996).
[Die93] B. Dietz and U. Smilansky. A scattering approach to the quantization of billiards – the inside–outside duality. *Chaos* **3**, 581 (1993).
[Dör98] U. Dörr, H.-J. Stöckmann, M. Barth, and U. Kuhl. Scarred and chaotic field distributions in three-dimensional Sinai-microwave resonators. *Phys. Rev. Lett.* **80**, 1030 (1998).
[Dor90] E. Doron, U. Smilansky, and A. Frenkel. Experimental demonstration of chaotic scattering of microwaves. *Phys. Rev. Lett.* **65**, 3072 (1990).
[Dor91] E. Doron, U. Smilansky, and A. Frenkel. Chaotic scattering and transmission fluctuations. *Physica D* **50**, 367 (1991).
[Dor92] E. Doron and U. Smilansky. Semiclassical quantization of chaotic billiards: a scattering theory approach. *Nonlinearity* **5**, 1055 (1992).
[Eck93] B. Eckhardt. Correlations in quantum time delay. *Chaos* **3**, 613 (1993).
[Eri63] T Ericson. A theory of fluctuations in nuclear cross sections. *Ann. Phys.* **23**, 390 (1963).
[Fyo97] Y.V. Fyodorov and H.-J. Sommers. Statistics of resonance poles, phase shifts and time delays in quantum chaotic scattering: Random matrix approach for systems with broken time-reversal invariance. *J. Math. Phys.* **38**, 1918 (1997).
[Guh98] T. Guhr, A. Müller-Groeling, and H.A. Weidenmüller. Random matrix theories in quantum physics: common concepts. *Phys. Rep.* **299**, 189 (1998).
[Haa92] F. Haake, F. Izrailev, N. Lehmann, D. Saher, and H.-J. Sommers. Statistics of complex levels of random matrices for decaying systems. *Z. Phys. B* **88**, 359 (1992).
[Haa96] F. Haake, M. Kuś, P. Šeba, H.-J. Stöckmann, and U. Stoffregen. Microwave billiards with broken time reversal invariance. *J. Phys. A* **29**, 5745 (1996).
[Jac62] J.D. Jackson. *Classical Electrodynamics*. John Wiley & Sons New York 1962.
[Jal90] R.A. Jalabert, H.U. Baranger, and A.D Stone. Conductance fluctuations in the ballistic regime: A probe of quantum chaos? *Phys. Rev. Lett.* **65**, 2442 (1990).
[Kud95] A. Kudrolli, V. Kidambi, and S. Sridhar. Experimental studies of chaos and localization in quantum wave functions. *Phys. Rev. Lett.* **75**, 822 (1995).
[Lan57] R. Landauer. Spatial variation of currents and fields due to localized scatterers in metallic conduction. *IBM J. Res. Develop.* **1**, 223 (1957).
[Lan87] R. Landauer. Electrical transport in open and closed systems. *Z. Phys. B* **68**, 217 (1987).
[Lee85] P.A. Lee and T.V. Ramakrishnan. Disordered electronic systems. *Rev. Mod. Phys.* **57**, 287 (1985).
[Lee97] Y. Lee, G. Faini, and D. Mailly. Quantum transport in chaotic and integrable ballistic cavities with tunable shape. *Phys. Rev. B* **56**, 9805 (1997).
[Leh95] N. Lehmann, D Saher, V.V. Sokolov, and H.-J. Sommers. Chaotic scattering: the supersymmetry method for large number of channels. *Nucl. Phys. A* **582**, 223 (1995).
[Lew91] C.H. Lewenkopf and H.A. Weidenmüller. Stochastic versus semiclassical approach to quantum chaotic scattering. *Ann. Phys.* **212**, 53 (1991).
[Mag66] W. Magnus, F. Oberhettinger, and R.P. Soni. *Formulas and Theorems for*

the Special Functions of Mathematical Physics. Springer-Verlag New York 1966.

[Mah69] C. Mahaux and H.A. Weidenmüller. *Shell-Model Approach to Nuclear Reactions*. North-Holland Amsterdam 1969.

[Mai52] L.C. Maier and J.C. Slater. Field strength measurements in resonant cavities. *J. Appl. Phys.* **23**, 68 (1952).

[Mar92] C.M. Marcus, A.J. Rimberg, R.M. Westervelt, P.F. Hopkins, and A.C. Gossard. Conductance fluctuations and chaotic scattering in ballistic microstructures. *Phys. Rev. Lett.* **69**, 506 (1992).

[McD79] S.W. McDonald and A.N. Kaufman. Spectrum and eigenfunctions for a Hamiltonian with stochastic trajectories. *Phys. Rev. Lett.* **42**, 1189 (1979).

[McD88] S.W. McDonald and A.N. Kaufman. Wave chaos in the stadium: Statistical properties of short-wave solutions of the Helmholtz equation. *Phys. Rev. A* **37**, 3067 (1988).

[Mor53] P.M. Morse and H. Feshbach. *Methods of Theoretical Physics*. McGraw-Hill New York 1953.

[Mül97] K. Müller, B. Mehlig, F. Milde, and M. Schreiber. Statistics of wave functions in disordered and in classically chaotic systems. *Phys. Rev. Lett.* **78**, 215 (1997).

[O'Con87] P. O'Connor, J. Gehlen, and E.J. Heller. Properties of random superpositions of plane waves. *Phys. Rev. Lett.* **58**, 1296 (1987).

[Per96] E. Persson, T. Gorin, and I. Rotter. Decay rates of resonance states at high level density. *Phys. Rev. E* **54**, 3339 (1996).

[Plu95] Z. Pluhař, H.A. Weidenmüller, J.A. Zuk, C.H. Lewenkopf, and F.J. Wegner. Crossover from orthogonal to unitary symmetry for ballistic electron transport in chaotic microstructures. *Ann. Phys.* **243**, 1 (1995).

[Por56] C.E. Porter and R.G. Thomas. Fluctuations of nuclear reaction widths. *Phys. Rev.* **104**, 483 (1956).

[Pri95] V.N. Prigodin, N. Taniguchi, A. Kudrolli, V. Kidambi, and S. Sridhar. Spatial correlation in quantum chaotic systems with time-reversal symmetry: Theory and experiment. *Phys. Rev. Lett.* **75**, 2392 (1995).

[Ric98] A. Richter. Playing billiards with microwaves – quantum manifestations of classical chaos. In D.A. Hejhal, J. Friedman, M.C. Gutzwiller, and A.M. Odlyzko, editors, *Emerging Applications of Number Theory* IMA Volume 109 page 479 New York 1998. Springer-Verlag.

[Sch95a] H. Schanz and U. Smilansky. Quantization of Sinai's billiard – a scattering approach. *Chaos, Solitons & Fractals* **5**, 1289 (1995).

[Sch95b] P. Schardt. *Mikrowellenexperimente zum chaotischen Verhalten eines supraleitenden Stadion-Billards und Entwicklung einer Einfangsektion am S-DALINAC.* Dissertation Technische Hochschule Darmstadt 1995.

[Sok88] V.V. Sokoloff and V.G. Zelevinsky. On a statistical theory of overlapping resonances. *Phys. Lett. B* **202**, 10 (1988).

[Sre96a] M. Srednicki. Gaussian random eigenfunctions and spatial correlations in quantum dots. *Phys. Rev. E* **54**, 954 (1996).

[Sre96b] M. Srednicki and F. Stiernelof. Gaussian fluctuations in chaotic eigenstates. *J. Phys. A* **29**, 5817 (1996).

[Sri91] S. Sridhar. Experimental observation of scarred eigenfunctions of chaotic microwave cavities. *Phys. Rev. Lett.* **67**, 785 (1991).

[Sri92] S. Sridhar and E.J. Heller. Physical and numerical experiments on the wave mechanics of classically chaotic systems. *Phys. Rev. A* **46**, R1728 (1992).

[Ste92] J. Stein and H.-J. Stöckmann. Experimental determination of billiard wave functions. *Phys. Rev. Lett.* **68**, 2867 (1992).
[Ste95] J. Stein, H.-J. Stöckmann, and U. Stoffregen. Microwave studies of billiard Green functions and propagators. *Phys. Rev. Lett.* **75**, 53 (1995).
[Stö90] H.-J. Stöckmann and J. Stein. 'Quantum' chaos in billiards studied by microwave absorption. *Phys. Rev. Lett.* **64**, 2215 (1990).
[Sto95] U. Stoffregen, J. Stein, H.-J. Stöckmann, M. Kuś, and F. Haake. Microwave billiards with broken time reversal symmetry. *Phys. Rev. Lett.* **74**, 2666 (1995).
[Stö98] H.-J. Stöckmann and P. Šeba. The joint energy distribution function for the Hamiltonian $H = H_0 - iWW^\dagger$ for the one-channel case. *J. Phys. A* **31**, 3439 (1998).
[Ver85] J.J.M. Verbaarschot, H.A. Weidenmüller, and M.R. Zirnbauer. Grassmann integration in stochastic quantum physics: The case of compound-nucleus scattering. *Phys. Rep.* **129**, 367 (1985).

Chapter 7

[Bal71] R. Balian and C. Bloch. Asymptotic evaluation of the Green's function for large quantum numbers. *Ann. Phys.* **63**, 582 (1971).
[Bal76] H.P. Baltes and E.R. Hilf. *Spectra of Finite Systems.* BI-Wissenschaftsverlag Mannheim 1976.
[Ber72] M.V. Berry and K.E. Mount. Semiclassical approximations in wave mechanics. *Rep. Prog. Phys.* **35**, 315 (1972).
[Ber77] M.V. Berry and M. Tabor. Level clustering in the regular spectrum. *Proc. R. Soc. Lond. A* **356**, 375 (1977).
[Ber81] M.V. Berry. Quantizing a classically ergodic system: Sinai's billiard and the KKR method. *Ann. Phys.* **131**, 163 (1981).
[Ber94] M.V. Berry and C.J. Howls. High orders of the Weyl expansion for quantum billiards: resurgence of periodic orbits, and the Stokes phenomenon. *Proc. R. Soc. Lond. A* **447**, 527 (1994).
[Bog88] E.B. Bogomolny. Smoothed wave functions of chaotic quantum systems. *Physica D* **31**, 169 (1988).
[Bra97] M. Brack and R.K. Bhaduri. *Semiclassical Physics.* Addison-Wesley New York 1997.
[Bun74] L.A. Bunimovich. On ergodic properties of certain billiards. *Funct. Anal. Appl.* **8**, 254 (1974).
[Cre95] P. Crehan. Chaotic spectra of classically integrable systems. *J. Phys. A* **28**, 6389 (1995).
[Dah95] P. Dahlqvist. Approximate zeta functions for the Sinai billiard and related systems. *Nonlinearity* **8**, 11 (1995).
[Eck88] B. Eckhardt. Quantum mechanics of classically non-integrable systems. *Phys. Rep.* **163**, 205 (1988).
[Eck89] B. Eckhardt and E. Aurell. Convergence of the semi-classical periodic orbit expansion. *Europhys. Lett.* **9**, 509 (1989).
[Ein17] A. Einstein. Zum Quantensatz von Sommerfeld und Epstein. *Verhandlungen der Deutschen Physikalischen Gesellschaft* **19**, 82 (1917).
[Fey65] R.P. Feynman and A.R. Hibbs. *Quantum Mechanics and Path Integrals.* McGraw-Hill New York 1965.

[Gas93] P. Gaspard and D. Alonso. $\hbar$ expansion for the periodic-orbit quantization of hyperbolic systems. *Phys. Rev. E* **47**, R3468 (1993).

[Gas95] P. Gaspard. $\hbar$ expansion for quantum trace formulas. In G. Casati and B. Chirikov, editors. *Quantum Chaos* page 385. Cambridge University Press 1995.

[Grä92] H.-D. Gräf, H.L. Harney, H. Lengeler, C.H. Lewenkopf, C. Rangacharyulu, A. Richter, P. Schardt, and H.A. Weidenmüller. Distribution of eigenmodes in a superconducting stadium billiard with chaotic dynamics. *Phys. Rev. Lett.* **69**, 1296 (1992).

[Gut67] M.C. Gutzwiller. Phase-integral approximation in momentum space and the bound states of an atom. *J. Math. Phys.* **8**, 1979 (1967).

[Gut69] M.C. Gutzwiller. Phase-integral approximation in momentum space and the bound states of an atom. II. *J. Math. Phys.* **10**, 1004 (1969).

[Gut70] M.C. Gutzwiller. Energy spectrum according to classical mechanics. *J. Math. Phys.* **11**, 1791 (1970).

[Gut71] M.C. Gutzwiller. Periodic orbits and classical quantization conditions. *J. Math. Phys.* **12**, 343 (1971).

[Gut89] M.C. Gutzwiller. The semi-classical quantization of chaotic Hamiltonian systems. In M.-J. Giannoni, A. Voros, and J. Zinn-Justin, editors. *Chaos and Quantum Physics* page 201. Les Houches Session LII North-Holland 1989.

[Gut90] M.C. Gutzwiller. *Chaos in Classical and Quantum Mechanics.* Springer-Verlag New York 1990.

[Hel89] E.J. Heller. Wavepacket dynamics and quantum chaology. In M.-J. Giannoni, A. Voros, and J. Zinn-Justin, editors, *Chaos and Quantum Physics* page 547. Les Houches Session LII North-Holland 1989.

[Kuś93] M. Kuś, F. Haake, and D. Delande. Prebifurcation periodic ghost orbits in semiclassical quantization. *Phys. Rev. Lett.* **71**, 2167 (1993).

[Mag66] W. Magnus, F. Oberhettinger, and R.P. Soni. *Formulas and Theorems for the Special Functions of Mathematical Physics*. Springer-Verlag New York 1966.

[Mas81] V.P. Maslov and N.V. Fedoriuk. *Semi-Classical Approximation in Quantum Mechanics.* Reidel Dordrecht 1981.

[Mes61] A. Messiah. *Quantum Mechanics, Volume I.* North-Holland Amsterdam 1961.

[Pav95] N. Pavloff and C. Schmit. Diffractive orbits in quantum billiards. *Phys. Rev. Lett.* **75**, 61 (1995).

[Pri96] H. Primack, H. Schanz, U. Smilansky, and I. Ussishkin. Penumbra diffraction in the quantization of dispersing billiards. *Phys. Rev. Lett.* **76**, 1615 (1996).

[Sie93] M. Sieber, U. Smilansky, S.C. Creagh, and R.G. Littlejohn. Non-generic spectral statistics in the quantized stadium billiard. *J. Phys. A* **26**, 6217 (1993).

[Ste92] J. Stein and H.-J. Stöckmann. Experimental determination of billiardwave functions. *Phys. Rev. Lett.* **68**, 2867 (1992).

[Ste95] J. Stein, H.-J. Stöckmann, and U. Stoffregen. Microwave studies of billiard Green functions and propagators. *Phys. Rev. Lett.* **75**, 53 (1995).

[Tom91] S. Tomsovic and E.J. Heller. Semiclassical dynamics of chaotic motion: Unexpected long-time accuracy. *Phys. Rev. Lett.* **67**, 664 (1991).

[Tom93] S. Tomsovic and E.J. Heller. Long-time semiclassical dynamics of chaos: The stadium billiard. *Phys. Rev. E* **47**, 282 (1993).

[Vat94] G. Vattay, A. Wirzba, and P.E. Rosenqvist. Periodic orbit theory of diffraction. *Phys. Rev. Lett.* **73**, 2304 (1994).

Chapter 8

[Aga93] O. Agam and S. Fishman. Quantum eigenfunctions in terms of periodic orbits of chaotic systems. *J. Phys. A* **26**, 2113 (1993).
[Aga94] O. Agam and S. Fishman. Semiclassical criterion for scars in wave functions of chaotic systems. *Phys. Rev. Lett.* **73**, 806 (1994).
[Arg93] N. Argaman, Y. Imry, and U. Smilansky. Semiclassical analysis of spectral correlations in mesoscopic systems. *Phys. Rev. B* **47**, 4440 (1993).
[Art90] R. Artuso, E. Aurell, and P. Cvitanović. Recycling of strange sets: I. Cycle expansions. *Nonlinearity* **3**, 325 (1990).
[Aur88] R. Aurich, M. Sieber, and F. Steiner. Quantum chaos of the Hadamard–Gutzwiller model. *Phys. Rev. Lett.* **61**, 483 (1988).
[Aur90] R. Aurich and F. Steiner. Energy-level statistics of the Hadamard–Gutzwiller ensemble. *Physica D* **43**, 155 (1990).
[Aur91] R. Aurich and F. Steiner. Exact theory for the quantum eigenstates of a strongly chaotic system. *Physica D* **48**, 445 (1991).
[Aur93] R. Aurich and F. Steiner. Statistical properties of highly excited quantum eigenstates of a strongly chaotic system. *Physica D* **64**, 185 (1993).
[Aur95] R. Aurich and F. Steiner. Periodic-orbit theory of the number variance $\Sigma^2(L)$ of strongly chaotic systems. *Physica D* **82**, 266 (1995).
[Bal86] N.L. Balazs and A. Voros. Chaos in the pseudosphere. *Phys. Rep.* **143**, 109 (1986).
[Ber72] M.V. Berry and K.E. Mount. Semiclassical approximations in wave mechanics. *Rep. Prog. Phys.* **35**, 315 (1972).
[Ber85] M.V. Berry. Semiclassical theory of spectral rigidity. *Proc. R. Soc. Lond. A* **400**, 229 (1985).
[Ber86] M.V. Berry. Riemann's zeta function: A model for quantum chaos? In T.H. Seligman and H. Nishioka, editors. *Quantum Chaos and Statistical Nuclear Physics* page 1 Berlin 1986. Springer-Verlag.
[Ber88] M.V. Berry. Semiclassical formula for the number variance of the Riemann zeros. *Nonlinearity* **1**, 399 (1988).
[Ber89] M.V. Berry. Quantum scars of classically closed orbits in phase space. *Proc. R. Soc. Lond. A* **423**, 219 (1989).
[Ber92] M.V. Berry and J.P. Keating. A new asymptotic representation for $\zeta(\frac{1}{2} + it)$ and quantum spectral determinants. *Proc. R. Soc. Lond. A* **437**, 151 (1992).
[Bog88] E.B. Bogomolny. Smoothed wave functions of chaotic quantum systems. *Physica D* **31**, 169 (1988).
[Bog92] E.B. Bogomolny, B. Georgeot, M.-J. Giannoni, and C. Schmit. Chaotic billiards generated by arithmetic groups. *Phys. Rev. Lett.* **69**, 1477 (1992).
[Bog93] E.B. Bogomolny, B. Georgeot, M.-J. Giannoni, and C. Schmit. Trace formulas for arithmetical systems. *Phys. Rev. E* **47**, R2217 (1993).
[Bog94] E.B. Bogomolny and P. Leboeuf. Statistical properties of the zeros of zeta functions – beyond the Riemann case. *Nonlinearity* **7**, 1155 (1994).
[Bog95] E.B. Bogomolny and P. Keating. Random matrix theory and the Riemann zeros I: three- and four-point correlations. *Nonlinearity* **8**, 1115 (1995).
[Bog96a] E.B. Bogomolny and J.P. Keating. Gutzwiller's trace formula and

spectral statistics: Beyond the diagonal approximation. *Phys. Rev. Lett.* **77**, 1472 (1996).

[Bog96b] E.B. Bogomolny and P. Keating. Random matrix theory and the Riemann zeros II: n-point correlations. *Nonlinearity* **9**, 911 (1996).

[Bog96c] E.B. Bogomolny, F. Leyvraz, and C. Schmit. Distribution of eigenvalues for the modular group. *Commun. Math. Phys.* **176**, 577 (1996).

[Boh86] O. Bohigas, M.-J. Giannoni, and C. Schmit. Spectral fluctuations of classically chaotic quantum systems. In T.H. Seligman and H. Nishioka, editors, *Quantum Chaos and Statistical Nuclear Physics* page 18 Springer-Verlag Berlin 1986.

[Bol92] J. Bolte, G. Steil, and F. Steiner. Arithmetical chaos and violation of universality in energy level statistics. *Phys. Rev. Lett.* **69**, 2188 (1992).

[Cas85] G. Casati, B.V. Chirikov, and I. Guarneri. Energy-level statistics of integrable quantum systems. *Phys. Rev. Lett.* **54**, 1350 (1985).

[Col85] Y. Colin de Verdiére. Ergodicité et fonctions propres du Laplacien. *Commun. Math. Phys.* **102**, 497 (1985).

[Cou95] M. Courtney, N. Spellmeyer, H. Jiao, and D. Kleppner. Classical, semiclassical, and quantum dynamics in the lithium stark system. *Phys. Rev. A* **51**, 3604 (1995).

[Cvi88] P. Cvitanović. Invariant measurement of strange sets in terms of cycles. *Phys. Rev. Lett.* **61**, 2729 (1988).

[Cvi89] P. Cvitanović and B. Eckhardt. Periodic-orbit quantization of chaotic systems. *Phys. Rev. Lett.* **63**, 823 (1989).

[Eck89] B. Eckhardt and E. Aurell. Convergence of the semi-classical periodic orbit expansion. *Europhys. Lett.* **9**, 509 (1989).

[Eck95] B. Eckhardt, S. Fishman, J. Keating, O. Agam, J. Main, and K. Müller. Approach to ergodicity in quantum wave functions. *Phys. Rev. E* **52**, 5893 (1995).

[Edw74] H.M. Edwards. *Riemann's Zeta Function*. Academic Press New York 1974.

[Fri89] H. Friedrich and D. Wintgen. The hydrogen atom in a uniform magnetic field – an example of chaos. *Phys. Rep.* **183**, 37 (1989).

[Fro94] T.M. Fromhold, L. Eaves, F.W. Sheard, M.L. Leadbeater, T.J. Foster, and P.C. Main. Magnetotunneling spectroscopy of a quantum well in the regime of classical chaos. *Phys. Rev. Lett.* **72**, 2608 (1994).

[Fro95a] T.M. Fromhold, A. Fogarty, L. Eaves, F.W. Sheard, M. Henini, T.J. Foster, P.C. Main, and G. Hill. Evidence for quantum states corresponding to families of stable and chaotic classical orbits in a wide potential well. *Phys. Rev B* **51**, 18029 (1995).

[Fro95b] T.M. Fromhold, P.B. Wilkinson, F.W. Sheard, L. Eaves, J. Miao, and G. Edwards. Manifestation of classical chaos in the energy level spectrum of a quantum well. *Phys. Rev. Lett.* **75**, 1142 (1995).

[Gal94] T.F. Gallagher. *Rydberg Atoms*. Cambridge University Press 1994.

[Gut67] M.C. Gutzwiller. Phase-integral approximation in momentum space and the bound states of an atom. *J. Math. Phys.* **8**, 1979 (1967).

[Gut83] M.C. Gutzwiller. Stochastic behavior in quantum scattering. *Physica D* **7**, 341 (1983).

[Gut90] M.C. Gutzwiller. *Chaos in Classical and Quantum Mechanics.* Springer-Verlag New York 1990.

[Guy94] R.K. Guy. *Unsolved Problems in Number Theory.* Springer-Verlag New York second edition 1994.

[Han84] J.H. Hanney and A.M. Ozorio De Almeida. Periodic orbits and a correlation function for the semiclassical density of states. *J. Phys. A* **17**, 3429 (1984).
[Har22] G.H. Hardy and J.E. Littlewood. Some problems of 'partitio numerorum'; III: On the expression of a number as a sum of primes. *Acta Mathematica* **44**, 1 (1922).
[Har60] G.H. Hardy and E.M. Wright. *An Introduction to the Theory of Numbers*. Oxford University Press fourth edition 1960.
[Has89] H. Hasegawa, M. Robnik, and G. Wunner. Classical and quantal chaos in the diamagnetic Kepler problem. *Prog. Theor. Phys. Suppl.* **98**, 198 (1989).
[Hej76] D.A. Hejhal. *The Selberg Trace Formula for PSL (2,R). Volume 1.* Lecture Notes in Mathematics 548. Springer-Verlag Berlin 1976.
[Hej83] D.A. Hejhal. *The Selberg Trace Formula for PSL (2,R). Volume 2.* Lecture Notes in Mathematics 1001. Springer-Verlag Berlin 1983.
[Hel84] E.J. Heller. Bound-state eigenfunctions of classically chaotic Hamiltonian systems: Scars of periodic orbits. *Phys. Rev. Lett.* **53**, 1515 (1984).
[Hön89] A Hönig and D. Wintgen. Spectral properties of strongly perturbed Coulomb systems: Fluctuation properties. *Phys. Rev. A* **39**, 5642 (1989).
[Hol86] A. Holle, G. Wiebusch, J. Main, B. Hager, H. Rottke, and K.H. Welge. Diamagnetism of hydrogen atom in the quasi-Landau regime. *Phys. Rev. Lett.* **56**, 2594 (1986).
[Hol88] A. Holle, J. Main, G. Weibusch, H. Rottke, and K.H. Welge. Quasi-Landau spectrum of the chaotic diamagnetic hydrogen atom. *Phys. Rev. Lett.* **61**, 161 (1988).
[Iu91] C. Iu, G.R. Welch, M.M. Kash, D. Kleppner, D. Delande, and J.C. Gay. Diamagnetic Rydberg atom: Confrontation of calculated and observed spectra. *Phys. Rev. Lett.* **66**, 145 (1991).
[Kea92] J.P. Keating. Periodic orbit resummation and the quantization of chaos. *Proc. R. Soc. Lond. A* **436**, 99 (1992).
[Li97] B. Li. Numerical study of scars in a chaotic billiard. *Phys. Rev. E* **55**, 5376 (1997).
[Mai86] J. Main, G. Wiebusch, A. Holle and K.H. Welge. New quasi-Landau structure of highly excited atoms: The hydrogen atom. *Phys. Rev. Lett.* **57**, 2789 (1986).
[Mar92] C.M. Marcus, A.J. Rimberg, R.M. Westervelt, P.F. Hopkins, and A.C. Gossard. Conductance fluctuations and chaotic scattering in ballistic microstructures. *Phys. Rev. Lett.* **69**, 506 (1992).
[Meh91] M.L. Mehta. *Random Matrices*. Academic Press San Diego second edition 1991.
[Mor53] P.M. Morse and H. Feshbach. *Methods of Theoretical Physics*. McGraw-Hill New York 1953.
[Nöc97] J.U. Nöckel and A.D. Stone. Ray and wave chaos in asymmetric resonant cavities. *Nature* **385**, 45 (1997).
[Odl87] A.M. Odlyzko. On the distribution of spacings between zeros of the zeta function. *Math. Comp.* **48**, 273 (1987).
[Odl89] A.M. Odlyzko. The 10^{20}-th zero of the Riemann zeta function and 70 million of its neighbors. Preprint, AT&T Bell Laboratories 1989.
[Rot86] H. Rottke and K.H. Welge. Photoionization of the hydrogen atom near the ionization limit in strong electric fields. *Phys. Rev. A* **33**, 301 (1986).
[Rud94] H. Ruder, G. Wunner, H. Herold, and F. Geyer. *Atoms in Strong Magnetic Fields.* Springer-Verlag Berlin 1994.
[Sch89] C. Schmit. Quantum and classical properties of some billiards on the

hyperbolic plane. In M.-J. Giannoni, A. Voros, and J. Zinn-Justin, editors, *Chaos and Quantum Physics* page 331. Les Houches Session LII North-Holland 1989.

[Sch96] H. Schanze. *Realisierung von nichteuklidischen Billards durch Wellenwannen.* Diplomarbeit Philipps-Universität Marburg 1996.

[Shn74] A.I. Shnirelman. Ergodic properties of eigenfunctions (in Russian). *Usp. Mat. Nauk.* **29**, 181 (1974).

[Sie91] M. Sieber and F. Steiner. Quantization of chaos. *Phys. Rev. Lett.* **67**, 1941 (1991).

[Sok64] I.S. Sokolnikoff. *Tensor Analysis*. John Wiley & Sons New York second edition 1964.

[Stö90] H.-J. Stöckmann and J. Stein. 'Quantum' chaos in billiards studied by microwave absorption. *Phys. Rev. Lett.* **64**, 2215 (1990).

[Stö95] H.-J. Stöckmann, J. Stein, and M. Kollmann. Microwave studies in irregularly shaped billiards. In C. Casati and B. Chirikov, editors, *Quantum Chaos Between Order and Disorder* page 661. Cambridge University Press 1995.

[Tak89] K. Takahashi. Distribution functions in classical and quantum mechanics. *Prog. Theor. Phys. Suppl.* **98**, 109 (1989).

[Tan91] G. Tanner, P. Scherer, E.B. Bogomolny, B. Eckhardt, and D. Wintgen. Quantum eigenvalues from classical periodic orbits. *Phys. Rev. Lett.* **67**, 2410 (1991).

[Tit51] E.C. Titchmarsh. *The Theory of the Riemann Zeta-Function.* Clarendon Press Oxford 1951.

[Vor88] A. Voros. Unstable periodic orbits and semiclassical quantization. *J. Phys. A* **21**, 685 (1988).

[Wei93] D. Weiss, K. Richter, A. Menschig, R. Bergmann, H. Schweizer, K. von Klitzing, and G. Weimann. Quantized periodic orbits in large antidot arrays. *Phys. Rev. Lett.* **70**, 4118 (1993).

[Win 87] D. Wintgen. Connection between long-range correlation in quantum spectra and classical periodic orbits. *Phys. Rev. Lett.* **58**, 1589 (1987).

[Win88] D. Wintgen. Semiclassical path-integral quantization of nonintegrable Hamiltonian systems. *Phys. Rev. Lett.* **61**, 1803 (1988).

[Win89] D. Wintgen and A. Hönig. Irregular wave functions of a hydrogen atom in a uniform magnetic field. *Phys. Rev. Lett.* **63**, 1467 (1989).

[Zel87] S. Zelditch. Uniform distribution of eigenfunctions on compact hyperbolic surfaces. *Duke Math. J.* **44**, 919 (1987).

Index

Printed in the United States
71241LV00001B

Paperback Re-issue

This book introduces the quantum mechanics of classically chaotic systems, or Quantum Chaos for short. The basic concepts of quantum chaos can be grasped easily by any student of physics, but the underlying physical principles tend to be obscured by the mathematical apparatus used to describe it. The author's philosophy, therefore, has been to keep the discussion simple and to illustrate theory, wherever possible, with experimental or numerical examples. The microwave billiard experiments, initiated by the author and his group, play a major role in this respect. The author assumes a basic knowledge of quantum mechanics.

Beginning with a presentation of the various types of billiard experiment, random matrix theory is described, including an introduction to supersymmetry techniques. There is a chapter that deals with systems with periodic time dependences, with special emphasis on dynamical localization. Another topic is the analogy between the dynamics of a one-dimensional gas with a repulsive interaction and spectral level dynamics, where an external parameter takes the role of time. In another chapter on scattering theory distributions and fluctuations, properties of scattering matrix elements are presented. In the final two chapters semiclassical quantum mechanics and periodic orbit theory are addressed, including a derivation of the Gutzwiller trace formula together with a number of applications.

This book will be of great value to anyone working in quantum chaos.

Cover Illustrations:

The front cover shows the field intensity of microwaves propagating through a Josephson junction (C. Krülle *et al., Physica D* 78, 214 (1994)). C. Krülle is thanked for the kind permission to reprint an unpublished photograph.

The back cover shows a pulse reconstruction in a microwave billiard of the shape of a quarter stadium billiard (J. Stein *et al., Phys. Rev. Lett.* 75, 53 (1995)).